12692832

WITHDRAWN

Lake and Reservoir Restoration

G. Dennis Cooke
Department of Biological Sciences
Kent State University
Kent, Ohio

Eugene B. Welch
Department of Civil Engineering
University of Washington
Seattle, Washington

Spencer A. Peterson
U.S. Environmental Protection Agency
Corvallis Environmental Research Laboratory
Corvallis, Oregon

Peter R. Newroth
Ministry of Environment
Province of British Columbia
Victoria, British Columbia

Butterworths
Boston London Durban Singapore Sydney Toronto Wellington

An Ann Arbor Science Book

Ann Arbor Science is an imprint of Butterworth Publishers.

Copyright © 1986 by Butterworth Publishers.
All rights reserved.

No part of this publication may be reproduced, stored in a retrieval system, or transmitted, in any form or by any means, electronic, mechanical, photocopying, recording, or otherwise, without the prior written permission of the publisher.

Library of Congress Cataloging-in-Publication Data
Main entry under title:

Lake and reservoir restoration.

Includes index.
1. Lake renewal. 2. Reservoirs. 3. Water quality management. 4. Eutrophication. I. Cooke, G. Dennis (George Dennis), 1937-0000.
TC409.L35   1986   627'.14   85-24318
ISBN 0-250-40643-8

Butterworth Publishers
80 Montvale Avenue
Stoneham, MA 02180

10 9 8 7 6 5 4 3 2 1

Printed in the United States of America

Contents

Preface · vii

I. OVERVIEW · 1

1. INTRODUCTION · 3

2. LIMNOLOGY, LAKE DIAGNOSIS, AND SELECTION OF RESTORATION METHODS · 9
Introduction · 9
Lakes and Reservoirs · 9
The Eutrophication Process · 10
Lake or Reservoir Evaluation · 22
Selection of Lake Restoration Alternatives · 40
The Lake Improvement-Restoration Plan · 46

II. PHYSICAL AND CHEMICAL METHODS TO REDUCE NUTRIENT CONCENTRATIONS · 53

3. ADVANCED TREATMENT AND DIVERSION OF WASTEWATER AND STORMWATER · 55
Mechanisms · 55
Extent of Recovery Expected · 56
Case Studies · 60
Costs · 67
Further Treatment Following Diversion · 70
Needed Research · 70

4. HYPOLIMNETIC WITHDRAWAL · 75
Case Studies · 75
General Considerations · 79

5. DILUTION AND FLUSHING · 83
Theory and Predictions · 84
Case Studies · 86
General Applications and Precautions · 96
Summary · 97

iii

6. PHOSPHORUS PRECIPITATION AND INACTIVATION — 101
The Reactions of Aluminum and Phosphorus in Lake Water — 102
Dose Determination and Application Techniques — 103
Case Studies — 111
Lake and Reservoir Pretreatment — 118
Negative Aspects — 119
Costs — 122
Needed Research — 122
Use of Fly Ash for Phosphorus Inactivation — 123

7. SEDIMENT OXIDATION — 133
Equipment and Application Rates — 133
Case Studies — 134
Costs — 134
Related Effects — 136

8. SEDIMENT REMOVAL — 139
Purposes of Sediment Removal — 139
Environmental Concerns — 141
Sediment Removal Depth — 143
Sediment Removal Techniques — 146
Suitable Lake Conditions — 156
Dredge Selection and Disposal Area Design — 157
Case Studies — 191
Costs — 197
Summary — 205

9. HYPOLIMNETIC AERATION — 213
Principle — 213
Sizing — 214
Case Studies — 216
Undesirable Side Effects — 220
Costs — 221

III. PROCEDURES TO CONTROL PLANT BIOMASS — 223

10. ARTIFICIAL CIRCULATION — 225
Theoretical Effects of Circulation — 225
Devices and Air Quantities — 233
Costs — 243
Effects of Circulation on Trophic Indicators — 243
Undesirable Effects — 245
Summary and Recommendations — 245

11. WATER-LEVEL DRAWDOWN — 253
Case Studies — 253
Responses of Aquatic Plants to Water-Level Drawdown — 266

		Positive and Negative Factors	268
		Summary	271
12.		**HARVESTING**	275
		Objectives and Benefits of Aquatic Plant Control	275
		Descriptions of Mechanical Control Technologies	277
		Case Studies	288
		Summary of Costs	297
		Environmental Effects	300
		Control of Nutrient Levels	302
		Effectiveness	304
		Summary	306
13.		**BIOLOGICAL CONTROLS**	315
		Phytophagous Fish	315
		Cyanophages and Bacteria	325
		Plant Pathogens and Insects	325
		Biomanipulation	328
		Summary	340
14.		**SURFACE AND SEDIMENT COVERS FOR MACROPHYTE CONTROL**	349
		Case Studies	350
		Sediment Covers—Summary	356
		Application Procedures for Sediment Covers	357
		Shading of Macrophytes with Surface Covers	358
IV.		**ACIDIFIED LAKES**	361
15.		**LIMING ACIDIFIED LAKES**	363
		Acidification of Lakes	363
		Ecological Effects	364
		Liming Practices	365
16.		**SUMMARY AND CONCLUSIONS**	377
		Techniques with Demonstrated Effectiveness	377
		Techniques Requiring More Research and Demonstration	380
		Acidified Lakes	383
		Combinations of Lake Restoration Techniques	383
		Conclusion	385
		INDEX	388

Preface

Eutrophication is perhaps the most evident and pervasive water quality problem in the world, and contributes significantly to impaired water use. Despite the infusion of substantial resources over the past decade or so to upgrade sewage treatment facilities, to curb industrial discharges, to enact land management, and to carry out basic and applied research, the quality of lakes and reservoirs continues to be a problem of increasing significance and scope. The industrialized world's demand for clean water is increasing, commensurate with world population increases and the expanding requirements for clean water for agriculture, domestic and industrial supply, and recreation.

In order to protect, manage, and restore the quality of lakes and reservoirs, basic and applied research necessary to understand them must be expanded. Lake and reservoir management and restoration is a new science, but its development could ultimately be essential to mankind's well-being. Progress in this field, however, is extremely hampered by restricted funding. It is imperative that these needs are recognized and supported, not only by researchers and others in this field, but by politicians, administrators, and others who influence governmental policy and decisions about appropriations.

Lake and Reservoir Restoration is about the eutrophication process and about the methods to protect, restore, and manage lakes and reservoirs, including a brief account of acidified lakes. We have written a text which we believe will be readable and very useful to the professional limnologist and engineer as well as to the interested layman and on-site lake or reservoir manager. This book is only a beginning, however. Our need to understand lakes and reservoirs and the methods to protect and restore them is great and much more must be learned.

We have described the eutrophication process and the development of a pre-restoration diagnosis and feasibility study in a simplified but hopefully easily understood manner. The chapters on the various restoration methods include an introduction to the theory of the problem and the restoration technique, a description of application procedures, an analysis of case studies and success stories, and a discussion of costs and potential negative impacts. The chapters contain extensive bibliographies, making them suitable not only to the person interested in the actual implementation of a method, but also to those who are entering the field and learning about the problems and management of lakes and reservoirs.

The content of the entire book is a product of the study, input, and concurrence of all of the authors. Specific chapter authorship is as follows: Chapters 1, 2, 6,

11, 13, 14, and 16 are by G. Dennis Cooke, Chapters 3, 4, 5, 7, 9, 10, and 15 by Eugene B. Welch, Chapter 8 by Spencer A. Peterson, and Chapter 12 by Peter R. Newroth and G. Dennis Cooke.

Limnology is a field very rich in scholars and practitioners who are stimulating and innovative. New ideas are rapidly emerging, particularly in the areas of biological controls and the use of complementary methods. This book is a review, a consolidation, and a synthesis of lake and reservoir restoration and management experiences. We hope it will serve as a resource to this community of scientists, engineers, and lake managers as well as a base from which the science of lake and reservoir restoration and management can further develop.

We have greatly benefitted from the productive and stimulating discussions and collaborations with our colleagues and students and we are grateful to them all. We especially wish to recognize those many persons, too numerous to name, who have given of themselves to make the North American Lake Management Society (NALMS) successful. The annual meetings and the proceedings are invaluable to everyone interested in this field. Dr. Robert Johnson and Ms. Judy Taggart of NALMS have been particularly encouraging and supportive to us and we are grateful to them.

The field of lake and reservoir management and restoration owes much to its pioneers, those persons who first put into practice the ideas which have lead to the development of our current understanding of these methods, and which are the basis of this book. We acknowledge the following as being among the leaders who took the first steps toward implementing or encouraging lake restoration techniques: T. Ahl, I. Ahlgren, G. Andersson, A. Bartsch, G. Bennett, H. Bernhardt, S. Björk, S. Born, R. Dunst, W.T. Edmondson, A. Fast, G. Fitzgerald, C. Forsberg, T. Freeman, W. Funk, R. Gächter, A. Jernelöv, R. Johnson, L. Klessig, D. Knauer, G. Lindmark, K. Mackenthun, S. Nichols, W. Ohle, J. Peterson, R. Pechlaner, C. Powers, W. Ripl, G. Rohlich, S. Rying, D. Schindler, J. Shapiro, R. Sylvester, E. Thomas, P. Uttormark, and T. Wirth.

We are deeply grateful to many persons for their support during the preparation of the manuscript. Marilyn Silvey, Jane Lybecker, Bobby Greer, and Linda Matz typed and re-typed the manuscript and we thank them for their skill and patience. We also thank Dr. Keith Ewing, Chairman, Department of Biological Sciences, Kent State University, and Dr. Ronald J. Buchanan, Water Management Branch, British Columbia Ministry of Environment for their interest and support of us during the preparation of the book. We also acknowledge Dr. Ray Montgomery, U.S. Army Corps of Engineers, for his extensive assistance and suggestions in the development of the chapter on sediment removal.

Finally, we acknowledge those skillful teachers and researchers, our mentors, who were willing to aid young investigators at the start of a career in aquatic ecology. We thank Dr. Ralph W. Dexter, Dr. Richard V. Bovbjerg, and Dr. Eugene P. Odum (GDC), Dr. Robert C. Ball and Dr. Max Katz (EBW), and Dr. Joe K. Neel (SAP).

G. Dennis Cooke
Eugene B. Welch
Spencer A. Peterson
Peter R. Newroth

I

Overview

Overview

1

Introduction

"Lakes seem, on the scale of years or of human life spans, permanent features of the landscape, but they are geologically transitory, usually born of catastrophes, to mature and die quietly and imperceptibly" (Hutchinson, 1957). This introductory sentence from Hutchinson's classic book alerts us to the fact that lakes (and reservoirs) are not static. It also suggests that changes in lake systems are very gradual. In 1985, nearly 30 years after Hutchinson wrote these words, it is clear that lakes and reservoirs exposed to the effects of human culture have in fact changed very rapidly and conspicuously.

One of the most obvious, pervasive, and world-wide water quality problems is eutrophication of lakes and reservoirs. These water bodies have deteriorated rapidly, through excessive additions of plant nutrients, organic matter, and silt, which combine to produce increased populations of algae and rooted plants and decreased lake or reservoir volumes (Likens, 1972). Lakes and reservoirs in this condition lose much of their beauty, their attractiveness for recreation, and their usefulness as industrial and domestic water supplies. Rooted and floating plant masses may become so dense that most uses of the water are curtailed. Symptoms of eutrophication, such as algal blooms (including surface scums), rapid loss of volume in reservoirs, noxious odors, tainted fish flesh and domestic water supplies, dissolved oxygen depletion, fish kills, and the development of nuisance animal populations (e.g., common carp) can bring about economic losses in the forms of decreased property values, high-cost treatment of drinking water, depressed recreation industries, expenditures for herbicide applications, and the need to build new reservoirs.

Protection of clean water systems and restoration of eutrophic water bodies is greatly needed because the supply of clean, fresh surface water is limited and because the already high demand for recreational, industrial, and domestic uses of this water appears to be increasing (Wetzel, 1983). The first and most obvious step toward protection and restoration of a lake or reservoir is to divert or treat excessive nutrient, organic, and silt loads. Even this expensive process, while necessary, may be insufficient to produce immediate and long-lasting effects, due to internal recycling of nutrients and the associated production of algae and macrophytes. Several authors (e.g., Rich and Wetzel, 1978; Carpenter, 1981) have identified feedback loops in eutrophic systems that can maintain the eutrophic state for some period after loading is curtailed. These include macrophyte growth-death-decay cycles, nutrient release under aerobic and anaerobic conditions, and bioturbation. In these lakes and reservoirs a second step, a technique to manipulate or alter an internal chemical, biological, or physical process, may be needed to promptly restore the water body.

Lake and reservoir restoration techniques are the subject of this book. Most of these procedures or manipulations are meant to be used after diversion, but obviously some are used in conjunction with diversion or to temporarily improve the lake without actual restoration. Techniques to restore or improve a eutrophic lake or reservoir must produce long-term improvement in the lake's trophic state through ecologically sound principles that change or retard plant growth, improve environmental conditions for lake organisms, or both. These goals may be met through harvesting of plant biomass or selected fish species, by alteration of the bottom substrate for rooted plants or internal nutrient supply, by oxygenation and circulation, or by altering the lake's fauna or flora to enhance the activities of pathogens, predators, or herbivores. In this book we examine the most common in-lake techniques or procedures, plus nutrient diversion. We review each with regard to its scientific basis, methods of application, known effectiveness, feasibility, drawbacks, and where known, its costs. We also identify areas for further research and development.

The history of this environmental technology is comparatively short, although there has been a long period of awareness of the problem (Hutchinson, 1973, provides a historical review). Hasler, in 1947, was among the first to recognize that restoration of eutrophic lakes would be difficult. He stated (p. 391): "The problem is especially serious because there is no way at present for reversing the process of eutrophy.... Unless some method can be discovered for oxidizing a large proportion of the additional deposits created by fertilizers, the lake's service to man is limited to a relatively short period." Twenty years later most of the techniques described in the present book had been suggested, some had been tried, and most were the subjects of symposia (National Academy of Sciences, 1969; American Association for the Advancement of Science, 1970) or literature reviews (Stewart and Rohlich, 1967; Hasler, 1969).

In the 1970s, stimulated and supported in part by the U.S. Environmental Protection Agency's (USEPA) Clean Lakes Program (PL 92-500, Section 314, 1972), research into and application of these techniques intensified. Lake restoration conferences, initiated at the University of Wisconsin, were held. The United States joined with other nations of the Organization for Economic Cooperation and Development (OECD) to study the problem (USEPA, 1977). Numerous reviews and conference proceedings were published, among them Tenney et al. (1972), Boyter and Wanielista (1973), Peterson et al. (1974), Dunst and others (1974), Born (1979), USEPA and North American Lake Management Society (NALMS) (1979; 1980 a,b; 1983), Rast and Lee (1978), Janik et al. (1980), Klapper (1980), Welch (1980), and Uhlmann (1982).

In 1980, as a product of these conferences and the intense concern about eutrophication and lake and reservoir protection and management, a new professional society, the North American Lake Management Society (P.O. Box 217, Merrifield, Virginia 22116), was established. This organization has continued to keep information flowing among lake scientists and the public on matters concerning eutrophication and lake management, while activities in the USEPA's Clean Lakes program have declined sharply.

Introduction 5

The history of lake restoration and protection, though brief, has not been without controversy. Based on limited evidence it was widely believed that phosphorus and, to a lesser extent, nitrogen were the elements most likely to limit the production of algal biomass in fresh waters. Efforts to improve lakes and reservoirs were therefore directed toward reducing the concentration of these elements in them through advanced waste treatment, diversion, and bans on the sale of phosphorus-containing detergents. The assertion that algal biomass was most often limited by phosphorus concentration was challenged by proponents (largely from the detergent industry) of the belief that carbon was the limiting element (the so-called "limiting nutrient controversy"). Considerable research was carried out to test this issue; it is now clear that phosphorus is most often the limiting nutrient and that carbon or nitrogen is only briefly or rarely the limiting nutrient (Vollenweider, 1968; Likens, 1972; Schindler, 1974).

These findings about phosphorus limitation of algal biomass led many lake managers to base their approach to lake and reservoir restoration and management on controlling the concentration of phosphorus. While this emphasis on phosphorus has been very useful, it has often restricted the view of eutrophication and its control to simply that of external nutrient income and algal biomass. The problem is more complicated because most lakes and many reservoirs are small or shallow, with extensive littoral zones, macrophyte development, and high ratios of bottom sediment to lake volume. We now know that lakes and reservoirs contain interacting food webs and dynamic stores of nutrients in their bottom sediments that interact with the water column. They are not simply reaction vessels containing nutrients and algae. Managing and restoring lakes and reservoirs must include a recognition of the significance of littoral zones and macrophyte development, as well as the roles of biological interactions and feedback processes, which might combine to maintain high nutrient concentrations and plant biomass long after nutrient diversion. We have attempted to emphasize these views throughout this book, and to discourage the erroneous belief that the effects of eutrophication are solely the result of excessive nutrients and algae.

An equally serious water quality problem, though perhaps more restricted geographically, is that of acid rain and the development of lakes with a low pH (Likens and Bormann, 1974). Soft-water lakes are poorly buffered and very sensitive to acid rain. A drop in pH may cause the loss of many plant and animal species. The technology to treat acid lakes lags behind that for eutrophication, because the problem is relatively new, because the long-term solution is reduction or elimination of the income of acid precipitation, and because of the complex political and social problems associated with control of the emissions from power plants and cars that cause acid rain. The state of this lake restoration technology is reviewed briefly in Chapter 15.

There are several very important topics in lake and reservoir management and protection that we have chosen not to address, in part because they have been covered extensively elsewhere. Control of stormwater and sediment income, except for the impact of nutrients, has been treated in several recent reports (e.g., Wanielista, 1978); we therefore consider only the nutrient-loading aspects of stormwater. The presence of toxic metals and organics in water bodies, surely one of the current and critical environmental issues, deserves a separate and lengthy treatment and is thus not discussed here.

One very relevant topic to lake management, the use of herbicides and algicides, has been omitted for other reasons. These chemicals are among the most widely used means of attempting to control nuisance aquatic vegetation, in part because until the last 10 to 20 years they were the only technique available to lake users. Most of these agents have been found to be at least briefly effective, especially when used in combination with newer methods (e.g., Goldsby et al., 1978); and some are neither directly toxic to nontarget species nor leave toxic residues.

We have not discussed the use of herbicides and algicides in this book on lake restoration techniques because, unlike any of the procedures we describe here, there is abundant evidence that herbicidal and algicidal chemicals have been associated with major adverse impacts on lake systems, and none of them is "restorative." These adverse impacts include nutrient releases to the water following plant death (e.g., Simsiman et al., 1972; Hestand and Carter, 1978; James, 1984); dissolved oxygen depletion following plant decay (e.g., Brooker and Edwards, 1975; Anderson, 1981; Carpenter and Greenlee, 1981); toxic effects on nontarget aquatic organisms at recommended doses (e.g., De Mayo et al., 1982); and rapid regrowth of plants following treatment (e.g., Conyers and Cooke, 1983). There are also conflicting and unresolved issues regarding mutagenic and carcinogenic effects of some of these chemicals on humans (e.g., Shearer, 1980; Mullison, 1981).

It is our belief that far more research is needed about the impacts of herbicides and algicides on aquatic communities — the actual level of biological organization to which they are applied, rather than additional studies of their toxic effects to individual organisms — before they can be endorsed for broad-scale application to lakes and reservoirs. There is also little research to compare the cost-effectiveness of chemical treatment with the methods we describe in this book. These critical studies of potential adverse effects and of cost versus effectiveness need to be carried out before herbicidal and algicidal chemicals can be considered for addition to any list of lake "restoration" techniques. Until these studies are completed, we do not recommend the use of chemicals as the only treatment of nuisance aquatic plants, except in cases where every other option has been considered and found inappropriate or ineffective.

We have organized lake and reservoir restoration procedures into three groups. The first, techniques for physical and chemical control of nutrients, includes diversion and advanced waste treatment, hypolimnetic withdrawal, dilution and flushing, phosphorus precipitation and inactivation, sediment oxidation, sediment removal, and hypolimnetic aeration. The second group, methods to control plant biomass, includes artificial circulation, water-level drawdown, harvesting, biological controls, and surface and sediment covers. We conclude with a chapter on liming acidified lakes.

This book can only begin to meet the needs in this field. We hope that it will stimulate additional research and applications that will improve our understanding of what does and does not work. At the same time, we hope the book will serve as a useful guide to limnologists, consultants, engineers, lake managers, and others who are trying to solve lake problems.

REFERENCES

American Association for the Advancement of Science. 1970. *Lake Restoration*, proceedings of a symposium, Washington, DC (tapes 67-70. Sessions 1 and 2, tapes 2-3730-2-3733). Sponsored by AAAS.

Anderson, L.W.J. 1981. Control of aquatic weeds with hexazinone. *J Aquatic Plant Manage* 19:9-14.

Born, S.M. 1979. Lake rehabilitation: A status report. *Environ Manage* 3:145-153.

Boyter, C.J., and Wanielista, M.P. 1973. Review of lake restoration procedures. *Water Res Bull* 9:499-511.

Brooker, M.P., and Edwards, R.W. 1975. Aquatic herbicides and the control of water weeds. *Water Res* 9:1-5.

Carpenter, S.R. 1981. Submersed vegetation: An internal factor in lake ecosystem succession. *Am Nat* 118:372-383.

Carpenter, S.R., and Greenlee, J.K. 1981. Lake deoxygenation after herbicide use: A simulation model analysis. *Aquatic Bot* 11:173-186.

Conyers, D.L., and Cooke, G.D. 1983. A comparison of the costs of harvesting and herbicides and their effectiveness in nutrient removal and control of macrophyte biomass. In *Lake Restoration, Protection, and Management*. USEPA 440/5-83-001, 317-321.

De Mayo, A., Taylor M.C., and Taylor, K.W. 1982. Effects of copper on humans, laboratory and farm animals, terrestrial plants, and aquatic life. *CRC Crit Rev Environ Control* 12:183-255.

Dunst, R.C., Born, S.M., Uttormark, P.D., Smith, S.A., Nichols, S.A., Peterson, J.O., Knauer, D.R., Serns, S.L., Winter, D.R., and Wirth, T.L. 1974. *Survey of Lake Rehabilitation Techniques and Experiences*. Tech. Bull. 75, Dept. of Natural Resources, Madison, Wisconsin.

Goldsby, T.L., Bates, A.L., and Stanley, R.A., 1978. Effect of water level fluctuation and herbicide on Eurasian watermilfoil in Melton Hill Reservoir. *J Aquatic Plant Manage* 16:34-38.

Hasler, A.D. 1947. Eutrophication of lakes by domestic drainage. *Ecology* 28:383-395.

Hasler, A.D. 1969. Cultural eutrophication is reversible. *BioScience* 19:425-431.

Hestand, R.S., and Carter, C.C. 1978. Comparative effects of grass carp and selected herbicides on macrophyte and phytoplankton communities. *J Aquatic Plant Manage* 16:43-50.

Hutchinson, G.E. 1957. *A Treatise on Limnology. Volume I. Geography, Physics and Chemistry*. John Wiley and Sons, Inc., New York.

Hutchinson, G.E. 1973. Eutrophication. *Am Sci* 61:269-279.

James, W.F. 1984. Effects of endothall treatment on phosphorus concentration and community metabolism of aquatic communities, misc. paper A-84-1, U.S. Army Corps of Engineers, Vicksburg, MS.

Janik, J.J., Taylor, W.D., and Barko, J.W. 1980. A compilation of common algal control and management techniques, tech. report E-80-1, U.S. Army Corps of Engineers, Vicksburg, MS.

Klapper, H. 1980. Experience with lake and reservoir restoration techniques in the German Democratic Republic. *Hydrobiol* 72:31-41.

Likens, G.E., ed. 1972. *Nutrients and Eutrophication*. American Society of Limnology and Oceanography, special symposium, Lawrence, KS: Allen Press.

Likens, G.E., and Bormann, F.H. 1974. Acid rain: A serious regional environmental problem. *Science* 184:1176-1179.

Mullison, W.R. 1981. Public concerns about the herbicide 2,4-D. Dow Chemical Co., Midland, Michigan.

National Academy of Sciences. 1969. *Eutrophication: Causes, Consequences, Correctives*, proceedings of a symposium, Washington, DC. Sponsored by NAS.

Peterson, J.O., Born, S.M., and Dunst, R.C. 1974. Lake rehabilitation techniques and experiences. *Water Res Bull* 10:1228-1245.

Rast, W., and Lee, G.F. 1978. Summary analysis of the North American (U.S. portion) OECD eutrophication project: Nutrient loading-lake response relationships and trophic state indices. USEPA 600/3-78-008.

Rich, P.H., and Wetzel, R.G. 1978. Detritus in the lake ecosystem. *Am Nat* 112:57-71.

Schindler, D.W. 1974. Eutrophication and recovery in experimental lakes: Implications for lake management. *Science* 184:897-899.

Shearer, R. 1980. Public health effects of the aquatic use of herbicides—2,4-D, dichlobenil, endothall, and diquat. In *Literature Reviews of Four Selected Herbicides: 2,4-D, Dichlobenil, Diquat, and Endothall*. Seattle, WA: Municipality of Metropolitan Seattle (METRO).

Simsiman, G.V., Chesters, G., and Daniel, T.C. 1972. Chemical control of aquatic weeds and its effect on the nutrient and redox status of water and sediments. *Proceedings of the 15th Annual Conference on Great Lakes Research*, 1972:166-180.

Stewart, K.M., and Rohlich, G.A. 1967. *Eutrophication—A Review*. Pub. 34, State Water Control Board, State of California, Sacramento, CA.

Tenney, M.W., Yaksich, S.M., and De Pinto, J.V. 1972. Restoration of water bodies. In B.G. Liptak, ed., *Environmental Engineers Handbook*. Philadelphia: Chelton Press.

Uhlmann, D. 1982. Evaluation of strategies for controlling eutrophication of lakes and reservoirs. *Int Rev Ges Hydrobiol* 67:821-835.

U.S. Environmental Protection Agency. 1977. *North American Project—A Study of U.S. Water Bodies*. EPA 600/3-77-086.

U.S. Environmental Protection Agency. 1979. *Lake Restoration*. EPA 440/5-79-001.

U.S. Environmental Protection Agency. 1980a. *Restoration of Lakes and Inland Waters*. EPA 440/5-81-010.

U.S. Environmental Protection Agency. 1980b. *Clean Lakes Program Guidance Manual*. EPA/400-5-81-003.

U.S. Environmental Protection Agency. 1983. *Lake Restoration, Protection and Management*, EPA 440/5-83-001.

Vollenweider, R.A. 1968. *Scientific Fundamentals of the Eutrophication of Lakes and Flowing Waters, with Particular Reference to Nitrogen and Phosphorus as Factors in Eutrophication*, OECD Rept. DAS/CSI/68.27, Organization for Economic and Cooperative Development, Paris.

Wanielista, M.P. 1978. *Stormwater Management. Quantity and Quality*, Ann Arbor: Ann Arbor Science Publishers.

Welch, E.B. 1980. *Ecological Effects of Wastewater*. London: Cambridge University Press.

Wetzel, R.G. 1983. *Limnology*, 2nd ed., New York: Saunders.

2

Limnology, Lake Diagnosis, and Selection of Restoration Methods

INTRODUCTION

Lake managers, consultants, and others need a basic understanding of limnology, at least as it pertains to the eutrophication process, in order to make cost-effective decisions about lake and reservoir management and restoration. Errors in understanding can lead to expensive and frustrating errors in the choice of management and restoration alternatives. This chapter provides, first, a basic outline of the main ideas of limnology that are significant to eutrophication. A brief review of the basic differences between lakes and reservoirs is presented, along with a description of the forces, both external and internal to the water body, that can promote problems in lakes and reservoirs. Procedures to diagnose a lake's condition and to obtain the data necessary to select a lake or reservoir restoration alternative are described. Finally, suggestions are presented for preparing a report about a lake restoration.

Many readers will be thoroughly familiar with the ideas presented in this chapter and may wish to proceed directly to the sections on the various restoration methods. This chapter is not meant to be a substitute for the in-depth study of limnology that is needed to make competent and effective decisions. Instead, we hope to provide an outline or guide to supplement the knowledge of persons in related fields, such as engineering, and to be a beginning for those attempting to become familiar with this field. The reader is referred to Ruttner (1963), Reid and Wood (1976), Cole (1983), Goldman and Horne (1983), and especially to Wetzel (1983).

LAKES AND RESERVOIRS

Natural lakes and man-made impoundments of streams may both become eutrophic from increases in the incomes of substances to them. Some newly constructed reservoirs may be almost instantly eutrophic from the decomposition of materials left in the flooded basin, or from rich soils. Similarly, many natural lakes have been at least moderately productive from their inception because of the fertile soils of their drainage basins.

Lakes and reservoirs have fundamental similarities in terms of basic biological and some physical processes. They can have identical fauna and flora, both can stratify thermally, and they can have the same types of habitats. Thus we can expect that aspects of the eutrophication process will be fundamentally the same in these two types of systems and that we can transfer some basic understandings gained in one system to the other.

There are, however, significant differences between natural lakes and reservoirs, and these differences may alter the choices of restoration procedures possible. Reservoirs usually have greater drainage basin areas, making control of income of material very difficult, especially if the drainage basin has several different municipal or agricultural discharges from politically separate units. Also, reservoirs usually have greater surface area, greater mean and maximum depth, shorter water residence time, and a greater areal water load. Thornton et al. (1980) have summarized these differences (Table 2-1). Kennedy et al. (1985) have emphasized the significant longitudinal gradients found in reservoirs, ranging from the light-limited, turbidity-dominated, upper riverine zone to the pool zone at the dam, which may behave like a natural lake. These differences between reservoirs and natural lakes, especially between reservoirs and the very common small (less than 100 ha) lakes, emphasize the need to recognize that the study and management of the eutrophication process in these systems will often be different as well.

THE EUTROPHICATION PROCESS

Lakes and reservoirs are usually consid[ered ...] [p]roduction of algae, rooted and floating plants, or [...] conditions such as taste and odor in potable water [...] recreation. It has recently been recognized that o[...] [m]atter from algal extracellular products, in raw wat[er ...] [treat]ment plants can produce organohalides or trihalomethanes [...] [treatm]ent process (e.g., Wachter and Andelman, 1984). Drinking water with these substances may have increased mutagenic potential. Excessive biological production in the forms of algae and macrophytes is caused by interactions of the following factors: characteristics of the drainage basin, such as land use and area; morphometric features of the lake or reservoir basin; the thermal history of the water body; and biological processes. These factors are briefly summarized and then combined in a pictoral model of the eutrophication process.

Definitions

We have modified the definition of eutrophication by Likens (1972) to include the income of silt. Eutrophication, then, is the process of excessive addition of inorganic

Table 2-1 Comparison of Geometric Means of Selected Variables of Natural Lakes and Army Corps of Engineers Reservoirs (modified from Thornton et al., 1980)

Variable	Natural Lakes (N = 309)	Reservoirs (N = 107)	Probability that Means are Equal
Drainage area (km^2)	222.00	3228.00	< 0.0001
Surface area (km^2)	5.60	34.50	< 0.0001
Maximum depth (m)	10.70	19.80	< 0.0001
Mean depth (m)	4.50	6.90	< 0.0001
Hydraulic residence time (yr)	0.74	0.37	< 0.0001
Areal water load (m yr^{-1})	6.50	19.00	< 0.0001
Drainage/Surface area	33.00	93.00	< 0.0001
P loading (gm m^{-2} yr^{-1})	0.87	1.70	< 0.0001
N loading (gm m^{-2} yr^{-1})	18.00	23.00	< 0.0001

nutrients, organic matter, and/or silt to lakes and reservoirs, leading to increased biological production and a decrease in volume.

Traditionally, eutrophication has referred only to the income of inorganic plant nutrients, their excessive concentration in the water, and the high production of algae that may occur as a consequence of that concentration. We wish to emphasize, following the views of Likens (1972) and Wetzel (1983), that organic matter income may also lead to a loss of volume and to the release of nutrients, through mineralization and through release from sediments, as dissolved oxygen is depleted. Silt, which can be rich in sorbed nutrients and organic matter, can also contribute to loss of volume through the development of nutrient-rich sediments. No matter how volume loss is produced, the development of substantial shallow areas can foster the growth of macrophytes, the rooted and floating vascular plants and their attendant epiphytic and filamentous algae. Ultimately these plants promote further losses of dissolved oxygen and release of organics and nutrients, as they decay (Carpenter, 1980; 1981). Many eutrophic lakes and reservoirs, therefore _____ water bodies with algal problems. It is common to find them to be _____ _____ sediments and extensive areas of macrophytes. The sign_____ _____ to the eutrophication process remains incompl_____ _____ and reservoirs also have colored water (green/br_____ _____ ygen in the deepest areas. Biological productivi_____ _____ ecific phytoplankton populations might exist. F_____ _____ igh it may be dominated by less desirable species. _____

An oligotrophic lake or reservoir is _____ _____ ctivity. This can occur because of low income of nutrient sub_____ _____ of that income by large basin volume and short water residence time, or it may occur because of extreme water hardness, which fosters coprecipitation of essential algal nutrients with calcium carbonate (e.g., marl lakes). Oligotrophic lakes are usually deep, with nutrient-poor sediments, few macrophytes, and large amounts of dissolved oxygen in the deepest water. Diversity of phytoplankton can be high.

Sources of Nutrients, Organic Matter, and Silt

The potential of a lake or reservoir to become eutrophic is greatly influenced by the nature of the drainage basin, because it is usual [...] urce of the nutrients, organic matter, and silt that c [...] hese materials are classified as either "point" or "[...] al in origin, usually pipes from sewage treatment p [...] stry. Nonpoint sources, which are far more difficu[...] trol, include inputs from bank erosion, fertilizer ar [...] ater.

[handwritten annotation: Watershed / Drainage / Input]

Materials enter a water body from biolo [...] gical sources (Likens and Bormann, 1974). Meteoro [...] ow, is rarely a major source except in remote area [...] soil is igneous (low nutrient output) and the ratio of the area of the drainage basin to area of the lake is 10:1 or less, direct precipitation can be a major part of the income of nutrients. Similarly, in the case of forested, sedimentary soils when the ratio is 3:1 or less, direct precipitation can be important (Rigler, 1974).

Biological sources are often significant as well. Animal feedlots produce a tremendous nonpoint load of nutrients in forms readily available for plant uptake. Discharges from sewage treatment plants (STP) following primary or secondary treatment are very high in nutrients. Untreated or primary-treated sewage contains much dissolved organic matter and creates high oxygen demand (Welch, 1980).

The hydrogeological loading of materials to a water body, while including cultural sources along the stream such as sewage effluent, contains mostly "nonpoint" loadings, from agriculture, marsh discharges, forest runoff, and groundwater. Some drainage basins yield more nonpoint loading than others. The amount of yield is a function of soil types, land use (e.g., urban versus forest), vegetation cover, slopes, and climate.

The effects of land use on stream quality and discharge to a lake or reservoir must be emphasized. Omernik (1977) analyzed drainage-basin characteristics and stream runoff data for a nationwide group of 928 watersheds characterized by nutrients from nonpoint sources. He found that agriculture produces much higher stream nutrient concentrations than forests. Table 2-2 illustrates the relationship between the percentages of drainage basin in urban and agricultural use combined and the mean total phosphorus and nitrogen concentrations for streams. There is a nearly ten-fold difference in concentration between streams without agricultural-urban drainage and those with 100 percent of their discharge coming from these sources. Figures 2-1 and 2-2 illustrate even more clearly that urban-agricultural land use, which is predominant in the eastern and midwestern areas of the United States, is a very significant nonpoint source of nutrients to streams.

Deforestation, perhaps through logging or subdivision development, sharply increases water and nutrient yield from the drainage basin. This was clearly demonstrated in a deforestation-reforestation project in the Hubbard Brook Valley ecosystem in New Hampshire (Bormann and Likens, 1979). In the deforested watershed, annual streamflow increased between 26 and 41 percent. Soils remained wet, due to an absence of evapotranspiration, which produced significant soil storage of water. Coincident with increased streamflow, loss of nutrients and particulate matter from the soil increased.

Table 2-2 Estimated Mean Total Phosphorus and Total Nitrogen Concentrations (mg l^{-1}) for the Eastern U.S. (modified from Omernik, 1977)

% Agricultural + % Urban	Average Total P	67% Limit*	Average Total N	67% Limit*
0	0.015	(0.008-0.026)	0.51	(0.33-0.77)
25	0.026	(0.014-0.048)	0.87	(0.57-1.32)
50	0.045	(0.024-0.083)	1.49	(0.98-2.26)
75	0.078	(0.042-0.144)	2.55	(1.68-3.88)
100	0.136	(0.073-0.251)	4.37	(2.87-6.64)

*There is a 67% probability that the true value falls within this range.

The loading of silt to a lake or reservoir can be very significant, since this increases the area of shallow water and the amount of nutrient-rich sediments. This may promote weed growth, release of dissolved and particulate organic matter, and loss of dissolved oxygen. Erosion of the Hubbard Brook forest was curtailed as long as the forest floor remained intact following deforestation. Plowing or the use of construction equipment to remove the forest floor produced a rapid loss of soil. The tonnage of silt or suspended solids from urban and agricultural areas can be very large. Weibel (1969) reported that the average concentration of suspended solids in urban stormwater runoff from an 11-hectare (27-acre) residential-light commercial area in Cincinnati, Ohio, was 641 kg ha^{-1} yr^{-1} (183 tons mi^{-2} yr^{-1}). Piest et al. (1975) estimated an average loss from an Iowa corn-cropped watershed over a six-year period to be 74,993 kg ha^{-1} yr^{-1} (21,410 tons mi^{-2} yr^{-1}), more than 100 times that of the urban-commercial area. In contrast, the yield of sediment from an undisturbed pre-Cambrian Shield forest was 0.005 kg km^{-2} yr^{-1} (19 tons mi^{-2} yr^{-1}) (Naiman, 1982).

The income of dissolved and particulate organic matter also has a decided effect of enhancing productivity in lakes and reservoirs. Organic substances contain phosphorus, nitrogen, and other elements that can become available to plants for growth, following mineralization. Also, dissolved organic molecules may stimulate photosynthesis (Francko and Wetzel 1981a; b). Undecomposed organic matter contributes to a loss of basin volume. Finally, imported organic substances add to the oxygen-demanding materials produced in the lake or reservoir. Sewage discharges, unless given extended aeration or other BOD-consuming processing, are extremely high in dissolved and particulate organics as well as in available nutrients. Leaves from the terrestrial community, as well as stream-produced biomass, can also add to the organic load.

Basin Morphometry

The response of a lake or reservoir to loadings of materials from the land is directly coupled to the area, depth, and shape of its basin, as well as to the rate of its nutrient and water loading (Wetzel, 1983). Figure 2-3 is a depth-area hypsograph of two

14 Overview

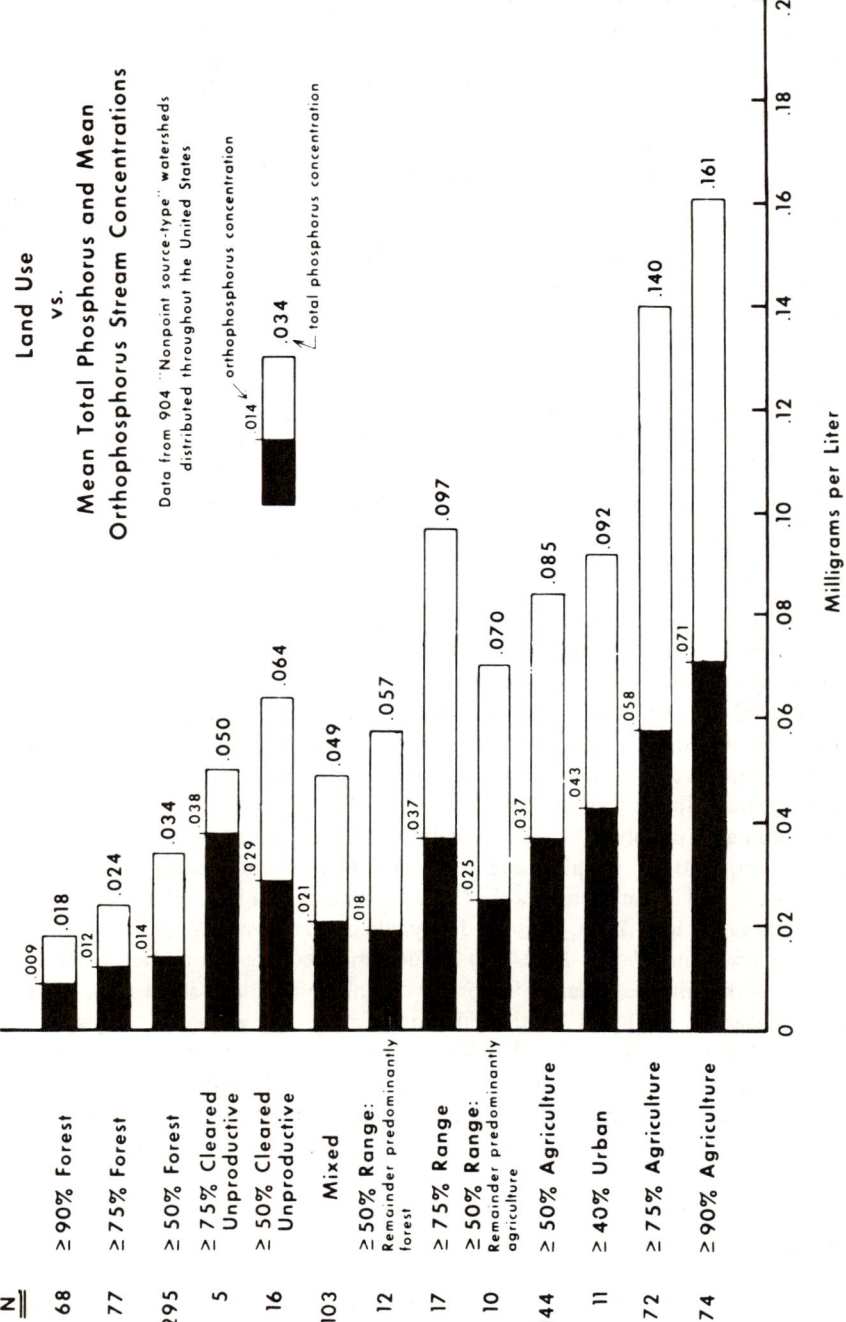

Figure 2-1 Relationships between general land use and total phosphorus and orthophosphorus concentrations in streams (from Omernik, 1977).

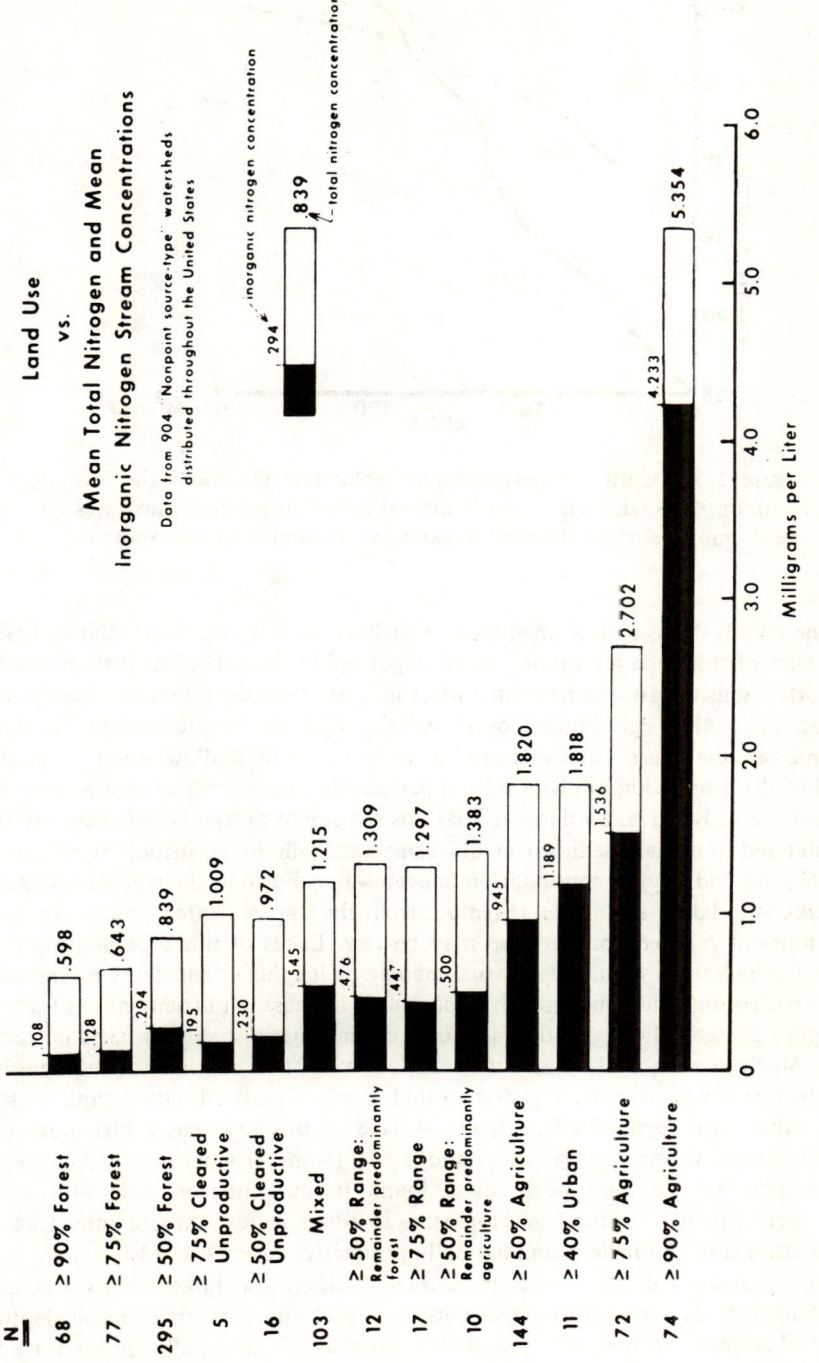

Figure 2-2 Relationships between general land use and total nitrogen and inorganic nitrogen concentrations in streams (from Omernik, 1977).

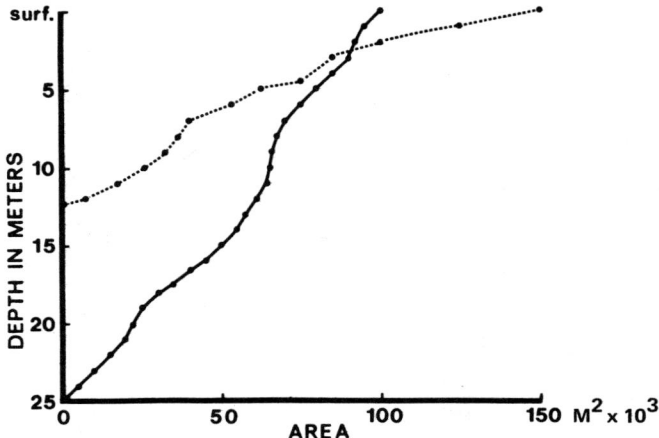

Figure 2-3 Depth–area hypsographs. Solid line illustrates the less common deep lake with a small littoral zone; dotted line illustrates the common shallow lake with extensive littoral area and volume.

lakes. One lake is deep, with a small area of shallow water and a large volume. Lakes or reservoirs of this type are often labeled oligotrophic. This lake has little potential to support extensive growths of rooted macrophytes ("weeds") because these plants would become either light-limited or limited by hydrostatic pressure in the deep water and because there usually is only a limited area of shallow water. This also means that the production of dissolved and particulate organic matter by macrophytes will be small relative to lake volume so that this subsidy of nutrients and energy to the open water and to the lake sediments will be small and will not contribute significantly to algal blooms and oxygen consumption in deep water. Further, the regenerative zone of the lake, the lake's sediments, is remote from the upper, lighted productive zone so that nutrient regeneration to algae may be low. Lakes of this type may support large algal populations when the external income or loading of nutrients is large and the water detention time long enough to permit an increase in nutrient concentration.

Figure 2-3 also illustrates the far more common basin shape for lakes in North America. Much of such a lake's area is in shallow water (<3 m) and the regenerative and productive zones are close together so that one is subsidized with organic matter and the other with regenerated nutrients. A lake of this type has a high potential to support extensive macrophyte areas and algal blooms. It would rarely be labeled as oligotrophic or nutrient poor, unless found in nutrient-poor soils with small drainage area and little cultural development. In other words, the common lake of North America is small, shallow, and naturally productive (Wetzel, 1983).

The "permanence" of a lake or reservoir basin is also linked to its area and shape. Many natural lake basins are essentially round and have small mean depths. Lakes of this type are not only productive but may fill in rapidly, in contrast to basins of identical shore lengths with larger volumes. Kerekes (1977) has called the ratio of basin volume to shoreline length the "index of basin permanence." A lake with an index ≤ 0.1 usually has abundant macrophytes and high productivity, and is senescent. Some reservoirs have vast shoreline lengths, several basins, and several entering streams. Their volumes, however, are small relative to shoreline lengths (low index of permanence) and many also have abundant macrophytes and high productivity. Lakes and reservoirs with low permanence index numbers, including the type illustrated in Figure 2-3, are basins that will be difficult to protect and restore.

Thermal Stratification and Dissolved Oxygen Distributions

One of the most significant determinants of the physical, chemical, and biological interactions of a lake or reservoir is its annual temperature cycle (Wetzel, 1983). In temperate climates, deep basins (≥ 7-10 meters deep) stratify thermally during summer months into a deep, dark, stagnant, cold layer (termed the hypolimnion) and an upper, lighted, circulating, warm layer (termed the epilimnion). These zones are separated by a band of water in which temperature declines rapidly with depth, the metalimnion (Figure 2-4). They remain thermally stratified through the summer. Thorough descriptions of how this normal thermal stratification occurs are found in Ruttner (1963) and Wetzel (1983).

The significance of thermal stratification to the eutrophication process and to lake or reservoir restoration lies in separation of the upper and lower lake zones during the summer through differences in water density (cold water being heavier).

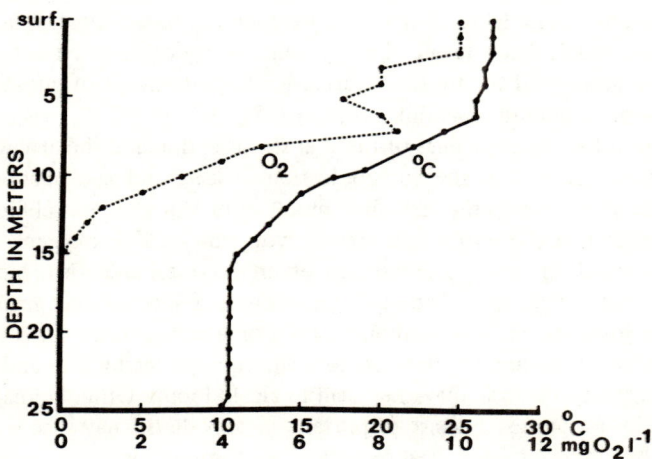

Figure 2-4 Distribution of temperature (solid line) and dissolved oxygen (dotted line) during summer thermal stratification of a eutrophic lake.

In eutrophic lakes and reservoirs the high content of organic matter in the lake sediments, originating from the production of algae and macrophytes as well as from income via streams, provides an excellent source of energy for bacteria, fungi, Protozoa, and some invertebrate animals. The respiration of such organisms consumes the oxygen that was stored in the hypolimnion at the time of thermal stratification in the spring and releases some of the elements bound in the organic matter. There may also be extensive chemical oxygen demand. As oxygen is reduced, at a rate called the lake's oxygen deficit rate, the ability of the sediments to retain elements such as phosphorus in iron-hydroxy complexes is altered and these elements are released to the hypolimnetic water. Figure 2-4 illustrates the typical summer distribution of dissolved oxygen by depth in an eutrophic lake. At a low dissolved oxygen concentration, phosphorus, usually the most limiting nutrient, is readily released from its association with iron (III) into the water column (see review by Bostrom et al., 1982).

As summer progresses, phosphorus, nitrogen, and other nutrients increase in concentration in the hypolimnion and frequently are transported vertically into the lighted, warm epilimnion where they subsidize algal growths, a process called internal nutrient loading. Stauffer and Lee (1973) found that cold summer rains depress the metalimnion as the cold, more dense water of the rainstorm sinks. A return to windy, warm weather brings the metalimnion back toward the surface, enriched with hypolimnetic phosphorus. All of the phosphorus necessary to support the series of summer algal blooms in phosohprus-limited Lake Mendota, Wisconsin could be accounted for through this process of vertical entrainment of sediment-released phosphorus during thermal stratification, coupled with low dissolved oxygen.

Thus one process of nutrient regeneration and recycling is governed directly by the existence of thermal stratification and the associated loss of dissolved oxygen in eutrophic reservoirs and lakes. A significant percentage of the increase in phosphorus content of a eutrophic lake during the dry summer months, when external loading is small, can be accounted for by the internal loading or release of phosphorus in the lake's oxygen-free hypolimnion (Cooke et al., 1977).

There are other consequences of thermal stratification and the attendant oxygen losses that are significant to the eutrophication process and lake restoration. Fish, particularly the predatory game fish that prefer cold water, are excluded from the hypolimnion and thus the entire lake or reservoir due to low oxygen. This loss of predators may result in great increases in other, less desirable fish, including zooplanktivorous and rough fish. Zooplankton, including species that graze the algae, cannot use the low-oxygen hypolimnion as a daytime refuge from these sight-feeding planktivorous fish. Elimination of these zooplankton species by fish and subsequent low grazing mortality of algae may then lead to algal blooms. Other animals, including some benthic invertebrates that are important in fish diets, may also be eliminated from the lake by absence of dissolved oxygen in the hypolimnion.

Water withdrawn from an anoxic hypolimnion for use as drinking water may have serious problems of taste, odor, and color. This is often caused by high concentrations of iron and manganese, which are released from the sediment in the reducing environment of an oxygen-free hypolimnion, as well as by algal excretory products.

Some shallow lakes and reservoirs circulate continuously from top to bottom. In these water bodies calm, warm weather conditions produce temporary thermal stratification which can be easily disrupted during summer wind storms. When temporary stratification occurs, dissolved oxygen may be rapidly consumed over the enriched organic sediments and phosphorus will be released into the water. A subsequent breakdown of stratification then circulates phosphorus to the upper, lighted water, to stimulate an algal bloom. In other shallow lakes, high pH or microbial decomposition controlled by temperature can promote phosphorus release from oxic sediments (Welch et al., 1982; Jacoby et al., 1982; Bostrom et al., 1982).

Phosphorus

The quantity or biomass of algae in a lake or reservoir, which is an indicator of the degree of eutrophication, is limited (usually) by the concentration of an essential element or nutrient (the limiting nutrient concept). It has been amply demonstrated, in experiments ranging from laboratory bioassays to fertilization of in situ enclosures to whole-lake experiments, that most often phosphorus is that limiting nutrient. In some lakes nitrogen, silicon, and carbon have been, at times, limiting nutrients. Sakamoto (1966) was the first to report the linear relationship of a log-log plot of phosphorus and algal chlorophyll (a measure of algal biomass). Figure 2-5, from Dillon and Rigler (1974a), illustrates this relationship. It should be noted that this is a log-log plot of the data. There is considerable scatter to the linear data, indicating effects from other factors in the pelagic environment, such as light, nitrogen, or zooplankton grazing, in limiting algal biomass. This also indicates that there can be wide variation in the expected chlorophyll from any given phosphorus concentration.

Phosphorus is an element with no gaseous component in its biogeochemical cycle. Its concentration in the water is therefore determined by its rate of income and sedimentation to the lake bottom, by the rate of water renewal (phosphorus loss through the outlet), by dilution in the lake or reservoir basin, and by the rate of release of phosphorus from sediments, macrophytes, and other sources inside the basin. The preceding sections of this chapter have outlined external sources of nutrients, organic matter, and silt, the effect basin morphometry may have on this loading, and the role of thermal stratification in internal phosphorus release. These relationships of loading, dilution, sedimentation, and wash-out are treated in more detail in a later section of this chapter, using mass balance models to describe the determinants of phosphorus concentration.

We will focus on phosphorus rather than other elements, not only because it is often the limiting nutrient for further growth of algae, but because its concentration can be more easily controlled than elements with gaseous phases to their biogeochemical cycles. Nitrogen, for example, can be obtained from the atmosphere by nitrogen-fixing blue-green algae and bacteria. Nitrogen-fixers are common among the nuisance species of algae in eutrophic lakes. Carbon has been shown to diffuse into the water column at rates sufficient to meet the needs of photosynthesizing cells. Phosphorus, on the other hand, cycles between living and nonliving particulate forms and the dissolved form, the only form of phosphorus cells can incorporate.

20 *Overview*

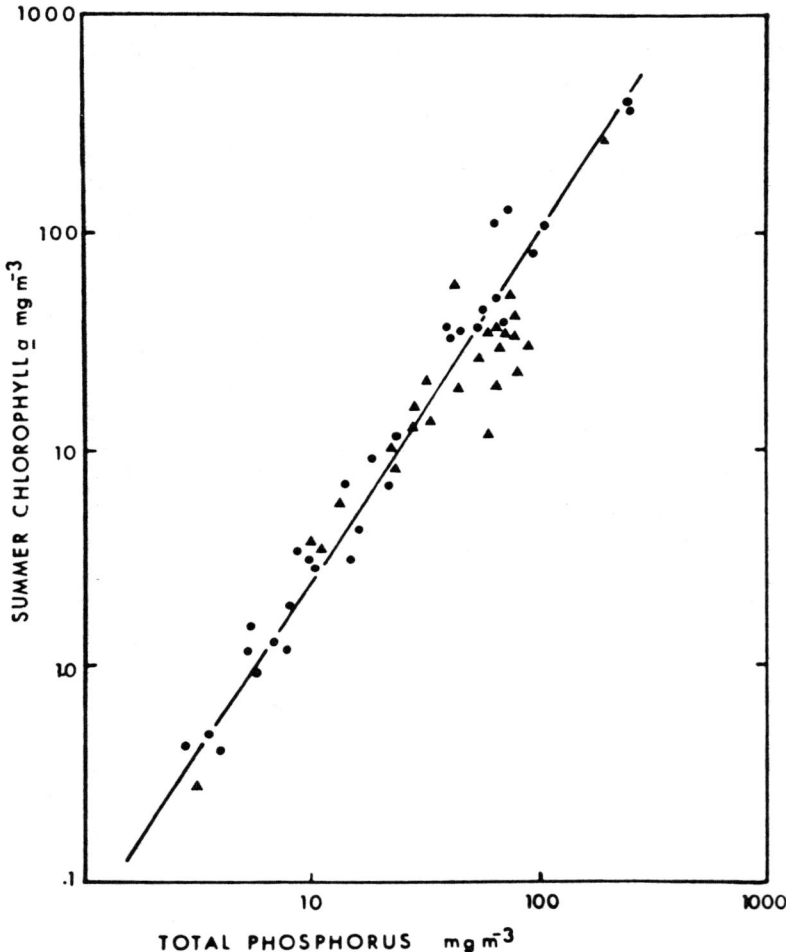

Figure 2-5 Summer average chlorophyll versus total phosphorus concentration at spring overturns log-log plot (from Dillon and Rigler,, 1974a).

Phosphorus income can be curtailed, for example, by precipitation with iron, aluminum, or calcium salts in an STP, and its concentration in the water can be lowered as a result of combining it with aluminum salts. The atmosphere as a source of phosphorus is not an important factor in lake management.

The Biological Communities of Lakes and Reservoirs

Lakes or reservoirs have three zones, which contain distinct but interacting biological communities—the littoral zone, the pelagic, and the benthic. The littoral zone, which contains what is known popularly as "weed beds," is actually the interface between the land and the lake. Its extent is governed by the area of fertile, shallow water sediments (recall the hypsographs of Figures 2-3), and the rates of silt deposition from the land and accumulation of partially decayed plant material. Many lakes have extensive littoral areas, and the production of organic matter by these zones is very significant in the cycling of nutrients and in the nutrition of lake organisms. In particular, the large annual production of rooted plants and attached algae may be a major source of decomposing organic matter, and thus nutrients, to the open water and to the sediments of the hypolimnion (Carpenter, 1980; 1981; 1983a;b). Not only are the sediments of the littoral zone the source of nutrients for these rooted plants, the plants may release nutrients to the water column via aerobic decomposition.

The pelagic zone, or open water area, is characterized by freely floating organisms—the zooplankton and phytoplankton—and by certain fish species. This plankton community obtains energy from sunlight and from dissolved and particulate organic matter transported from the littoral zone by water movements. Nutrients enter the pelagic zone from the littoral zone and by vertical entrainment from the nutrient-rich hypolimnion. The pelagic community can be dominated by one or a few highly adapted species of algae, which in eutrophic lakes are often the scum-forming blue-green algae. In some cases it appears that these algal blooms may be due as much to the absence of zooplankton grazing on them, following planktivorous fish feeding, as to the abundance of nutrients in the water.

The benthic community, the organisms living in and on the bottom sediments, is subsidized by organic matter from the littoral and pelagic communities. The deep-water benthos may have a substantial community of insects, molluscs, and oligochaetes, although in eutrophic lakes these are reduced to a few low-oxygen-tolerant species. The dominant organisms are microbes that convert dissolved and particulate organic matter into energy for themselves and release nutrients that can be reused by plants when transported to upper, lighted waters.

A Pictorial Summary of the Eutrophication Process

Figure 2-6 is a pictorial model that summarizes the interactions of the eutrophication process. Circles A and B represent the incomes of organic matter and silt and the income and concentration of nutrients. Box C is the central, in-lake compartment containing nonliving dissolved and particulate organic matter from external loading and from the excretion and death of biota. Boxes D, E, and F illustrate that the biotic components are stimulated by, consume, and produce organic matter. Algal growth is stimulated by nutrients (B), while macrophytes tend to become extensive where organic matter and nutrient-rich silt have deposited. Both plant groups, but particularly planktonic algae, may be grazed upon and controlled by animals. Boxes G, H, and I represent lake or reservoir problems (including problems with recreational use of the water). Finally, the model illustrates that internal processes such as nutrient regeneration from sediments (I to B) and the deposition of undecomposed organic matter (C to G) can promote algal and macrophyte production without substantial incomes from the land. Once a lake or reservoir becomes nutrient-rich and shallow enough

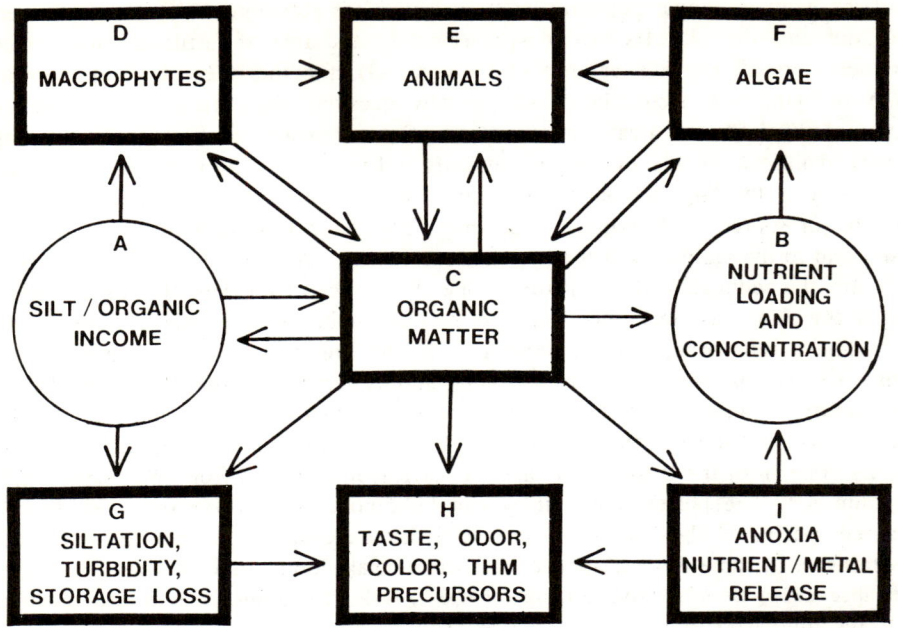

Figure 2-6 Primary interactions and effects of eutrophication process. THM-trihalomethane.

or develops an extensive anoxic hypolimnion which regenerates nutrients, it may remain eutrophic long after diversion of the excess nutrient incomes.

Eutrophication, then, is a far more complicated process in most lakes and reservoirs than just the income of nutrients and excessive production of algae. The simple pictoral model of Figure 2-6 illustrates the processes and compartments that are the targets of lake restoration or lake improvement techniques. One group of procedures, which we place under the heading of procedures to control nutrient concentration, aims at ways to reduce the impact of Circles A and B or the regeneration of nutrients to Circle B. The other group of techniques, procedures to control biomass, aims at Boxes D, E, and F. Both sets of procedures, where successful, will reduce the significance of Box C to nutrient regeneration and promotion of macrophytes.

LAKE OR RESERVOIR EVALUATION

Before any lake or reservoir restoration method can be implemented, the uses of the lake and the causes of its problems must be evaluated. Four questions are asked prior to any field evaluation. This initial examination of the lake and its problems should be done with the lake owners, the homeowner association, or some person knowledgeable about the lake or reservoir. The questions are:

1. What are the primary uses or desired uses of the lake? The approach to understanding and solving the lake's problems may vary with the answer to this question. For example, a single-purpose lake (e.g., fishing) may need an entirely different solution to its problems than a comparable multipurpose lake.

2. What are the problems of this lake (e.g., algal blooms, turbidity, excessive macrophytes, etc.), and what condition or trophic state do the users desire? Is the desired state realistic?

3. What is the history of this lake? When was the dam constructed, homes built, fish stocked, treatment plant constructed, etc., and how have the symptoms of eutrophication been treated in the past? Are there previous reports or other data available?

4. Is restoration or lake improvement feasible? For example, a small lake community may receive extensive sewage discharge to their lake from an upstream municipality. The lake users may be economically or politically powerless to have this source diverted or treated. If restoration is not possible, can any improvements be made to alleviate the symptoms?

The answers to these questions will help in the design of a diagnosis and feasibility study that will be sufficiently comprehensive, and yet will emphasize an analysis of this lake's particular problems and desired uses.

Design of a Diagnosis and Feasibility Study

The first step in prescribing a restoration or lake improvement technique is to perform a diagnosis and feasibility study of the lake or reservoir and its watershed. The purposes of such a study are (1) to determine the sources and rates of nutrient-silt-organic matter income or loading; (2) to determine the state or condition of the lake or reservoir; and (3) to determine the most effective technique(s) for correcting the problem and for providing long-term protection and restoration of the lake.

What variables should be measured in the study, how often, and why? We have divided these measurements into those for the lake (physical, chemical, and biological) and those for the watershed, or the loading determination part of the study. The reader is referred to Walker (1985) for general and specific guidelines for development of a monitoring, data reduction, and data analysis program. Three computer programs, FLUX, PROFILE, and BATHTUB are described for reduction and analysis of tributary and reservoir discharge monitoring data, in-reservoir water quality data, and application of empirical eutrophication models.

The In-Lake Study

Physical Measurements. Accurate measurements of lake area and depth contours are essential, because from these data the variables of area, mean depth, and volume are calculated for modeling and for the development of a hypsograph and an index of basin permanence. Depth-contour maps may be made by manual soundings of with a recording fathometer (see Lind, 1979, and Wetzel and Likens, 1979, for details on how to map a lake).

Water temperature should be determined at a deep-water station at one-meter intervals over the year, with determinations made at least twice per month during spring and summer. Secchi disc transparency should be determined at least twice monthly during the summer. In a reservoir or multibasin lake, determine the temperature profile at each location established for determination of water chemical characteristics.

Water Chemistry Determinations. A deep-water station should be established. In large lakes, and particularly in reservoirs, there is a gradient in the content of chemical characteristics from inlet to outlet, so several stations are necessary. The purpose of the study is to estimate a mean concentration for the whole lake volume on each date, plus annual and seasonal means, depending upon the variables and specific needs for the data. One way of selecting station locations in reservoirs is to place at least one station in each reservoir zone (riverine, transition or mixing, and pool), or if a multibasin lake or reservoir is under study, a station may be established near the deep area in each basin. A single station is insufficient in a reservoir or in many lakes.

The following represent the minimum combination of variables to be determined at least twice monthly over the summer and at monthly intervals over the year:

1. Dissolved oxygen at meter intervals to bottom (avoid inclusion of bottom sediments).
2. All phosphorus fractions (at least total and soluble reactive phosphorus) at a minimum of three intervals from surface to bottom in shallow lakes and at 3-meter to 5-meter intervals in deep lakes.
3. Nitrogen, at least nitrate, at the same depth intervals.
4. Surface pH and alkalinity, plus monthly determinations at depth intervals as for phosphorus fractions.
5. Lake sediment samples (cores) for determination of available, organic, and refractory phosphorus fractions (see methodology outline in Bostrom et al., 1982).

There are many ways in which these data will be used. The following are examples. The chapters on lake restoration methods will suggest other uses. Determination of phosphorus concentration is essential, since these data will be used to validate model predictions, to compute a lake trophic-state index before and after restoration, to calculate internal phosphorus loading rate from mass balance equations, and to assess the potential nutritional state of algae. Sediment phosphorus fractionation may indicate which mechanisms are important for internal loading (e.g., Bostrom et al., 1982). The total nitrogen-total phosphorus ratio is one way to assess whether phosphorus is the limiting nutrient. According to Smith (1979), an $N:P$ ratio greater than 21 strongly suggests that phosphorus is the limiting element, whereas when the ratio is less than 13 nitrogen may be the limiting factor. Lakes with intermediate ratios could be limited from time to time by either element, but by controlling phosphorus, it could be made the limiting factor. Dissolved oxygen data can be used to assess a lake's trophic state, to calculate an oxygen-deficit rate, and to help explain nutrient dynamics. The pH–alkalinity data will be needed to compute a proper dose of aluminum sulfate if nutrient inactivation is chosen as a lake restoration method, as well as to understand sediment phosphorus release and the factors that promote dominance by blue-green algae.

Biological Determinations. The biological portion of a diagnostic/feasibility study is the portion most likely to be inadequate, even though eutrophication is a biological problem and may have biological solutions. The books by Lind (1979), Wetzel and Likens (1979), and APHA (1980) are among the several useful guides to biological methods in limnology. The following determinations should be made at each of the stations established for water chemical variables and taken at least twice monthly:

1. Surface phytoplankton, reported as cell volume for each species or genus. Taxonomic identifications to genus have far less value. At least monthly during the summer season, samples at depth should be taken using a Van Dorn bottle.
2. Chlorophyll *a* in surface samples, as well as at the depths within the epilimnion sampled for other variables.
3. Zooplankton samples at the surface and in the metalimnion and hypolimnion, with a transparent Van Dorn bottle. Determine the abundance of each species of microcrustacea. Counts of zooplankton genera have far less interpretive value.
4. Species of macrophytes present and their individual areal coverage.
5. Fish species and their relative abundance. This can be done by talking with fishermen, examining creels, or by a gill net catch.
6. Benthos sampling at least twice during the summer, with a minimum of three dredge hauls per station. List the genera and determine their abundance.

The biological portion of the diagnosis and feasibility study can be the most formidable to consultants and managers since few engineering firms employ people with the necessary training. An adequate study is mandatory, however, and limnological consultants should be hired.

The following examples of the use of these data will indicate their importance. In the chapters on each type of in-lake restoration technique, considerable additional detail will be given about the uses of the biological data. The cell-volume data and phytoplankton species list will document the nature and severity of the problem. The nature of the seasonal succession of species can provide significant information about the importance of zooplankton grazing and about the probability of success with artificial circulation. The chlorophyll *a* data will be used to compute a trophic state index, and as part of a predictive model of lake recovery. In combination with phytoplankton and fish data, an analysis of zooplankton species composition, abundance, and distribution can indicate the impact (or lack of impact) of grazing on algae, and can form the basis for developing a recommendation for restoration through biological controls. Lake-level drawdown is often recommended as a means of macrophyte control. Since only some species are susceptible to this method and others are enhanced, an accurate species list is needed. In other lakes, knowledge of the area of macrophytes that could be harvested may be important. The benthos samples, along with the chemical characteristics of the sediment, will be needed to assess the impact of a sediment removal project.

The Watershed Study

Watershed Mapping. The first step is to obtain the relevant government geological survey topographic maps of the area and delineate the drainage basin. This is done by studying the contours on the map and by a field check of the actual drainage area to be certain that it is identical to that chosen by examination of the map. Significant hydrological changes, such as the construction of an upstream real estate lake, may have occurred since the topographic map was made.

Identify all of the point-source discharges, particularly storm and sanitary sewer discharges. Storm drainage pipes are often installed to discharge below the lake surface, making detection difficult. Storm-sewer tracing often must be done by walking the perimeter of the urbanized area of the lake and looking for street drains. Local or state officials can be another source of information about the location of storm drains. Significant nonpoint sources, such as construction sites, feedlots, and mining areas, should also be noted.

The areas of the watershed classified as urban, rural or agricultural, wetlands, and forest must be carefully delineated. Uttormark et al. (1974) have concluded that, with the degree of our current knowledge of the relationship between land use and nutrient discharge, there is little reason for a more detailed delineation.

Nutrient-Water Budgets. A nutrient–water budget for a lake or reservoir is a measurement of the annual input and output of these substances. Suspended and dissolved solids may also be determined, particularly if the lake or reservoir appears to be silting in.

Budgets of substances are developed for a lake because they are powerful management tools. Eutrophication, as defined earlier, is due to an income of sediments, organic matter, and nutrients at rates faster than the lake can assimilate or discharge them. The establishment of the rates of input and output allows the lake manager or consultant to estimate the impact of any changes (whether increase or decrease) and to assess which sources of input and output are of greatest significance. A predicted change in nutrient concentration from the loading models permits an estimate of the change in the degree of eutrophication that can occur. There may be proposed changes in the lake's mean depth or water residence time, as from sediment removal or flushing, factors that can significantly alter nutrient concentration.

Estimates of silt and organic matter income have seldom been attempted, despite the significance of these materials to the eutrophication process. Determination of the load of dissolved organics, the most abundant fraction of organic matter loading (Wetzel, 1983), is difficult and time-consuming. A determination of suspended solids, however, taken from the same samples used for the phosphorus budget and followed by a determination of ash weight after ignition in a furnace (APHA, 1980), is relatively simple. These data will identify the sources that are significant in the loading of oxygen-consuming materials and in the deposition of substances leading to a loss of lake or reservoir volume and the creation of macrophyte habitats. These data may be of particular value in determining the efficacy of a sediment removal project versus a plan to curtail erosion on the watershed. An addition of these two determinations to a phosphorus budget will greatly enhance the quality of many diagnostic studies at little additional cost.

Uttormark and Hutchins (1980) offer a particularly lucid description of the three phosphorus and water input–output models most widely used as management tools in lake restoration work. These are the models of Dillon and Rigler (1974b) and of Vollenweider (1975; 1976). They are described here briefly and presented in detail in Reckhow and Chapra (1983). The reader is also urged to consult the original literature.

Both models are based on the same assumptions:

1. The lake is phosphorus limited.
2. The lake behaves, on a long-term basis, as a completely mixed system.
3. Lake volume remains constant.
4. The influx of phosphorus is constant, losses occur through deposition and outflow, and net internal loss is proportional to phosphorus concentration in the lake.

Phosphorus concentration at steady state is described as

$$[P] = \frac{L}{\bar{Z}(\rho + \sigma)}, \qquad (2.1)$$

where $[P]$ = in-lake total phosphorus concentration, mg P m^{-3}
L = areal loading of total phosphorus, gm P m^{-2} lake surface yr^{-1}
$\bar{Z}$ = mean depth, meters
ρ = flushing rate, years; $\rho = Q/V$
Q = annual water outflow, m^3 yr^{-1}
V = lake volume, m^3
σ = specific sedimentation rate, yr^{-1}.

The specific sedimentation rate of phosphorus cannot be measured practically. The three models vary only in the way in which equation (2.1) was modified to estimate sedimentation. Dillon and Rigler modified equation (2.1) to include a phosphorus retention coefficient R, where

$$R = \frac{P_{in} - P_{out}}{P_{in}}. \qquad (2.2)$$

Equation (2.1) then becomes

$$[P] = \frac{L(1-R)}{\bar{Z}\rho}. \qquad (2.3)$$

Vollenweider (1975) plotted sedimentation rate data, determined by knowing all variables in equation (2.1) except σ, versus mean depth and found that the rate was estimated by

$$\frac{10}{\bar{Z}}. \qquad (2.4)$$

Equation (2.1) then becomes

$$[P] = \frac{L}{10 + \overline{Z}\rho} . \qquad (2.5)$$

Vollenweider (1976), and independently Larsen and Mercier (1976), then found a relationship between sedimentation and the hydraulic residence time, expressed as

$$\sigma = \sqrt{\rho} \qquad (2.6)$$

Equation (2.1) was rewritten as

$$[P] = \left(\frac{L}{\overline{Z}\rho}\right) \cdot \left(\frac{\rho}{\sqrt{\rho}+\rho}\right) = \left(\frac{L}{\overline{Z}\rho}\right) \cdot \left(\frac{1}{1+\sqrt{\rho}}\right), \qquad (2.7)$$

where $L/\overline{Z}\rho$ is the average inflow concentration.

General observations suggest that a concentration of 10 mg P m^{-3} (total phosphorus) or less during spring circulation is likely to be associated with few problems of algal blooms and low transparency, whereas concentrations above 20 mg P m^{-3} are often associated with high amounts of algae. Vollenweider (1975; 1976) and Dillon and Rigler (1974b) therefore used as the maximum "acceptable" loading of phosphorus the amount that would produce a total steady-state phosphorus concentration of 10 mg P m^{-3} or less; loadings which produced a concentration of twice that or more were called "unacceptable."

A water-phosphorus budget may be obtained in either of two ways. Potentially the least accurate but the least expensive is the method proposed by Dillon and Rigler (1975) as modified by Reckhow and Simpson (1980) and Reckhow and Chapra (1983). Phosphorus export coefficients for the various land-use categories are employed, along with measured lake morphometry and water outflow data, to estimate the annual loading. The second method involves actual measurement of all sources of water and phosphorus input and output over a year. Both procedures are briefly described here, but the reader is urged to consult the original literature, including the above papers, USEPA (1980), and Rast and Lee (1983).

Export Coefficients. The procedure of Dillon and Rigler (1975) was originally developed to estimate the capacity of a lake to assimilate development with houses and cottages around the shoreline. Their approach allowed the consultant or lake manager to estimate the current mean lake phosphorus concentration and compare it to predicted concentration, and thus to predict new algal biomass (measured as water transparency) after various amounts of shore development. Lake phosphorus concentration is obtained by summing the expected export from each of the various land-use areas (urban + agricultural + forest), the amount added by direct rainfall, and the amount expected from artificial sources such as septic tanks. Water flow is estimated from runoff maps, and lake volume and area from topographic maps or direct measurement.

Reckhow and Simpson (1980), in agreement with Dillon and Rigler, point out that the potential for error with this method is very large. They developed a procedure for estimating the uncertainty involved in each estimation of phosphorus export and provided improved phosphorus export coefficients. Reckhow (1983) describes a method of error analysis appropriate when a new steady-state phosphorus concentration must be predicted for a change in land use. Existing lake quality data are used, eliminating the need to project all land-use impacts with the model.

Rast and Lee (1978) developed a set of phosphorus export coefficients for the three land-use types (wetlands are considered to have no net annual output) plus precipitation, based upon data from 473 subdrainage areas in the eastern United States (USEPA, 1974) and the data of Uttormark et al. (1974) and Sonzogni and Lee (1974). These coefficients are listed in Table 2-3. Rast and Lee believe these coefficients approximate the measured load by a factor of ± 100 percent.

At present there is little evidence that the coefficients of Reckhow and Simpson (1980) are either superior or inferior to those of Rast and Lee (1978). Reckhow and Simpson present a means of estimating the uncertainty of the computed loading value. Rast and Lee (1983) found good agreement between the loading computed from their export coefficients and the loading rate empirically determined for 38 United States waterbodies.

Export coefficients can produce a reasonable estimate of phosphorus (and nitrogen—see Rast and Lee, 1978) loading to a lake, and at comparatively low cost. The degree of uncertainty must be computed, and the export rates should be validated with field measurements of several storm inputs. To use the models to predict a new steady-state phosphorus concentration in the lake after a land-use change, knowledge of the annual water budget is required. This should be determined directly (see next section). Using runoff maps, as in Dillon and Rigler (1975), or calculating annual discharge by the difference between estimated annual precipitation to and evaporation from the watershed leads to large errors (Scheider et al., 1979). Sewage treatment plant discharges are not included in export coefficients, but measurements of flow and concentration are usually available from plant records.

Water–Nutrient Budget Determination. The actual determination of an annual water–phosphorus budget can be a costly procedure, and as error-prone as the method

Table 2-3 Watershed Phosphorus Export Coefficients (from Rast and Lee, 1978)

Source	Watershed Export Coefficient $(gm\ P\ m^{-2}\ yr^{-1})$
Urban land	0.10
Rural/agricultural land	0.05
Forest land	0.01
Rainfall	0.02
Dry fallout	0.08

of estimation just described. An estimate of the actual loading can be obtained, however. Scheider et al. (1979) provide a highly useful outline of the procedures to follow to obtain an accurate water-phosphorus budget, and the following discussion is based in part upon their report. Winter (1981) describes in detail the methods, uncertainties, and problems of estimating a lake's water budget.

The generalized equation for the water budget of a lake is

$$\text{input} = \text{output} \pm \Delta \text{ storage}. \tag{2.8}$$

Input is the sum of stream and groundwater inflow, direct precipitation on the lake's surface, nonchannelized overland flow, and flow from point sources such as sewage treatment plant discharges. Output is the sum of stream and groundwater outflow, evaporation, and industrial or domestic withdrawal. The following paragraphs briefly describe how each component of the budget is determined.

Stream discharge (in or out) is measured in one of two ways. In the first method a flow meter is used to measure the flow rate of each stream over the range of expected annual flow. A staff gauge is often used. A regression of measured flow versus the product of stream depth and width is used to estimate unmeasured flow:

$$\text{discharge } (m^3) = \text{flow } [m \text{ sec}^{-1} \times \text{cross-section of stream area } (m^2)]. \tag{2.9}$$

The regression must include the entire range of flow and of stream depths and widths. A regression may be developed for each season and for both rising and falling hydrographs. An alternative and far more accurate method is to place continuous flow recorders on each stream. Flow versus stream depth is determined as before and a linear regression equation computed. In this case the relationship is used to convert continuous stream-depth determinations into continuous stream discharge. This is far more accurate than the method of visiting the stream at intervals to read the staff gauge, because precipitation events can be sudden and of very high volume. Unless the sampling team were in place to measure the hydrologic event, its entire contribution to the budget will be unmeasured. Scheider et al. (1979) compared values for discharge obtained from discrete measurements of stream discharges, and estimations from runoff maps or precipitation and evaporation, with the continuous flow recorder method. Table 2-4, in which continuous flow is considered to have zero error, lists their results.

Precipitation is determined with a collector installed at the lake, preferably on the water rather than on shore. The collector is left open constantly so that dry fallout is obtained. Each precipitation event should be collected because volumes and phosphorus concentration can vary enormously between storms. At a large lake or reservoir several collectors will be needed. The significance of precipitation as a direct source of water and phosphorus varies with the ratio of watershed area to lake area. For Ontario lakes, 23 percent of the total phosphorus income is from precipitation when the ratio is 10:1, but only 9 percent when the ratio increases to 30:1 and 3 percent when the ratio is 100:1 (Rigler, 1974).

Table 2-4 Comparison of Hydraulic Input as Calculated by Five Commonly Used Methods (seven streams on Harp Lake, Ontario, January–December 1977) (from Scheider et al., 1979)

Data	Stream Discharge Calculation Method	Mean Absolute Percent Error	Range in Percent Error
Discharge calculated from continuous stage records	Integration of continuous discharge vs. time plot	0	0
Discharge measured at discrete time intervals	Integration of discrete discharge vs. time plot	12	−19 to +35
	Three-point running mean of discrete discharge	35	−15 to +130
No measured discharge	Long-term unit runoff (Pentland, 1968)	18	− 2 to +68
	Precipitation–evapotranspiration (Morton, 1976)	36	+12 to +91

The groundwater contribution to the budget is difficult to determine but cannot be ignored. Detailed descriptions of this problem are presented by Winter (1978; 1980; 1981). Some lakes apparently receive little, if any, groundwater discharge. This cannot be assumed, however. One method of measurement is to treat groundwater discharge as the residual term in the hydrologic equation (2.8). This approach assumes that the other terms of the equation have been determined very accurately. La Baugh and Winter (1984) have found in their study of two Colorado reservoirs that the residual term, rather than being a contribution of an unmeasured portion of the budget, can be accounted for almost entirely by measurement errors. Another measurement method is to calculate groundwater in a flow net, using the equation

$$Q = KIA, \qquad (2.10)$$

where Q = groundwater discharge
 K = hydraulic conductivity
 I = hydraulic gradient
 A = cross-sectional area through which groundwater flows.

This procedure requires establishing nests of piezometers to determine the hydraulic gradient of the water table, measuring hydraulic conductance through pump tests, and establishing hydrogeologic boundaries. Piezometers are also used to measure the phosphorus content of groundwater.

Lee (1977), Lee and Hynes (1978), and Barwell and Lee (1981) have introduced the use of seepage meters made from steel barrel halves, inverted over the lake bottom so that groundwater flows into an attached collecting bag, as a means of estimating groundwater. Large errors can occur with these meters, since they determine flow at discrete sites; therefore a sampling design is required to improve accuracy of estimates of flow. Further, the meters are installed and sampled using scuba gear, which limits the practical portion of the sampling in northern latitudes to good weather periods. Despite these problems, seepage meters are very valuable for determining the direction of groundwater flow and its phosphorus content.

There is no direct, simple, accurate method of measuring the groundwater component of the hydraulic equation. Since a reliable value is required, it is suggested that at least two methods be attempted. One method could be to solve for groundwater as the residual in the hydraulic equation. The other could be establishing a flow net. In either case, samples of the groundwater are needed to measure phosphorus concentration.

Measurements of overland or nonchannelized flow are essentially nonexistent in water budget studies. This source may be very minor, but a walk around some or all of a lake's perimeter during a rainstorm may be needed to confirm that conclusion. Some workers have assigned this component of the hydrologic equation to the residual term (along with groundwater), but this increases the potential for error in the overall budget. Others have used the Soil Conservation Service technique described by Mockus (1971), in which soil infiltration characteristics, five-day antecedent precipitation, and land use are used to calculate overland flow. As a practical matter, there probably is little alternative but to include this term in the residual of the hydrologic equation.

Point sources entering the lake, such as urban stormwater pipes, can be very significant in the water–phosphorus budget. Discharge to the lake is intermittent but often very large and of exceedingly poor water quality. Great care must be given to an accurate measurement of each storm event. Arnell (1982) provides a method of estimating runoff volume from paved urban areas, but development of the necessary regressions of rainfall versus runoff volumes requires the use of recording devices on storm-flow streams or pipes whose flows span the range of storm flows in the area. Brater and Sherrill (1975) review the literature regarding precipitation and urban runoff and develop a predictive model to estimate runoff. The alternative to using continuous-flow measurement devices is discrete sampling of flow rate, a procedure that can produce substantial error if large storms or snowmelts are missed.

Lake water outflow must also be measured. Normally discharge is through a single stream or over a dam, simplifying measurement to that of an accurately measured, channelized flow. Evaporation is a water-loss term that can be measured in several ways, all with significant possibilities for error. Winter (1981) suggests that calculation through an energy budget is most accurate. The use of an evaporation pan is the more common procedure, but no standard pan technique appears to exist, and there are problems in translating the pan evaporation rate to the lake evaporation rate. Many limnologists have simply obtained an evaporation rate from the nearest National Weather Service station and used the coefficient of 0.7 to estimate lake evaporation from evaporation from a class A pan. As Siegel and Winter (1980) have pointed out, this coefficient was determined by use of annual averages but is incorrectly used for monthly values. If phosphorus export coefficients are being used to estimate the lake's phosphorus budget, then these outflow determinations and their phosphorus content are the only ones required to compute the annual water-phosphorus budget.

Finally, the lake level, or the storage term of the water budget, must be measured. A water-level recording device is installed in the lake, or frequent observations of a staff gauge are made. In some reservoirs these data, along with outflow, are regularly recorded. Errors in lake-level measurement are largely attributable to errors in measurement of lake area and volume and to seiches in large lakes and reservoirs.

Determining a lake's phosphorus budget requires measuring concentration in all inflows and outflows. Concentration can change drastically between sampling dates and from source to source on any given lake. Table 2-5 from Scheider et al. (1979) compares the error in eight methods of measurement, with an assumed zero error for Method 1, which uses discrete samples of total phosphorus concentration and continuous samples for water flow. If flow recorders are not available, the authors conclude that Method 7 provides the best estimate of stream phosphorus input.

Discrete samples for phosphorus measurements must be taken frequently. Sampling at regular intervals, such as weekly, may miss significant runoff events and produce a phosphorus budget that underestimates loading. It is better to concentrate the sampling effort on high-discharge periods such as snowmelt and spring rains. The concentration of phosphorus in groundwater will usually not vary greatly.

Water-Nutrient Budget Data Evaluation. The data from the annual budget, along with measurements of lake volume, area, and mean depth, may be used to compute the steady-state phosphorus concentration (eq. 2.3, 2.5, or 2.7). Uttormark and Hutchins (1980) have evaluated these three models and found that all can be used to "reliably predict the general trophic state likely to occur in lakes as a result of phosphorus abatement." The Vollenweider (1976) model (eq. 2.7) gave accurate predictions in 80 percent of the study lakes; the other models were accurate in 70 percent of the lakes tested.

Reckhow (1979a; b) describes a method of estimating the total uncertainty or reliability of a prediction in which Vollenweider's (1976) model (eq. 2.7) was used. Calculation of uncertainty in the prediction is important, since this provides a measure of the value of the information contained in the prediction and thus allows the user to weigh the results in any decision-making process. Further discussion of this calculation is found in Reckhow and Chapra (1983).

To estimate the new trophic state or condition of a lake from the predicted phosphorus concentration after management or diversion of phosphorus loading, the quantity of algal chlorophyll, or the Secchi disc transparency, must be calculated for that future time. Several investigators have produced equations to describe the relationship between total phosphorus concentration and chlorophyll *a*, and there is wide variance among them (see Nicholls and Dillon, 1978, for a review). Differences are due to variations in sampling periods and technique, analytical procedures, grazing, and factors such as incident radiation, types of algae, and supplies of nutrients other than phosphorus that affect the amount of chlorophyll in an algal cell. The

Table 2-5 Comparison of Phosphorus Input Calculation by Nine Commonly Used Methods (seven streams on Harp Lake, Ontario, January–December 1977) (from Scheider et al., 1979)

Data		Phosphorus Input Calculation Method	Mean Absolute Percent Error	Range in Percent Error
Discharge calculated from continuous stage records; [P] measured at discrete time intervals	1.	Product of integrated discharge vs. time plot and [P] at midpoint of time interval	0	0
	2.	Product of integrated discharge vs. time plot and mean of [P] at end points of time interval	3	−4 to +5
Discharge and [P] measured at discrete time intervals	3.	Product of integrated discharge vs. time plot and [P] at midpoint of time intervals	11	−19 to +11
	4.	Product of integrated discharge vs. time plot and [P] at endpoints of time interval	14	−25 to +16
	5.	Product of discharge as calculated by three-point running mean and [P] at midpoint of time interval	30	−19 to +92
	6.	Integration of the plot of the product of discharge and [P] vs. time	10	−19 to +8
	7.	Three-point running mean of product of discharge and [P]	27	−14 to +57
No measured discharge and [P] measured monthly	8.	Product of total monthly discharge (Pentland, 1968) and [P]	49	+4 to +85
	9.	Product of total monthly discharge (precipitation−evapotranspiration) and [P]	71	+19 to +111

relationships given by Carlson (1977) are chosen here because they are based primarily on summer samples. The equations are

$$\ln \text{Chl} = 1.449 \ln \text{TP} - 2.442, \tag{2.11}$$

$$\ln \text{SD} = 3.876 - 0.98 \ln \text{TP}, \tag{2.12}$$

where Chl = chlorophyll *a* (mg m^{-3})
TP = total phosphorus (mg m^{-3})
SD = Secchi disc depth (meters)

As an example of the use of these data, assume these determinations of a hypothetical lake or reservoir:

phosphorus loading (L) = 0.8 gms P m^{-2} yr^{-1}
mean depth $(\bar{Z})$ = 8.0 meters
flushing rate = 0.9 years
mean epilimnetic [P] = 54 mg P m^{-3}
mean Secchi disc transparency = 0.96 meters

Assume a municipality that adds treated sewage effluent to this lake decides to ban the sale and use of phosphorus-containing detergents and that all persons comply. This will eliminate about 25 percent of the current phosphorus income (Jones and Lee, 1982). The Vollenweider (1976) model (eq. 2.7) predicts a new steady-state phosphorus concentration and Carlson's equation predicts the new Secchi disc transparency that would occur if chlorophyll concentration controls transparency and if algal biomass is limited by phosphorus:

(800 mg P m^{-2} yr^{-1}) (0.25) = 200

(800 − 200) = 600 mg P m^{-2} yr^{-1} (new P loading):

$$[P] = \left(\frac{L}{\bar{Z}\rho}\right)\left(\frac{\rho}{\sqrt{\rho}+\rho}\right) \tag{2.7}$$

$$= \left(\frac{600}{(8)(0.9)}\right)\left(\frac{0.9}{\sqrt{0.9}+0.9}\right)$$

$$= (83.33)(0.49)$$

$$= 40.57 \text{ mg P m}^{-3};$$

$$\ln SD = 3.876 - 0.98 \ln TP \qquad (2.12)$$

$$= 3.876 - 0.98 \ln 40.57$$

$$= 3.876 - 3.629$$

$$= 0.25$$

$$SD = 1.28 \text{ meters.}$$

The users of the lake would not likely be able to consistently distinguish the difference in transparency between the original mean of 1.0 meters and the new predicted mean of about 1.30 meters. Thus the effectiveness of the detergent ban alone would probably be of small significance. To achieve a noticeable change in water transparency, a larger fraction of the load must be removed. The phosphorus-containing detergent ban is a low-cost first step, but in this case it would have to be coupled with phosphorus removal at the treatment plant or some other significant lowering of external phosphorus load.

Lakes with persistent anoxic hypolimnia release phosphorus from sediments at high rates, and the phosphorus concentration predicted for the lake could be an underestimate of the actual concentration. The recovery of some lakes after sewage diversion has apparently been delayed by such internal phosphorus loading [e.g., Shagawa Lake, Minn. (Larsen et al., 1981); Lake Sammamish, Wash. (Welch et al., 1980)]. The rate of net internal phosphorus loading can be calculated from the nutrient budget data. High rates suggest that response to diversion could be slow and that an in-lake procedure to treat the release, such as phosphorus inactivation, sediment removal, or hypolimnetic withdrawal, may be needed. Net internal phosphorus loading may be calculated as follows (from Cooke et al., 1977), where loading to the lake during the stratified period = $\Sigma\ P_{L_{ext}}$; the change in the phosphorus content of the lake (sum of each depth stratum's volume $\times$ concentration at each stratum) from the beginning to the end of the stratified period = $\Delta\ P_{lake}$; and the sum of all phosphorus outflows = P_{out}. Then net internal phosphorus loading, $P_{L_{net}}$, is

$$\Sigma\ P_{L_{ext}} - P_{out} = P_{L_{net}}, \qquad (2.13)$$

$$\Delta\ P_{lake} - P_{L_{net}} = P_{L_{int}}. \qquad (2.14)$$

Net rates of sediment phosphorus release in eutrophic lakes have ranged from 6 to 28 mg P m^{-2} day^{-1}, using the area of the anoxic sediments to express release rates (Nurnberg, 1984).

Nurnberg (1984) has proposed a model, based upon a detailed analysis of internal loading in lakes with oxic and anoxic hypolimnia, that predicts total phosphorus concentration by accounting for internal loading. The equation is

$$[P] = L_{ext/q_s}(1 - R_{pred}) + (L_{int/q_s}), \qquad (2.15)$$

where

$$L_{int} = L_{ext}(R_{obs} - R_{pred}) \tag{2.16}$$

$$R_{obs} = (P_{in} - P_{out})/P_{in} \quad \text{(measured P retention)} \tag{2.17}$$

$$R_{pred} = 15/(18 + q_s), \tag{2.18}$$

where q_s = annual outflow/lake area. This model could be substituted for the other three predictive models described here (Dillon and Rigler, 1974a; Vollenweider, 1975; 1976).

Monitoring Data Evaluation

The determinations of physical, chemical, and biological characteristics of the lake are used to quantitatively describe the degree of eutrophication, or the trophic state, through the calculation of an index number. The data, after proper reduction, are also used to choose the in-lake restoration technique.

The calculation of a trophic state index (TSI) number for a lake is of great value to the consultant or lake manager, and to the lake users. An index number allows comparison of the lake's present condition with the past, and can illustrate the possible future states of the lake by assuming change in loading and associated change in phosphorus or chlorophyll concentrations or in Secchi disc transparency. The use of an index also permits easy communication with lake users by emphasizing that there are degrees of eutrophication. Their lake may well be oligotrophic according to the many published tables that assign the states of "oligotrophic," mesotrophic," and "eutrophic" on the basis of a range of phosphorus or chlorophyll concentrations. These labels, however, lack a sense of degree in conveying *how* oligotrophic, mesotrophic, or eutrophic the lake may be. Also, to tell lake users that their lake is eutrophic is meaningless if an adjacent lake is also eutrophic but has half the algal biomass and perhaps far greater potential for restoration. This topic has been reviewed by Carlson (1979).

Carlson (1977) has developed a useful trophic state index for lakes that have few macrophytes and little nonalgal turbidity. Carlson based his index on algal biomass in surface water, using a scale of 0-110. The scale uses a log transformation of Secchi disc transparency values as a measure of algal biomass. The scale is so constructed that each increase of ten units represents a doubling of biomass. Since chlorophyll and total phosphorus concentrations are often highly correlated with transparency, an index number can also be calculated from these measurements. The accuracy of Carlson's index is diminished by water coloration or suspended materials other than algae (Carlson, 1980; Lorenzen, 1980). The equations for computing the Carlson TSI are:

$$\text{TSI} = 60 - 14.41 \ln \text{Secchi disc (meters)} \tag{2.19}$$

$$\text{TSI} = 9.81 \ln \text{Chlorophyll (mg m}^{-3}) + 30.6 \qquad (2.20)$$

$$\text{TSI} = 14.42 \ln \text{Total Phosphorus (mg m}^{-3}) + 4.15. \qquad (2.21)$$

Table 2-6 lists the TSI numbers from 0 to 100 with the associated Secchi depths and phosphorus and chlorophyll concentrations. The range between 40 and 50 is most often associated with mesotrophy. Between these two points algal biomass doubles or halves (4 m at TSI 40 and 2 m at 50), a change that would be obvious to lake users through changes such as blue-green algal blooms and oxygen deficits. On the other hand, if a management strategy proposed for a phosphorus-limited lake with a TSI of 70 will only cut the concentration in half, then the lake users may not see any appreciable improvement (Table 2.6).

The Carlson index has been misused, particularly in lakes with high nonalgal turbidity or with extensive macrophyte infestations. It makes no sense to locate the sampling boat over the only macrophyte-free patch of water and measure trophic state from transparency, phosphorus, and algal chlorophyll. The lake could be classified as oligotrophic from these measurements, while anyone could see that it is severely infested with rooted and floating plants.

Canfield et al. (1983) have proposed an index for classification of lakes in this condition. The total biomass of submersed macrophytes is determined, and the total biomass estimate of each species is then multiplied by its phosphorus content, as determined by tissue analysis; this gives an amount of phosphorus associated with all the macrophytes. Then the phosphorus content of the water column (concentration in each stratum × strata volumes) and the amount associated with macrophytes are summed to give a total water-column value. Canfield et al. found that macrophytes have little effect on a trophic state evaluation when less than 25 percent of the phosphorus in the water column is associated with them and the mean macrophyte concentration is less than 1 gm dry wt m^{-3}. Use the total water column phosphorus

Table 2-6 Completed Trophic State Index (TSI) and Its Associated Parameters (from Carlson, 1977)

TSI	Secchi Disc (m)	Surface Phosphorus (mg m^{-3})	Surface Chlorophyll (mg m^{-3})
0	64	0.75	0.04
10	32	1.5	0.12
20	16	3	0.34
30	8	6	0.94
40	4	12	2.6
50	2	24	6.4
60	1	48	20
70	0.5	96	56
80	0.25	192	154
90	0.12	384	427
100	0.062	768	1183

concentration (water column phosphorus plus macrophyte phosphorus) in the Carlson index.

An alternative trophic state index, which includes transparency, phosphorus, nitrogen, chlorophyll *a*, dissolved oxygen, and macrophytes, has been proposed by Porcella et al. (1980). This index, termed the lake evaluation index (LEI), provides a single number based on determinations of these variables and is useful in assessing the change in lake state following a restorative treatment. Walker (1980, 1984) has noted that some lakes and many reservoirs may deviate in several ways from Carlson's equations, perhaps due to nitrogen limitation or non-algal turbidity. Walker (1984) has developed a two-dimensional classification system which appears to more completely describe reservoir water quality than the Carlson-type index. The consultant or manager must choose the index that most closely allows an accurate description of the lake or reservoir in question. The Carlson TSI has limited value in turbid lakes (Lorenzen, 1980) or in macrophyte-dominated lakes. The LEI was designed for use with either phosphorus or nitrogen, depending upon which is the limiting nutrient. The two should not be used simultaneously in the same equation. The limiting nutrient must be identified.

Lakes, and especially reservoirs with considerable horizontal heterogeneity, should be sampled along their length. A gradient of trophic state may be found. The TSI for a situation of this nature is not found by averaging the values over the season or along the reservoir's length. Rather, plot all of the index values against time and use a separate graph for each sampling station.

The remainder of the lake evaluation data must be assembled into meaningful graphs, tables, and calculations. A useful guide for this work is Wetzel and Likens (1979).

The vertical temperature–dissolved oxygen data should be plotted on a depth–time graph, rather than on a separate graph for each sampling date. This will clearly illustrate the establishment of thermal stratification and any loss of oxygen in the hypolimnion. The computation of a hypolimnetic oxygen deficit, defined as the difference in amount of oxygen per unit area below the top of the hypolimnion at the beginning of thermal stratification and on some later date just before zero oxygen is recorded, divided by the interval in days, is an indication of the lake's productivity. Determine the amount of oxygen in the hypolimnion just after thermal stratification is established in the spring and again just before the lowermost strata have been depleted of oxygen. In small, eutrophic lakes, this interval will be less than one month. Express the data as the rate of change per cm^2 of the top of the hypolimnion's surface. A valuable reference for data interpretation is Cornett and Rigler (1979), and an equation describing the computation is presented in the chapter of this book on Hypolimnetic Aeration and in Wetzel (1983).

The other water chemical data may be presented graphically and in tabular form. Recall that one of the purposes of these determinations is to assess the trophic state of the lake prior to any diversion or in-lake treatments and to serve as a benchmark against which future states of the lake may be compared. Data analysis should reflect that approach.

Plot the Secchi disc transparency changes against time. Phosphorus concentration measurements are presented in several meaningful ways. Plot the vertical concentrations of total phosphorus, temperature, and dissolved oxygen for several dates over the summer, to illustrate the effects of stratification and any oxygen depletion. This may also be done by plotting concentrations of surface and deep-water samples against time. The total phosphorus content of each stratum is obtained in order to compute a volume-weighted concentration. This may be expressed as the volume-weighted concentration of the three thermal zones and plotted over time. Soluble reactive phosphorus (SRP), a fraction that may be closely associated with orthophosphate (the only form of phosphorus assimilated by cells), should also be plotted on time and depth graphs. Calculate seasonal and thermal zone means and standard deviations.

The ratio of total nitrogen to total phosphorus is sometimes used to evaluate which nutrient is limiting. A ratio of about 20:1 is often considered to be evidence of phosphorus limitation to algal growth, and low ratios (below 13:1) are often given as evidence of nitrogen limitation (Smith, 1979). As pointed out by Lee et al. (1981), these ratio numbers do not necessarily indicate nitrogen or phosphorus limitation of growth per se, but only which nutrient will likely be used up first and become limiting if growth is not inhibited by light or another factor. If the soluble reactive phosphorus values are less than 5 mg m^{-3} and the N:P ratio is 20 or greater, then there is good reason to believe that algae are phosphorus limited.

Alkalinity and pH data are particularly significant for the phosphorus inactivation and artificial circulation procedures. A table of values should be prepared.

Phytoplankton samples, preserved in Lugol's solution, are normally counted in sedimentation chambers using an inverted microscope. Wetzel and Likens (1979) discuss alternatives if this microscope is unavailable. Data are expressed as cell volume, which should be measured for common species. An approximation of cell volumes for some lake species is given in Table 8-1 of Wetzel (1983). Plot changes in cell volumes of the dominant species over time. Also prepare a graph of the percentage of total algal biomass represented by each phylum over time and a graph of total algal cell volume over time. Chlorophyll *a* measurements, also an expression of algal biomass, are presented as vertical profile measurements, as surface values plotted over time, and as mean epilimnetic chlorophyll. Hallegraeff (1977) presents an excellent summary of the various ways to provide meaningful summaries of phytoplankton data.

Zooplankton, at least the microcrustacea, are counted by species. Plot densities of total microcrustacea, Cladocera, and Copepoda in each thermal zone. The relative abundance of the dominant species may be significant in the choice of a biological control technique, so a plot of the changes in relative species density is needed. One technique is the use of "pie" diagrams, as illustrated by Pennak (1957).

The fish species and the sediment fauna should be listed, with estimates of relative abundance. The distribution of macrophytes should be mapped to illustrate the area of greatest infestation, and a list of species and their relative abundances should be presented. The values for the present study should also be compared to any data available from the past.

SELECTION OF LAKE RESTORATION ALTERNATIVES

The diagnosis and feasibility study was designed to determine the causes of the lake's problems and to evaluate its current state or the severity of the problem. The consultant or lake manager will use these data to select the appropriate in-lake procedures to improve the lake.

We divide lake restoration techniques into two groups: methods that emphasize nutrient control and methods that emphasize direct control or removal of biomass. Each is briefly described below, along with some guidelines to aid in selection. The chapters that follow are intended to describe the state of the art for each technique.

Nutrient Control Methods

The consultant or lake manager must evaluate the loading and in-lake data to determine whether nutrient abatement, either internal or external, is needed and is feasible. High external loading as evaluated with nomographs, such as those from Vollenweider (1976), that give "permissible" and "dangerous" rates, or high internal loading and high in-lake concentration would make it obvious that nutrient control is needed. We discuss six methods, including diversion. More than one method could be used, and methods of biomass removal may be recommended along with nutrient control.

Nutrient Diversion: A reduction in nutrient income that is sufficient to produce a recognizable reduction in algal biomass is an essential step in lake restoration. While it is possible to manipulate algal biomass through enhanced grazing or by removing macrophytes, the basic cause of excessive plant growth—high concentrations of nutrients, and shallow, rich sediments—must ultimately be corrected.

Dilution and Flushing. Dilution is the addition of nutrient-poor water; flushing involves the addition of large volumes of water (which could be high in nutrients) to wash out algal cells. While these treatments are effective, availability of a suitable water supply may limit their application.

Phosphorus Inactivation. The internal release of phosphorus may be a significant source of nutrients and could delay lake recovery after diversion. Release can be controlled by adding aluminum salts to the sediments. Phosphorus is bound to aluminum hydroxide and is not released, even after the onset of anoxia. Inactivants could also be added to incoming waters. Phosphorus inactivation is a powerful and effective technique when coupled with diversion.

Sediment Oxidation. The large quantity of organic matter in the sediments of eutrophic lakes serves as a substrate for microbial respiration, and hence leads to oxygen depletion. Phosphorus bound to iron (III) is released when iron is reduced. Nitrate serves as an alternate electron receptor and is reduced to nitrite and N_2. The reduction of iron (III) is delayed in nitrate-rich waters. This observation has led lake managers to add nitrate to hypolimnetic waters to control sediment phosphorus release. This method has been successful, but the number of treatments has been very small to date.

Sediment Removal. This effective procedure is among the most frequently recommended methods for removing internal nutrient sources and for improving very shallow lakes. It has a significant advantage over nutrient inactivation in that the source of internal nutrient loading is removed rather than left in place. This treatment usually also has an impact on macrophyte density, whereas nutrient inactivation has not been shown to have an effect on macrophytes. Consultants and lake managers must recognize, however, that many lake situations are not amenable to sediment removal due to high cost or to disposal problems.

Hypolimnetic Withdrawal. The nutrient-enriched hypolimnetic waters may be removed by siphon, pumping, or deep discharge at the dam, thereby increasing hypolimnetic oxygen concentrations in deep water. This may reduce the impact of sediment-released nutrients on surface-water algae. There is little documentation of the efficacy of this procedure.

Methods for Control or Removal of Biomass

As with nutrient control methods, the results of biomass control may be temporary unless the source of loading is controlled. Several of these procedures, such as drawdown and sediment covers, work well together.

Hypolimnetic Aeration. Hypolimnetic aeration is a procedure to increase the dissolved oxygen concentration of the hypolimnion without destratifying the lake. This is usually accomplished with an air lift device which brings cold hypolimnetic water to the surface, where gases are exchanged, and then returns it to deep water. This procedure can improve reservoir discharge quality, allow the re-establishment of a cold-water fishery, provide a daytime refuge for zooplankton, and eliminate problems with iron and manganese in potable water supplies.

Artificial Circulation. Artificial circulation is a technique used to prevent or eliminate thermal stratification through the mixing action of a column of air bubbles. This procedure will improve dissolved oxygen conditions in the lake and in reservoir discharges, and it may control algal biomass in some situations. Oxygenation should also eliminate taste and odor problems in potable water supplies.

Lake Level Drawdown. This is a truly multipurpose technique for impoundments. Exposure of rooted plants to freezing or hot conditions eliminates some species, does not affect others, and stimulates a third group of species. If the lake is drawn down, other techniques such as sediment removal, emplacement of screens, or fish management can take place.

Harvesting. Removing macrophyte biomass from the lake is an effective, though sometimes symptomatic, treatment for nuisance macrophyte conditions. Nutrients and organic matter that could stimulate algal blooms through the decay of the macrophytes, causing further oxygen depletion, are removed. The fraction of the annual net external income of nutrients that could be removed is one of the criteria used in deciding about the use of harvesting. Harvesting can have negative effects, including dispersal of plant fragments to uninfested areas, and the appearance of algal blooms.

Biological Controls. Dispersing phytophagous insects and fish and manipulating food webs to enhance grazing on algae are among the newest lake improvement techniques. In all such techniques, significant questions remain, either about undesirable side-effects or about efficacy. Undoubtedly, some of these procedures will become standard tools to control eutrophication symptoms, but at present their use remains largely experimental.

Sediment Covers. Screening materials to stop rooted plant growth are an expensive, symptomatic, and highly effective means of control. Expense prohibits their use over large areas, forcing the consultant or lake manager to choose another means to deal with macrophytes lake-wide.

Guidelines for Choosing Lake Restoration Alternatives

A consultant or lake manager who is directing a lake project supported with USEPA Clean Lakes Program funds must follow the guidelines of that program in choosing and defending methods. These guidelines should also form the basis for choosing alternatives in lake restoration projects funded from sources other than the federal government. The reader should consult the USEPA Clean Lakes Program Guidance Manual (USEPA, 1980), especially Section 8 and Appendix F.

The diagnostic portion of the feasibility study provides the data to select restoration alternatives. Two fundamental questions are asked of the data: How can sufficient nutrient diversion (sufficient to protect the lake from further deterioration or sufficient to accomplish a significant change in trophic state) be accomplished? What in-lake procedures can be used to accelerate recovery or to accomplish further improvement? For each in-lake procedure evaluated, four questions are asked: How effective is it projected to be? How much restoration will be accomplished? How much will it cost? How effective and costly are the alternatives? The alternative of "no action" should also be described.

The appropriate technique, or techniques, to apply to a specific lake requires a decision based largely on judgment. Although cost may be the principal criterion, reliability and longevity of the technique(s) will also be important. To make the proper decision, the lake manager will probably go through a decision process in which one or more of the sixteen techniques described in this book will be chosen. Such a decision process may take the form of Figure 2-7, for algal problems.

With the aid of a sound nutrient budget, the first consideration should be the principal source of the limiting nutrient. Is the major load coming from external sources or internal recycling? If from external sources, are they point sources or non-point (diffuse) sources? If principally from point sources, diversion of sewage or stormwater would probably be considered before advanced treatment because it has usually been less costly and operation and maintenance costs are relatively small. If one of those techniques are employed and there is little or no recovery, i.e., lake nutrient concentration does not decrease sufficiently, then dilution/flushing, artificial circulation and/or biomanipulation could be the next logical consideration. These techniques could provide controls on biomass and nuisance species of algae where controls on the limiting nutrient are not possible. Or, as in the case of dilution, the concentration of limiting nutrient could be controlled without reducing the total load. Dilution/flushing is listed first because there would be control on the causative limiting nutrient concentration. However, costs and scarcity of low-nutrient water may make that technique an unlikely choice.

The problems associated with definition, estimation, and control of non-point nutrient loads, which is often urban stormwater runoff, have been discussed only

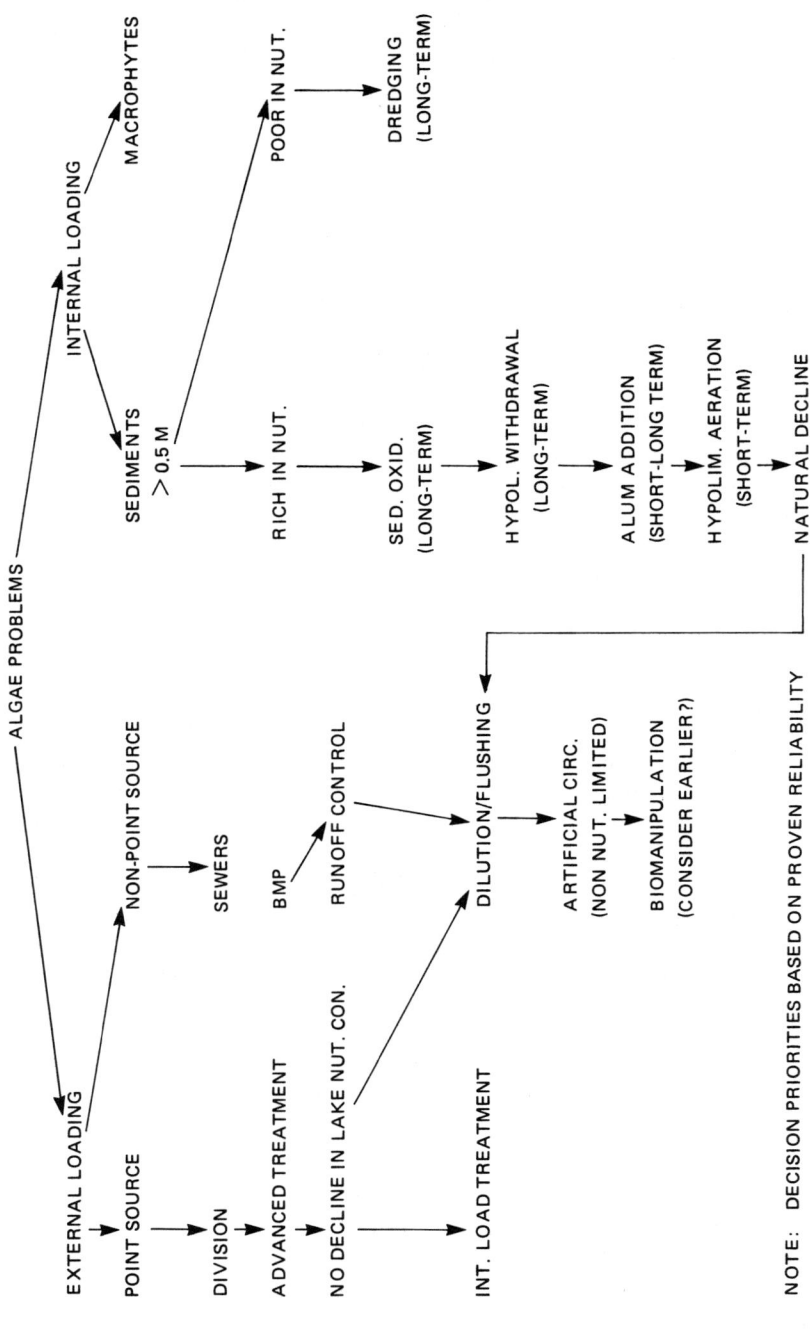

Figure 2-7 Decision tree for choice of best restoration procedure for control of algae problems.

lightly in the book. This is because that subject is large and is worthy of a separate treatment. Nevertheless, stormwater runoff is a principal cause for degradation of urban lakes (and rural lakes in the case of agricultural runoff). The techniques for reduction of non-point loads could be sewers to intercept stormwater and/or septic tank leachate, retention ponds, grassy swales, deep-well injection and chemical treatment in retention ponds, and best management practices (BMPs), such as fertilizer controls and minimizing impervious surfaces. If any one or a combination of these techniques do not result in improvements (nutrient decline), and the principal source of limiting nutrient is still external, then one of the three techniques previously mentioned could be considered.

The most common reason why lakes or reservoirs do not respond to controls on external inputs of nutrient is due to excessive internal loading or recycling of nutrients, particularly phosphorus. In that case, one can proceed to the right side of Figure 2-7. If sediments are the source of internal loading and the bulk of nutrients are located in the top 0.3-0.5 m of a sediment core, then removal of that layer by dredging should provide the most reliable and permanent solution, although it will be the most costly. If sediments are rich in nutrients below that depth, then dredging would result in only exposing more sediment with the same high nutrient content providing little or no expected decrease in internal loading. In that case there are seven techniques that could be considered. These are arranged in sequence of their reliability and expected longevity for control of the nutrient source itself.

Dilution/flushing, artificial circulation and biomanipulation are, again, aimed at control of algal biomass or nutrient concentration and do not interrupt the source of loading. Sediment oxidation, although not tested in a wide array of cases, does have as its major goal the actual restoration of the upper sediment layer and therefore should provide the most long-term solution with a single treatment. However, it is probably the most costly. Although the record for hypolimnetic withdrawal as a control for internal loading has not been as dramatic as that for alum addition, it nevertheless has the potential to deplete the sediment of nutrients. Alum, on the other hand, simply covers the sediment with a floc layer and while its reliability at interrupting sediment phosphorus release has been excellent, the layer has been observed to sink through the sediment presumably exposing P-rich sediment that is available for release. Alum addition is the least costly, with hypolimnetic aeration being the next most economic technique of the five. Hypolimnetic aeration has not been as effective as alum or sediment oxidation in controlling sediment release, although it provides a more direct and effective reaeration.

If the principal source of internal loading is suspected to be macrophytes then measures for their control must be undertaken. While enclosure and mass balance analyses have indicated the potential significance of macrophyte senescence to internal loading, there are as yet no demonstrations of lake-water P control through macrophyte control practices. Nevertheless, macrophytes clearly satisfy most of their nutrient demand from the sediment via their roots and, therefore, affecting lake-water P content through their control would seem to be a logical procedure.

A different sequence than that in Figure 2-7 may be needed for a given lake, depending on economic, political and social demands. However, a similar procedure should probably be used to consider each of the available techniques. There are other benefits and detriments associated with each of these techniques and their success/failure record is more equivocal than implied in this discussion. The reader is thus referred to the individual chapters to gain insight and judgment that will be more pertinent to an individual lake.

When an in-lake procedure is chosen, it should be reviewed against this checklist (USEPA, 1980, Table 8-4):

Will the project displace people?

Will the project deface existing residences or residential areas?

Will the project be likely to lead to changes in established land use pattern or an increase in development pressure?

Will the project adversely affect prime agricultural land or activities?

Will the project adversely affect parkland, public land, or scenic land?

Will the project adversely affect lands or structures of historic, architectural, archaeological, or cultural value?

Will the project lead to a significant long-range increase in energy demands?

Will the project adversely affect short-term or long-term ambient air quality?

Will the project adversely affect short-term or long-term noise levels?

If the project involves the use of in-lake chemical treatment, will it cause any short-term or long-term adverse effects?

Will the project be located in a floodplain?

Will structures be constructed in the flood plain?

If the project involves physically modifying the lake shore, its bed, or its watershed, will the project cause any short-term or long-term adverse effects?

Will the project have a significant adverse effect on fish and wildlife or wetlands or other wildlife habitat?

Will the project adversely affect endangered species?

Have all feasible alternatives to the project been considered in terms of environmental impacts, resource commitment, public interest, and cost?

Are there other measures not previously discussed that mitigate adverse impacts resulting from the project? Describe.

THE LAKE IMPROVEMENT-RESTORATION PLAN

A technical report is usually required of the consultant or lake manager. The report is a vital part of the diagnosis and feasibility study procedure, since it will be used by lake users or homeowners in choosing a course of action and the data will form the benchmark against which future studies will be compared. The report should follow the standard format of a scientific investigation. It should include:

1. description of the nature of the eutrophication process and a description of the lake's specific problems
2. listing of the particular questions asked in the diagnosis and feasibility study
3. description of the area, including maps and a table of morphometric-hydrologic data, and an accurate summary of all measurement methods and sampling locations
4. compilation of all results in tabular, graphical, and narrative form and an analysis or discussion of the implications of the findings
5. discussion of the recommendations, including their costs and environmental impacts
6. brief summary
7. citation of literature used in the study

Since few of the homeowners-lake users have technical backgrounds, most will not wish to read the technical report. A companion report that is brief and nontechnical should therefore also be prepared. This second report should include sections on the nature of the problem, questions asked, general findings, and recommendations and costs.

Finally, a public meeting is usually desirable to discuss the results with those who may have to pay for the project and those who will enjoy the benefits. The consultant or lake manager should therefore document the diagnostic field work thoroughly with color slides and should be prepared to make a lucid and brief presentation of the work, the recommendations, and the consequences of the "no action" alternative.

REFERENCES

American Public Health Association (APHA). 1980. 15th Ed. *Standard Methods for the Examination of Water and Wastewater*, Washington, DC: The Association.

Arnell, V. 1982. Estimating runoff volumes from urban areas. *Water Res Bull* 18:383-387.

Barko, J.W., and Smart, R.M. 1981. Sediment-based nutrition of submersed macrophytes. *Aquatic Bot* 10:339-352.

Barwell, V.K., and Lee, D.R. 1981. Determination of horizontal-to-vertical hydraulic conductivity ratios from seepage measurements on lake beds. *Water Res Bull* 17:565-570.

Bormann, F.H., and Likens, G.E. 1979. *Pattern and Process in a Forested Ecosystem*, New York: Springer-Verlag.

Bostrom, B., Jannson, M., and Forsberg, C. 1982. Phosphorus release from lake sediments. *Arch Hydrobiol Beih Ergebn Limnol* 18:5-59.

Brater, E.F., and Sherrill, J.D. 1975. *Rainfall-runoff Relations on Urban and Rural Areas*, EPA-670/2-75-046.

Canfield, D.E., Jr., Langeland, K.A., Maceina, M.J., Haller, W.T., Shireman, J.V., and Jones, J.R. 1983. Trophic state classification of lakes with aquatic macrophytes. *Can J Fish Aquatic Sci* 40:1713-1718.

Carlson, R.E. 1977. A trophic state index for lakes. *Limnol Oceanogr* 22:363-369.

Carlson, R.E. 1979. A review of the philosophy and construction of trophic state indices. In, T.E. Maloney, ed., *Lake and Reservoir Classification Systems*, EPA-600/3-79-074.

Carlson, R.E. 1980. More complications in the chlorophyll–Secchi disk relationship. *Limnol Oceanogr* 25:379–382.
Carpenter, S.R. 1980. Enrichment of Lake Wingra, Wisconsin, by submersed macrophyte decay. *Ecology* 61:1145–1155.
Carpenter, S.R. 1981. Submersed vegetation: An internal factor in lake ecosystem succession. *Am Nat* 118:372–383.
Carpenter, S.R. 1983a. Lake geometry: Implications for production and sediment accretion rates. *J Theor Biol* 105:273–286.
Carpenter, S.R. 1983b. Submersed macrophyte community structure and internal loading: Relationship to lake ecosystem productivity and succession. In *Lake Restoration, Protection and Management*, EPA-440-5/83-001, 105–111.
Cole, G.A. 1983. *Textbook of Limnology*. 3rd Ed. St. Louis: Mosby.
Cooke, G.D., McComas, M.R., Waller, D.W., and Kennedy, R.H. 1977. The occurrence of internal phosphorus loading in two small, eutrophic glacial lakes in northeastern Ohio. *Hydrobiology* 56:129–135.
Cornett, R.J., and Rigler, F.H. 1979. Hypolimnetic oxygen deficits: Their prediction and interpretation. *Science* 205:580–581.
Dillon, P.J., and Rigler, F.H. 1974a. The phosphorus–chlorophyll relationship in lakes. *Limnol Oceanogr* 19:767–773.
Dillon, P.J., and Rigler, F.H. 1974b. A test of a simple nutrient budget model predicting the phosphorus concentration in lakewater. *J Fish Res Board Can* 31:1771–1778.
Dillon, P.J., and Rigler, F.H. 1975. A simple method for predicting the capacity of a lake for development based on lake trophic status. *J Fish Res Board Can* 32:1519–1531.
Francko, D.A., and Wetzel, R.G. 1981a. Dynamics of cellular and extracellular cAMP in *Anabaena flos-aquae* (Cyanophyta): Intrinsic culture variability and correlation with metabolic variables. *J Phycol* 17:129–134.
Francko, D.A., and Wetzel, R.G. 1981b. Synthesis and release of cyclic adenosine 3′:5′-monophosphate by aquatic macrophytes. *Physiol Plant* 52:33–36.
Gabrielson, J.O., Perkins, M.A., and Welch, E.B. 1984. The uptake, translocation and release of phosphorus by *Elodea densa*. *Hydrobiology* 11:43–48.
Goldman, C.R., and Horne, A.J. 1983. *Limnology*. New York: McGraw-Hill.
Hallegraeff, G.M. 1977. A comparison of different methods used for the quantitative evaluation of biomass of freshwater phytoplankton. *Hydrobiology* 5:145–165.
Jacoby, J., Lynch, D., Welch, E.B., and Perkins, M.A. 1982. Internal phosphorus loading in a shallow eutrophic lake. *Water Res* 16:911–919.
James, W.F. 1984. Effects of endothall treatment on phosphorus concentration and community metabolism of aquatic communities. Misc. Paper A-84-1. U.S. Army Corps of Engineers, Vicksburg, MS.
Jones, R.A., and Lee, G.F. 1982. Recent advances in assessing impact of phosphorus loads on eutrophication-related water quality. *Water Res* 16:503–515.
Kennedy, R.H., Thornton, K.W., and Ford, D.E. 1985. Characterization of the reservoir ecosystem. In D. Gunnison, ed., *Microbial Processes in Reservoirs*, The Hague, Netherlands: Junk Publishers, Chapter 2.
Kerekes, J. 1977. The index of basin permanence. *Int Rev Ges Hydrobiol* 62:291–293.
LaBaugh, J.W., and Winter, T.C. 1984. The impact of uncertainties in hydrologic measurement on phosphorus budgets and empirical models for two Colorado reservoirs. *Limnol Oceanogr* 29:322–339.

Larsen, D.P., and Mercier, H.T. 1976. Phosphorus retention capacity of lakes. *J. Fish Res Board Can* 33:1742-1750.

Larsen, D.P., Schults, D.W., and Malueg, K.W. 1981. Summer internal phosphorus supplies in Shagawa Lake, Minnesota. *Limnol Oceanogr* 26:740-753.

Lee, D.R. 1977. A device for measuring seepage flux in lakes and estuaries. *Limnol Oceanogr* 22:140-147.

Lee, D.R., and Hynes, H.B.N. 1978. Identification of groundwater discharge zones in a reach of Hillman Creek in southern Ontario. *Water Pollut Res Can* 13:121-133.

Lee, G.F., Jones, R.A., and Rast, W. 1981. Index to evaluate lake restoration. *J Environ Eng Div ASCE* 107:1334-1336.

Likens, G.E. 1972. Eutrophication and aquatic ecosystems. In *Nutrients and Eutrophication, Proceedings of the Symposium on Nutrients and Eutrophication: The Limiting-Nutrient Controversy*, Lawrence, KS. Sponsored by the American Soc. Limnol. Oceanogr., 3-13.

Likens, G.E., and Bormann, F.H. 1974. Linkages between terrestrial and aquatic ecosystems. *BioScience* 24:447-456.

Lind, O.T. 1979. *Handbook of Common Methods in Limnology*. St. Louis: Mosby.

Lorenzen, M.W. 1980. Use of chlorophyll-Secchi disk relationships. *Limnol Oceanogr* 25:371-372.

Mockus, V. 1971. Estimation of direct runoff from storm rainfall. US Dept. Agriculture, Soil Conservation Serv., National Engineering Handbook. Sect. 4, Hydrology, Chap. 10.

Morton, F.I. 1976. Climatological estimates of evapotranspiration. *J. Hydraul Div ASCE* 2:275.

Naiman, R.J. 1982. Characteristics of sediment and organic carbon export from pristine boreal forest watersheds. *Can J Fish Aquatic Sci* 39:1699-1718.

Nicholls, K.H., and Dillon, P.J. 1978. Evaluation of phosphorus-chlorophyll-phytoplankton relationships for lakes. *Int Rev Ges Hydrobiol* 63:141-154.

Nurnberg, G.K. 1984. The prediction of internal phosphorus load in lakes with anoxic hypolimnia. *Limnol Oceanogr* 29:111-124

Omernik, J.M. 1977. *Nonpoint Source-Stream Nutrient Level Relationships: A Nationwide Study*. EPA 600/3-77-105.

Pennak, R.W. 1957. Species composition of limnetic zooplankton communities. *Limnol Oceanogr* 2:222-232.

Pentland, R.L. 1968. Runoff characteristics in the Great Lakes Basin. *Proc 11th Conf Great Lakes Res* 11:326-359.

Piest, R.F., Kramer, L.A., and Heinemann, H.G. 1975. Sediment movement from loessial watersheds. In *Present and Prospective Technology for Predicting Sediment Yields and Sources, Proc. of Workshop*, Agri. Res. Serv., ARS-S 40, 130-141.

Porcella, D.B., Peterson, S.A., and Larsen, D.P. 1980. Index to evaluate lake restoration. *J Environ Eng Div ASCE* 106:1151-1169.

Rast, W., and Lee, G.F. 1978. *Summary Analysis of the North American (U.S. Portion) OECD Eutrophication Project: Nutrient Loading-Lake Response Relationships and Trophic State*. EPA-600/3-78-008.

Rast, W., and Lee, G.F. 1983. Nutrient loading estimates for lakes. *J. Environ Eng Div ASCE* 109:502-517.

Reckhow, K.H. 1979a. *Quantitative Techniques for the Assessment of Lake Quality*. EPA-440/5-79-015.

Reckhow, K.H. 1979b. Uncertainty analysis applied to Vollenweider's phosphorus loading criterion. *J. Water Pollut Control Fed* 51:2123-2128.

Reckhow, K.H. 1983. A method for the reduction of lake model prediction error. *Water Res* 17:911-916.

Reckhow, K.H., and Chapra, S.C. 1983. *Engineering Approaches for Lake Management: Vol. I. Data Analysis and Empirical Modeling*, Boston: Butterworth.

Reckhow, K.H., and Simpson, J.T. 1980. A procedure using modeling and error analysis for the prediction of lake phosphorus concentration from land use information. *Can J Fish Aquatic Sci* 37:1439-1448.

Reid, G.K., and Wood, R.D. 1976. *Ecology of Inland Waters and Estuaries*. New York: Van Nostrand.

Rigler, F.H. 1974. Phosphorus cycling in lakes. In F. Ruttner, *Fundamentals of Limnology*. 3rd ed. Toronto: Univ. Toronto Press, 263-273.

Ruttner, F., ed. 1963. *Fundamentals of Limnology*. 3rd ed. Toronto: Univ. Toronto Press.

Sakamoto, M. 1966. Primary production by phytoplankton community in some Japanese lakes and its dependence on lake depth. *Arch Hydrobiol* 62:1-28.

Scheider, W.A., Moss, J.J., and Dillon, P.J. 1979. Measurement and uses of hydraulic and nutrient budgets. In *Lake Restoration* EPA 440/5-79-001, 77-83.

Siegel, D.I., and Winter, T.C. 1980. Hydrologic setting of Williams Lake, Hubbard County, Minnesota. U.S. Geol. Surv. open-file report 80-403.

Smith, V.H. 1979. Nutrient dependence of primary productivity in lakes. *Limnol Oceanogr* 24:1051-1064.

Sonzogni, W.C., and Lee, G.F. 1974. Nutrient sources for Lake Mendota—1972. Trans Wisc Acad Sci 62:133-164.

Stauffer, R.E., and Lee, G.F. 1973. The role of thermocline migration in regulating algal blooms. In E.J. Middlebrooks, D.H. Falkenborg, and T.E. Maloney, eds., *Modeling the Eutrophication Process*, Utah State Univ. 73-82.

Thornton, K.W., R.H. Kennedy, J.H. Carroll, W.W. Walker, R.C. Gunkel, and S. Ashby. 1980. Reservoir sedimentation and water quality—a heuristic model. In Proceedings of Symposium on Surface Water Impoundments. Amer. Soc. Civil Engineers, Minneapolis, Minn. p. 654-661.

United States Environmental Protection Agency. 1974. The relationship of nitrogen and phosphorus to the trophic state of Northeast and North-Central lakes and reservoirs. NES Working Paper 23, Corvallis Environmental Research Laboratory.

United States Environmental Protection Agency. 1980. *Clean Lakes Program Guidance Manual*, EPA 440/5-81-003.

Uttormark, P.H., and Hutchins, M.L. 1980. Input/output models as decision aids for lake restoration. *Water Res Bull* 16:494-500.

Uttormark, P.D., Chapin, J.D., and Green, K.M. 1974. *Estimating Nutrient Loadings of Lakes from Non-Point Sources*. EPA 660/3-74-020.

Vollenweider, R.A. 1975. Input-output models with special reference to the phosphorus loading concept in limnology. *Schweiz Z Hydrol* 37:53-84.

Vollenweider, R.A. 1976. Advances in defining critical loading levels for phosphorus in lake eutrophication. *Mem Ist Ital Idrobiol* 33:53-83.

Wachter, J.K., and Andelman, J.B. 1984. Organohalide formation on chlorination of algal extracellular products. *Environ Sci Technol* 18: 811-817.

Walker, W.W. Jr. 1980. Variability of trophic state indicators in reservoirs. In *Restoration of Lakes and Reservoirs*. EPA 440/5-81-010. 334-348.

Walker, W.W. Jr. 1984. Trophic state indices in reservoirs. In *Lake and Reservoir Management*. EPA 440/5-84-001. 435-440.

Walker, W.W. Jr. 1985. Empirical methods of predicting eutrophication in impoundments, Applications manual. EWQOS Program. U.S. Army Corps of Engineers. Vickburg, MS.

Weibel, S.R. 1969. Urban drainage as a factor in eutrophication. In *Eutrophication: Causes, Consequences, Correctives*. Wash., D.C: National Academy of Sciences, 383-403.

Welch, E.B. 1980. *Ecological Effects of Wastewater*. London: Cambridge Univ. Press.

Welch, E.B., Michaud, J.P., and Perkins, M.A. 1982. Alum control of internal phosphorus loading in a shallow lake. *Water Res Bull* 18:929-936.

Welch, E.B., Rock, C.A., Howe, R.C., and Perkins, M.A. 1980. Lake Sammamish response to wastewater diversion and increasing urban runoff. *Water Res* 14:821-828.

Wetzel, R.B. 1983. *Limnology* 2nd ed. Philadelphia: Saunders.

Wetzel, R.G., and Likens, G.E. 1979. *Limnological Analysis* Philadelphia: Saunders.

Winter, T.C. 1978. Ground-water component of lake water and nutrient budgets. *Verh Int Ver Limnol* 20:438-444.

Winter, T.C. 1980. Survey of errors for estimating water and chemical balances of lakes and reservoirs. *Symposium on Surface Water Impoundments*, sponsored by American Society of Civil Engineering, Minneapolis, 224-233.

Winter, T.C. 1981. Uncertainties in estimating the water balance of lakes. *Water Res Bull* 17:82-115.

II

Physical and Chemical Methods to Reduce Nutrient Concentrations

3

Advanced Treatment and Diversion of Wastewater and Stormwater

A well-recognized first step in restoring eutrophic lakes is to remove or treat direct inputs of wastewater, stormwater, or both that contain high concentrations of nitrogen and phosphorus. The purpose is to reduce external loading. Without this any long-term benefits of in-lake treatment will be minimal. There have been cases in which diversion was the only treatment required to restore lakes, but most have required in-lake treatments to control internal cycling of nutrients to realize significant improvements in water quality.

Many lakes have been observed to degrade as anthropogenic sources of wastewater (mainly sewage effluent), containing high nutrient content, increased over time. The increased algal and nutrient content, the appearance of surface scums, and decreasing water transparency, among more subtle changes, prompted corrective action. For lakes where the principal source(s) of nutrients causing the problem are well defined numerically and spatially, the corrective action has been to divert the wastewater or stormwater to a different receiving water where nutrient content does not limit growth of plankton algae. For example, if light is limiting in the receiving water, such that nutrients are not fully utilized, then additional nutrient increase should not cause increased algal abundance.

The waste source may already be collected in a wastewater system and represent a "point" source. Where individual household septic tank systems or stormwater runoff constitute nonpoint sources, a collection system may be a necessary part of the diversion project. In either case, relatively large pipes are usually installed to transport the diverted wastewater or stormwater, and this portion often represents the greatest cost. Advanced wastewater treatment (AWT), in which treated water with lowered nutrient content is returned to the lake, is an alternative to most of the piping and may represent a reduction in capital costs over that for diversion.

MECHANISMS

Diversion or AWT has worked successfully to recover lake quality in cases where external loading dominates the nutrient cycle. The concentration of the most limiting nutrient, usually phosphorus (P) determines the lake's trophic state. The maximum concentration of algae in the lake is set by the concentration of the most limiting nutrient. The lake concentration of nutrient, in turn, is dependent primarily on external loading, sedimentation rate, and outflow of nutrient. This is illustrated by the following steady-state equation for lake P concentration (Vollenweider, 1976; Chapra and Reckhow, 1979):

$$P = \frac{L}{\overline{Z}\rho} \cdot \frac{1}{1 + \frac{1}{\sqrt{\rho}}} \qquad (3.1)$$

where L is external loading, $\overline{Z}$ is mean depth, and ρ is flushing rate. The term $L/\overline{Z}\rho$ is the average inflow concentration, and $1/(1 + 1/\sqrt{\rho})$ is the correction for sedimentation loss where $1-1/(1 + 1/\sqrt{\rho})$ is the P retention coefficient. Thus the average concentration of P in the lake would not exceed the average inflow concentration if the lake content were dependent upon external loading, and would in fact be some fraction of the inflow defined by the sedimentation loss term. Diversion of wastewater or stormwater would result in a decrease in L and therefore a decrease in the inflow concentration of nutrient. Following an exponential decrease in lake concentration as old lake water is replaced with new, lower-nutrient water, a lower steady-state concentration would eventually be reached. This expectation is not always realized, as shown below.

The analysis of error using the above equation to estimate the P content of lakes is discussed in detail by Chapra and Reckhow (1979) and Reckhow and Chapra (1983). As an example interpretation, for a population of 117 lakes a model prediction of 20 μg l^{-1} would have approximately a 4, 46, 49, and 1 percent chance of actually occurring in, respectively, the concentration ranges (trophic states): < 10 μg l^{-1}, 10-20 μg l^{-1}, > 20 μg l^{-1} and > 50 μg l^{-1}.

EXTENT OF RECOVERY EXPECTED

Results from several diversion projects have shown that there is a high probability for lake P to decrease and for the lake to recover. Whether the lake returns to a "lower" trophic state, similar to what existed prior to degradation, however, depends upon the level to which P decreases. Of course other factors, such as predator-prey relations, can be important. Uttormark and Hutchins (1980) reviewed results from 23 lakes following diversion (or AWT), and while 83 percent responded in a predictable manner regarding change in eutrophic state (> 20 μg P l^{-1} for eutrophy, 10-20 μg l^{-1} for mesotrophy, and < 10 μg l^{-1} for oligotrophy), only 30 percent actually recovered to a "lower" trophic state. Thus half of the lakes showed positive and predictable responses to diversion in terms of decreased P concentration, but the amount of P diverted (actual decrease in L) was often insufficient to cause the lake concentration to reach 20 μg l^{-1} or less. Three of the lakes that did not respond according to the model actually recovered to "lower" trophic states than predicted. Thus, it seems clear that lakes might improve following diversion or AWT, but they might not recover to a "lower" trophic state if the fraction of nutrient diverted is insufficient.

Even if diversion is not likely to cause a shift from eutrophy to mesotrophy (i.e., < 20 μg l⁻¹), the project may still be a success. In concert with the P limits for trophic state, average summer chlorophyll a content should be less than about 8 μg l⁻¹ for mesotrophy (Chapra and Tarapchak, 1976). A decrease in mean summer chl a from 50 to 25 μg l⁻¹ would not qualify as a change in trophic state, but should still show a relatively large improvement in lake quality. Conditions of surface scum formation, odors surrounding the lake, and taste and odors in fish would no doubt decrease, and transparency of the water should show some improvement.

Improvement in transparency of the water is not linearly related with a reduction in chl a and P concentrations. For example, when 87 percent of the external P input to Lake Norrviken was diverted, the lake remained eutrophic. The summer total P concentration decreased from 263 to 94 and chl a decreased from 90 to 49 (all in μg l⁻¹), but Secchi transparency only increased from 0.8 to 1.0 m. The degree of improvement in transparency, for an equal amount of P diverted, would become greater as a mesotrophic state (< 20 μg l⁻¹ P) is approached. This is largely due to the nonlinear relationship between algal concentration and transparency, as reported by Carlson (1977):

$$\text{Secchi depth} = 7.7/\text{chl }a^{0.68}. \quad (3.2)$$

The graphical display of this relationship in Fig. 3-1 shows that larger and larger increases in transparency occur at each successive decrease in chl a content. A given decrease in P and chl a will be more apparent in terms of water clarity improvement in mesotrophic or lower eutrophic lakes than in higher eutrophic or hypereutrophic lakes.

Thus, to restore a lake to a particular trophic state the fraction of loading diverted is not as important as the final level of loading and consequently the lake concentration. No relation is apparent between the fraction diverted and the change

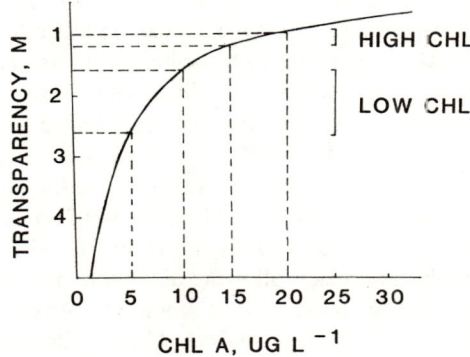

Figure 3-1 Chl a versus transparency (after Carlson, 1977), showing greater absolute benefits to transparency for an incremental change at low versus high chl a content.

in trophic state in the lakes reviewed by Uttormark and Hutchins (1980). Six lakes that had three-fourths or more of the external P diverted did not change trophic state, yet three of eleven lakes that changed trophic state had one-half or less of the external P diverted.

Although sediments are an ultimate sink for P, there is often a large interchange of P between sediments and water. The net effect is that more P goes into and is permanently trapped by sediments than is released back to the water. Because the sediments in most lakes do permanently trap part of the net annual input of P, Vollenwieder's equation (3.1), in which the sedimentation rate coefficient σ is estimated as $\sqrt{\rho}$, usually predicts in-lake P concentration with reasonable consistency. Jones and Bachmann (1976) showed that, for 143 lakes surveyed, the mean value calculated for σ was 0.65. Reckhow and Chapra (1983) have described the rationale for various approaches to estimating and generally applying sedimentation loss terms.

Notwithstanding the general applicability of a constant term for sedimentation loss, there are many lakes in which large releases from sediments do occur (Ahlgren, 1977; Bengtsson, 1975; Jacoby et al., 1982; Kamp-Nielsen, 1975; Larsen et al., 1981; Ryding and Forsberg, 1980; Welch, 1977). In some lakes such internal releases may not greatly alter the annual mean P concentration, although seasonal peaks, usually in summer, may occur and produce high concentrations of algae (Larsen et al., 1979). In other lakes the seasonal peaks may result in much higher annual means than predicted by Vollenwieder's model (eq. 3.1; Welch et al., 1979). In both instances, internal loading may significantly retard the recovery rate of a lake following treatment.

Although internal loading may initially retard recovery, it is reasonable to believe that the rate of internal loading will eventually decrease. This has been hypothesized by Lorenzen (1973) for lakes in general and by Kamp-Nielsen (1980) for hypereutrophic Lake Glumsø. A decrease in the rate of sediment P release has apparently occurred in two lakes, which before and after diversion had relatively high internal loading—Lake Norrviken (Ahlgren, 1977) and Lake Sammamish (Welch et al., 1980). Although the results are not entirely clear, the rate of increase of hypolimnetic P in summer decreased in both lakes during the years following diversion. This is indicated in Fig. 3-2, for Lake Sammamish, by the more gradual rate of increase in whole-lake P in the years following compared to those before diversion. Recent results show that during 1975 to 1984 (7-16 years after diversion) sediment release rate as estimated from changes in hypolimnetic P content has declined to 3.4 ± 2.7 mg M^{-2} day^{-1} from the pre- and post-diversion (1964-1974) rate of 5.8 ± 3.2 mg M^{-2} day^{-1}. In Norrviken the total P content of the sediment declined and the release rate, determined in the same way as for Lake Sammamish, declined from 9.2 to 1.6 mg M^{-2} day^{-1} over a period of six years. Larsen et al. (1979), however, did not observe a decline in sediment release rate in Shagawa Lake following reduction in P input.

An eventual decrease in sediment release should result from sediment, depositing on the lake bottom, that contains progressively less P following diversion or AWT. Although that phenomenon was shown in Lake Norrviken, it apparently did not completely explain the declining rate of hypolimnetic P increase. Nevertheless, as older, richer sediments are gradually removed from contact with overlying water, as a

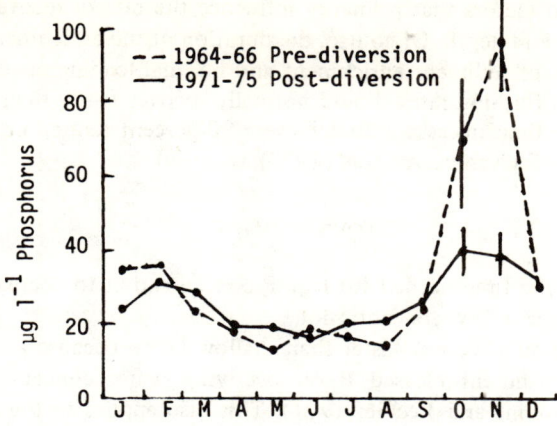

Figure 3-2 Mean volume-weighted total phosphorus concentration for three years before and five years after diversion in Lake Sammamish, Washington. Each point represents all data for that month, usually two observations, and vertical lines represent standard errors. Reprinted with permission from *Water Research*, vol. 14, Welch et al., Lake Sammamish, response to wastewater diversion and increasing urban runoff, copyright 1980. Pergamon Press Ltd.

result of continual covering with sediment poorer in P, the rate of release should decrease. Because each lake has an unique loss rate for internally loaded P, and because the release rates vary so greatly for any given sediment content of P, it is probably impossible to predict a general rate of decrease in internal loading applicable for all lakes. Aside from the examples mentioned above, the problem has not been researched sufficiently to provide a reasonable range in expected rates of decrease. Based on experiences in the three lakes mentioned, ten years may be necessary to observe a significant decrease in internal loading. Kamp-Nielsen (1980) has predicted recovery in three to four years for rapidly flushed, hypereutrophic, and internally loaded Lake Glumsø following the planned diversion of wastewater.

For the purpose of predicting improvements in lake quality that will result from diversion, it is recommended that a model be used that is modified to consider internal loading and seasonal changes in P, such as that developed by Larsen et al. (1979). Representing seasonal changes is important to the prediction of lake recovery in lakes with large internal releases because dealing with annual means may minimize the summer condition, which is highly visible to water users. At that time most of the P loading may be internal. Because the rate of decline in sediment release is uncertain, and because of other mechanisms that contribute to internal loading, it is further recommended that the recovery rate of P in the lake be predicted using a constant internal sediment release rate, as a conservative estimate, unless more specific information exists for the lake in question.

There are two factors that primarily influence the rate of recovery with respect to P—flushing rate and depth. Of course, the duration of the enrichment period is also important in terms of sediment enrichment and internal loading, as discussed above, but lakes with high flushing rates should normally recover faster than lakes with low flushing rates. The time necessary to achieve a 90-percent reduction in lake content from washout alone is given by Ahlgren (1979) as

$$t_{90} = \ln 10 / \rho. \qquad (3.3)$$

This also refers to the time needed for lake P concentration to decrease to within 10 percent of the average inflow concentration.

Deep lakes should recover faster than shallow lakes, because as depth increases the influence of sediment-released P on overlying water concentration decreases (Ahlgren, 1979; Stumm and Leckie, 1971). This also applies to the idea of internal loading from littoral regions (Cooke et al., 1977; Fee, 1979). Normally, as depth increases, less of the lake volume covers the littoral region. The effects of wind as an entrainment force for sediment-released P, in either stratified or unstratified lakes (Ahlgren, 1980; Larsen et al., 1981; Ryding and Forsberg, 1980; Stauffer and Lee, 1973), also diminishes as depth increases.

CASE STUDIES

Six examples of diversion and AWT will be discussed in some detail here. Although wastewater has been diverted from a sizeable number of lakes in the world, complete data sets that describe conditions from before and after treatment, even for P, are scarce. Uttormark and Hutchins (1980) compiled data sets for 46 treated lakes, but before-and-after data were available for only 23 of which 20 are listed in Table 3-1. By including six additional lakes from Sweden (Boren, Ekoln, Norrviken, Vallentunsajön, Edssjön, and Oxundasjön), one from Norway, and one from Switzerland there is a total of 30 for which before-and-after data exist (Table 3-1). Recently, Rast et al. (1983) indicate that sufficient data to evaluate response to diversion and AWT exist for only ten lakes in the world. Five of their ten (Gravenhurst Bay, Haley's Pond, Lake Minnetonka, Lake Ryssbysjön, and Lake Wahnbach) are not listed in Table 3-1. Unfortunately, data were not available on some of the 28 listed so only a change in trophic state is indicated rather than specific trophic state variables.

Lake Washington

The diversion of domestic wastewater from Lake Washington from 1963 to 1967 by the Municipality of Metropolitan Seattle is undoubtedly the most celebrated diversion and recovery project in the world. The primary reason for its fame was its rapid and predictable recovery at a time when there existed considerable doubt about the prospects for restoring lakes once they had become eutrophic. Using the Vollenweider

Table 3-1 Lakes for Which Complete Before-and-After Diversion Data Sets Exist.

Lake	Country	% Reduction External P Load	Trophic State*	Source
Boren	Sweden	70	+	3
Brielse Meer	Netherlands	52	0	1
Edssjön	Sweden	80	0	2
Ekoln	Sweden	70	0	3
Faakersee	Austria	85	+	1
Gjersjøn	Norway	78	0	3
Inkwill	Switzerland	75	0	1
Kegonsa	USA	72	0	1
Klopeiner	Austria	95	+	1
Mascoma	USA	34	0	1
Mendota	USA	20	0	1
Millstättersee	Austria	54	+	1
Norrviken	Sweden	87	0	2
Ossiachersee	Austria	47	+	1
Oxundasjön	Sweden	89	0	2
Sammamish	USA	35	+	1
Shagawa	USA	85	0	1
Skatutakee	USA	67	+	1
Stone	USA	97	0	1
Turnersee	Austria	83	+	1
Uttran	Sweden	53	−	1
Vallentunasjön	Sweden	92	0	2
Washington	USA	88	+	1
Waubesa	USA	73	0	1
Weissensee	Austria	87	+	1
Wörthersee	Austria	41	+	1
Zellersee	Austria	81	+	1
Zürich	Switzerland	54	0	4

*Trophic state change from eutrophic to mesotrophic or oligotrophic indicated by +, no change by 0, and no data by −.
1. Uttormark and Hutchins (1980)
2. Ahlgren (1979)
3. Ryding (1980)
4. Schantz (1978)

model (eq. 3.1) the P content of Lake Washington would have been expected to decline from 64 $\mu g \, l^{-1}$ (mean annual total P 1961-1964) prior to diversion to 21 $\mu g \, l^{-1}$ at equilibrium. TP actually did decrease to 21 $\mu g \, l^{-1}$ by 1969, two years after the diversion was completed. The 1969-1975 7 year mean was 19 $\mu g \, l^{-1}$ (Edmondson and Lehman, 1981; Table 3-2). TP changed little after that; the 1976-1979 4 year mean was 17 $\mu g \, l^{-1}$ (Table 3-2). The response is shown graphically in Fig. 3-3.

Phytoplankton chl *a* also decreased in direct proportion to TP. Although the lake had approached a nitrogen-limiting condition prior to diversion, primarily because the ratio of N to P in sewage effluent is rather low (2:1-3:1), P was quickly reestablished

Table 3-2 Characteristics of Four Lakes, Averaged Over Indicated Years Before and for Successive Periods Following Diversion or Wastewater Treatment (P Removal). $\bar{Z}$ = mean depth, m, ρ = flushing rate, yr^{-1}; L_{int} = internal loading; SD = Secchi transparency, m, as summer mean; TP = total phosphorus, μg/l^{-1}, as annual mean except June–Sept. for Norrviken; chl a, μg/l^{-1}, as summer mean.

Lake	$\bar{Z}$	ρ	L_{int}	Years pre/post	SD pre/post	TP pre/post	Chl a pre/post
Washington[1]	37	0.40	no	4 / 7	1.0 / 3.1 6.9	64 / 19 17	36 / 6 3
Sammamish[2]	18	0.55	yes	2 / 5 4	3.2 / 3.4 4.9	33 / 27 19	5 / 7 2.7
Norrviken[3]	5.4	1.2	yes	2 / 6 3	0.8 / 0.7 1.0	263 / 174 94	90 / 85 49
Shagawa[4]	5.6	1.6	yes	2 / 3		51 / 30	28 / 24

1. Edmondson and Lehman (1981)
2. Welch et al. (1980), Municipality of Metropolitan Seattle, WA.
3. Ahlgren (1980)
4. Larsen et al. (1979)

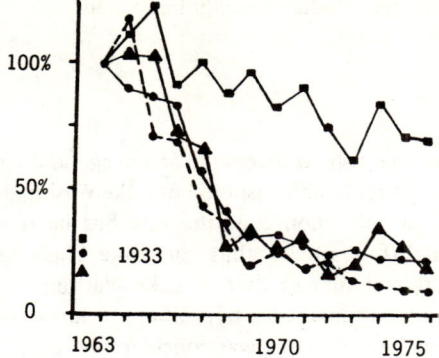

Figure 3-3 Summary of changes in relative nutrient and chlorophyll a content in Lake Washington following diversion of treated sewage effluent from 1963 (100 percent) through 1976. Circles, dashed line = chl a; circles, solid line = TP; triangles = phosphate P; squares = nitrate N (from Edmondson, 1978).

following diversion as the most limiting nutrient (Edmondson, 1970). Summer chl a content had reached a level of 7 μg l^{-1} by 1969. Chl a remained about the same through 1975, with a seven-year mean of 6 μg l^{-1} (Table 3-2; Fig. 3-3). Transparency improved threefold, from a summer mean of 1.0 m to one of 3.1 m, during the same period.

The lake showed yet another marked improvement after 1975. Transparency more than doubled during the subsequent four-year period (to 6.9 m), while chl a declined by half to 3 μg l^{-1} (Table 3-2). The improvement in lake quality has been related to the recurrence of *Daphnia* and its dominance of the zooplankton in 1976 (Edmondson and Litt, 1982). *Daphnia* populations increased at that time apparently because *Neomysis mercedis*, a predator, had decreased in the mid 1960s and blue-green algae (especially *Oscillatoria*) became markedly less abundant. Because blue-green algae, especially *Oscillatoria*, accumulate in the filtering mechanism of *Daphnia*, a relatively high abundance of blue-greens in the phytoplankton reduces the efficiency of *Daphnia*'s food consumption.

Lake Washington is deep (mean depth 37 m), flushes rapidly (0.4 yr^{-1}), and was enriched at its maximum loading for a relatively short time (lake P did not exceed 30 μg l^{-1} until 1959). These three factors are no doubt of primary importance in explaining the rapid and complete recovery of the lake. The great depth and the fact that its period of maximum enrichment was short combined to minimize the potential for internal loading, which would have delayed recovery. The hypolimnion always remained oxic, even in its most eutrophic period.

A flushing rate of 0.4 yr^{-1} suggests that phosphorus washout due to water exchange alone would have resulted in the lake concentration approaching 90 percent of its equilibrium level in 5.8 years (eq. 3.3). If sedimentation is included as $\sqrt{\rho}$

Lake Sammamish

in the denominator of equation 3.3, the 90-percent change should have occurred in 2.2 years, which is much closer to what actually happened.

Lake Sammamish

The response of Lake Sammamish to diversion of sewage and dairy plant effluent is a striking contrast to the very predictable response of Lake Washington. Lake Sammamish has responded favorably to diversion, but the rate has been much slower and the explanation not as straightforward as that for Lake Washington. Although Lake Sammamish flushes at a rate similar to that of Lake Washington, it is only about half as deep (Table 3-2), it has an anaerobic hypolimnion during late summer and early autumn, the decrease in extreme loading was considerably less (35% versus 88%, Table 3-1), and it received its treated-sewage P load via its principal inflow stream. The effluent entered the stream about 3 km from the lake, whereas treated sewage effluent was discharged directly into Lake Washington.

These differences meant that (1) Lake Sammamish had a much greater internal P loading (about 50% of total) than Lake Washington, due to the anaerobic conditions and the effect of iron at the sediment-water interface, as well as to the much greater ratio of sediment area to lake volume; (2) Lake Washington probably received a much greater fraction of its sewage P in a dissolved form, whereas sewage P released to Lake Sammamish had a greater opportunity to be converted to particulate form in the 3 km of stream before it entered the lake. Stream monitoring showed further that most of the P load to Sammamish entered in the winter, during high flow, and two-thirds to three-fourths was particulate, most of which probably sedimented and was not available for plankton uptake in the spring.

The annual mean whole-lake P concentration in Lake Sammamish is not predicted closely by equation 3.1 either before or after diversion. Predicted values are higher than those observed by 33 percent and 45 percent before and after diversion, respectively. If the sediment-loss term is adjusted to predict the prediversion level, then the apparent equilibrium level of about 20 μg l^{-1} is predictable. That concentration was reached in 1975, with even lower concentrations persisting in the four years from 1981 and 1984. The lake was slow to reach that level; levels remained much higher for six years after diversion (Fig. 3-4, Table 3-2). The slow recovery to lower P content is probably due to declining release rates from sediments and resultant decreased internal loading, as indicated earlier.

Unlike the response in Lake Washington, but consistent with the lack of a significant decrease in lake P, neither chl a or Secchi transparency in Sammamish changed during the first seven years after diversion. Further, there was little evidence that the lake had deteriorated as a result of sewage input in the first place. Various indices suggest that the lake did not respond to sewage enrichment in an expected pattern. The stream pathway for sewage P input, together with the high iron content in the hypolimnion, which effectively removes P at overturn, may have resulted in sedimentation of a very high proportion of the external loading to Sammamish. These factors tend to render P relatively unavailable for plankton algae in spring and summer.

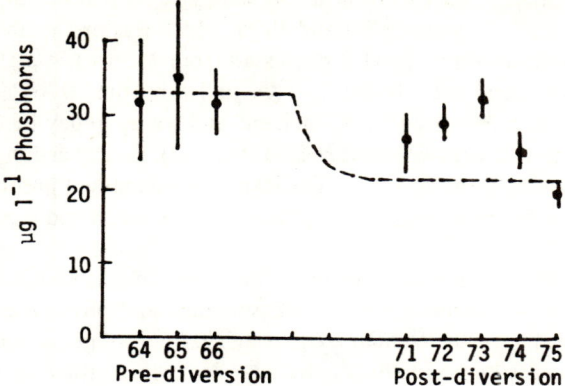

Figure 3-4 Mean volume-weighted total phosphorus concentration for prediversion and postdiversion years in Lake Sammamish, Washington. The dashed line represents predicted response from diverting one-third of the external phosphorus load in 1968. Reprinted with permission from *Water Research*, vol. 14, Welch et al., Lake Sammamish response to wastewater diversion and increasing urban runoff, copyright 1980, Pergamon Press Ltd.

Such a result is suggested by Fig. 3-2, where the spring-summer concentration was about the same before and after diversion, as well as by the overestimation of the prediversion and postdiversion lake P content by equation 3.1. Lake Sammamish may have a rather high capacity to assimilate changes in nutrient loading over the range recently experienced, as long as it is added via a stream and at least with respect to the spring and summer plankton biomass.

The lake did ultimately respond favorably. The fraction of the phytoplankton composed of blue-green algae decreased by an average of nearly 50 percent during the first seven years after diversion. As shown in Table 3-2, by 1981-1984 the mean summer chl *a* had decreased by nearly 50 percent and mean Secchi transparency in summer increased by nearly the same magnitude from prediversion levels. While stratification may prevent hypolimnetic P, released from sediment in late summer and early autumn, from reaching phytoplankton in the photic zone normally, the results from Lake Sammamish suggest that the apparent decrease in the rate of sediment P release by the 1980s may, in fact, have caused the eventual decrease in phytoplankton.

Lake Norrviken

This Swedish lake contrasts with both Lake Washington and Lake Sammamish in that it is much shallower and was hypereutrophic prior to treatment (Table 3-2). As a result of its anaerobic hypolimnion, Norrviken had a large internal loading, which has declined steadily since diversion of 87 percent of the P input (Ahlgren, 1977, 1979, and 1980). The maximum concentration in the lake at overturn declined from about 450 μg l^{-1} to 150 μg l^{-1} between 1970 and 1975, which was predictable largely from dilution alone. Summer levels of total P decreased from 263 to 174 μg l^{-1} during that same period, but decreased even further, to 94 μg l^{-1} by the 1976-1979 three-year period. As a result, chl a improved by 45 percent and transparency by 25 percent for the three-year period over prediversion levels (Table 3-2). Postdiversion summer mean transparency has been as great as 1.2 m. The lake showed marked improvement but is still eutrophic due to its initial hypereutrophic state and continued high loading from diffuse sources.

Although Norrviken is still eutrophic, the diversion is considered a success because of the degree of improvement in transparency and phytoplankton biomass. Also, decreased dominance by *Oscillatoria*, from nearly a monoculture in 1969, has greatly improved the aesthetic quality of the lake. Norrviken, then, is an example, as noted earlier, of the "success" of diversion even when the overall trophic state remains unchanged. Trophic state helps define lake conditions but should not be used too rigidly to interpret restoration success (i.e., restoration need not be seen as successful only if the trophic state changes).

Shagawa Lake

This lake was studied thoroughly during the period 1971 through 1976. The external P loading was reduced by 85 percent by P removal through AWT. The lake has shown some improvement, but has not reached expectations (Larsen et al., 1979).

The average annual total P concentration declined from 51 μg l^{-1} before treatment to 30 μg l^{-1} after, a 40 percent reduction (Table 3-2). Soluble P showed a greater reduction, 80 percent to 4.5 μg l^{-1}, which probably indicates P has become more consistently the limiting nutrient. The average chl a content in spring decreased by 50 percent, but little change was observed in summer, except that bloom duration decreased. The average lake total P content should have declined to 12 μg l^{-1} in 1.5 years, according to the steady-state model of Larsen et al., (1979). The reason the lake has not reached expectations is internal loading from anaerobic sediments during summer (Larsen et al., 1981). Net release rates have averaged 5.3 mg M^{-2} day^{-1} over the whole lake area (nearly double that for the profundal zone only) during summer. As much as 6-12 mg P M^{-2} day^{-1} may be transported vertically by turbulent diffusion during summer, when steep P gradients develop across the thermocline (Larsen et al., 1979). That internal supply and availability to the photic zone would account for the continuing high chl a content in summer, as well as the less-than-expected decline in annual mean total P. Stauffer and Lee (1973) suggested that entrainment of hypolimnetic P during summer storms accounted for plankton blooms in Lake Mendota.

From results in Lake Norrviken and Lake Sammamish, it may be reasonable to expect the internal loading in Shagawa to decline over several years. If that happens, improvements could be expected to occur in the summer, especially since the soluble

P content has been significantly reduced. The model developed for P by Larsen et al. (1979) fits the observations in Shagawa quite well during the first three years after treatment and assumes a constant rate of internal loading. Thus there was no indication that internal loading was declining.

Lake Waubesa

The results from this lake at Madison, Wisconsin, are similar to those for Lake Norrviken. Diversion of Waubesa's treated sewage effluent in 1958 decreased the external P load by 88 percent, resulting in a predictable decline in P from more than 500 μg 1^{-1} to about 50 μg 1^{-1}. Rural runoff has kept Lake Waubesa in an eutrophic state. Fitzgerald (1964) reported that algal biomass in Waubesa did not change from prediversion levels, but the nearly complete dominance by *Microcystis* changed to more mixed assemblages of species. (A similar change occurred in Lake Norrviken for *Oscillatoria*, as indicated previously.) With a more complete data record, changes in biomass might have been noted in Waubesa, although Stewart (1976) suggested that no significant change in either transparency or oxygen was detectable over and above normal year-to-year fluctuations.

Lake Zürich

Over a ten-year period beginning in about 1965, P removal was installed in all treatment plants within the immediate drainage to Lake Zürich. Conventional wastewater treatment plant construction was initiated in 1955. About 54 percent of the P load entering the lake prior to 1955 has been removed by AWT. Substantial decrease in hypolimnetic oxygen deficit has occurred, with the most improvement occurring in the ten years after installation of conventional treatment facilities in 1955 (Shantz, 1977). Oxygen deficit averaged 27 percent less during 1971-1976 than in 1953-1959. Although P and N concentrations in the lake have not been reduced, neither have they increased.

The transparency of Lake Zürich has increased about 50 percent as an annual mean between 1966 and 1976 as compared with 1953 to 1965 (Shantz, 1977). While it was unusual for maximum transparency to exceed 6 m in the years before treatment, it is now common for the maximum to reach 10 m.

The greatest improvement has occurred in winter and autumn. During those times of the year the lake is much clearer than it was before the turn of the century, which has been correlated with the disappearance of *Oscillatoria rubescens*. Summer transparency, however, has improved only slightly.

COSTS

The cost of diversion obviously varies greatly from site to site, primarily according to the required distance for wastewater transport. Also, from the standpoint of inflation, the sooner the project is completed the less will be the expense. With AWT the cost per capita would be more equal from site to site, and the total cost should depend largely on the volume of wastewater to be treated and on local charges for chemicals, transport, and sludge disposal. The cost for AWT (P removal) is considered to be roughly double the cost for secondary treatment.

Wastewater entering Lake Washington from eleven small secondary treatment plants located around the lake (Edmondson and Lehman, 1981) was collected and piped to a large primary treatment plant about 3 km from the lake. Effluent from that plant is discharged at depth in Puget Sound. Wastewater, including dairy waste as well as sewage, was diverted from Lake Sammamish through a collection system that transported it about 20 km to a secondary treatment plant. The effluent from that plant enters a river and eventually also reaches Puget Sound. Light limits primary productivity in Puget Sound, and as a result the added nutrients have not caused adverse effects of eutrophication in that water body.

The costs vary greatly among the cases described (Table 3-3). Because effluent was diverted from Lakes Washington and Sammamish between 1963 and 1967 and in 1968, respectively, the costs 16 years later are low. The costs corrected for inflation indicate that the Washington and Sammamish projects would have probably cost in the neighborhood of three times more by 1983. This is assuming that recovery would have occurred at the same rate. An additional 16 years of sediment enrichment could have caused a greater delay in recovery—possibly necessitating employment of additional measures—and in that sense would have been considered an additional cost if diversion had been delayed. Thus, the benefits accrued from the recovery of Lake Washington from eutrophy to near oligotrophy and the prevention of Lake Sammamish from becoming eutrophic, and in fact recovering to near oligotrophy, were obviously an exceptional bargain when viewed from today's costs. This is especially true because these lakes lie in a metropolitan area of over one million people and receive intense recreational use. The State Park on Lake Sammamish is the most heavily used in the State of Washington.

The same cost-effectiveness statement can be made for Lake Norrviken, because effluent was diverted from that lake in 1969-1970. There are three other lakes in a chain with Lake Norrviken (Vallentunasjön, Edssjön, and Oxundasjön) that benefitted from the same diversion. In this case the wastes from both domestic and industrial (yeast factory) uses were diverted to Stockholm's treatment system, from which effluent enters the Baltic Sea.

Advanced wastewater treatment (AWT) was installed at Ely, Minnesota, to remove P from sewage effluent entering Shagawa Lake. The treatment system was paid for by the U.S. Environmental Protection Agency as a test case to determine if AWT would alleviate the effects of eutrophication. Treatment has worked well, through a two-stage process of lime clarification and dual-media filtration, to maintain an effluent P content of about 50 μg l^{-1}. The relatively high costs of operating the AWT plant were due in part to the research nature of the project (Vanderboom et al., 1976).

Beginning in about 1955, the communities around Lake Zürich started building wastewater treatment plants. Between 1965 and 1975 all treatment plants were equipped with P elimination (AWT). The construction costs per capita for Lake Shagawa were about 50 percent higher than for Zürich (Table 3-3), while the per

Table 3-3 Estimated Costs for Diversion and Advanced Treatment of Sewage to Restore Five Lakes [Values in parentheses adjusted to 1983 dollars (USDL, 1983).]

Lake	Treatment	Year	Construction		Operation (per yr)	
			Total $ × 10⁶	$ per capita	$ × 10⁶	$ per capita
Washington[1]	diversion	1967	94.9 (296)	171 (534)	2,138	4
Sammamish[1]	diversion	1968	4.5 (14)	370 (1154)	146	12
Norrviken[2]	diversion	1970	44.5 (113)	106 (269)	6,736	16
Shagawa[3]	AWT	1973	1.9 (4.2)	380 (844)	389	77
Zürich[4]	AWT	1975	36.0 (66)	252 (461)	1,500	115

1. Municipality of Metropolitan Seattle, G. Farris (personal communication).
2. Käppalaforbundet, Årsredovisning 1980, Lidingö, Sweden: Käppalaverket
3. Vanderboom et al., 1976; R.M. Brice, USEPA (personal communication).
4. Shantz (1977). Conventional treatment included; PO_4 removal only is $2–3.30 per capita.

capita operating costs are much higher (50%) for Zürich than for the Minnesota treatment plant. There are 17 treatment plants in the Lake Zürich drainage below the upper lake, serving an average of about 8,400 inhabitants per plant. The Ely, Minnesota, plant serves 5,000 people. The AWT process used in plants around Lake Zürich is mostly ferric chloride, simultaneous with activated sludge, while removal at Ely is accomplished with lime. Although costs for AWT are higher than for diversion, the hidden cost of exporting the pollutants to another receiving water must be considered.

FURTHER TREATMENT FOLLOWING DIVERSION

Several lakes from which sewage effluent was diverted were subsequently treated by other techniques. Either the lakes showed no improvement after diversion, or internal loading was recognized as a problem that would retard the recovery process. Two of these lakes are discussed elsewhere in this review — Lake Trummen in Sweden and Twin Lakes in Ohio.

Sewage effluent was diverted from shallow Lake Trummen in 1959 and, because no improvement was apparent ten years later, the top one meter of sediment was dredged from the lake in 1970-1971. Removal of the rich surface sediment layer was sufficient to greatly reduce the internal loading and allow the lake to make a dramatic recovery (Björk, 1972; 1974). Details of the Lake Trummen dredging are covered in Chapter 8. The recovery of West Twin Lake, Ohio, was greatly accelerated by a hypolimnetic treatment with alum (Cooke et al., 1978). Wastewater that had been treated by septic tank systems was collected and diverted away from the West and East Twin Lakes. Recognizing that large internal sources of phosphorus (anaerobic hypolimnia) would probably slow West Twin Lake's recovery process, alum was added shortly after the diversion. The treated lake improved quickly as a result of the alum addition. The detailed results of the alum-treated lake's response, relative to the untreated lake, are presented in Chapter 6 on nutrient inactivation.

As in the case of Twin Lakes, it is important to recognize whether the lake in question receives a relatively large internal P loading. If internal loading is significant, it may take 10 or 15 years for recovery to occur, so in-lake treatment should be planned at the outset, rather than waiting five or ten years (with little or no observable improvement) before considering further treatment. Not only is the public initially informed of the total cost to improve the lake, but the costs will ultimately be less because inflation is avoided during the years that the residents wait for their lake to recover through diversion.

Diversion may not be practical because of cost, and it may not even be feasible. Stormwater is often so diffuse that complete sewering of the lake would be necessary and the fraction of nutrient load removed may not be significant in terms of benefits received. In such instances diversion may not be cost-effective and an in-lake treatment alone may be sufficiently cost-effective.

NEEDED RESEARCH

Obviously, better models are needed to predict the rate and extent of recovery across a range of lake types and trophic states, without in-depth studies on each lake. Most of what we know about lake recovery has been learned in the last 15 years. What is known comes from lakes that have been thoroughly studied, not only to observe the changes in characteristics with time, but also to understand through experimentation or modeling why changes did or did not occur. This effort is still needed, particularly on shallow lakes that have high internal loading.

Simple seasonal models for P need to be developed for shallow lakes. How much does internal loading vary from lake to lake and upon what factors does it depend? Such models could follow the format of that by Larsen et al., (1979), with terms for declining sediment release rates, as proposed by Lorenzen (1973) and Kamp-Nielsen (1980), if specific information is available.

REFERENCES

Ahlgren, I. 1977. Role of sediments in the process of recovery of a eutrophicated lake. In H.L. Golterman, ed., *Interactions Between Sediments and Fresh Water*, The Hague: Dr. W. Junk, 372-377.

Ahlgren, I. 1979. Lake metabolism studies and results at the Institute of Limnology in Uppsala. *Arch Hydrobiol Beih* 13:10-30.

Ahlgren, I. 1980. A dilution model applied to a system of shallow eutrophic lakes after diversion of sewage effluents. *Arch Hydrobiol* 89:17-32.

Bengtsson, L. 1975. Phosphorus release from a highly eutrophic lake sediment. *Verh Int Ver Limnol* 19:1107-1116.

Björk, S. 1972. Ecosystem studies in connection with the restoration of lakes. *Verh Int Ver Limnol* 18:379-387.

Björk, S. 1974. *European Lake Rehabilitation Activities*. Inst. Limnol. report, Univ of Lund.

Carlson, R.E. 1977. A trophic state index for lakes. *Limnol Oceanogr* 22:361-368.

Chapra, S.C., and Reckhow, K.H. 1979. Expressing the phosphorus loading concept in probabilistic terms. *J Fish Res Board Can* 36:221-229.

Chapra, S.C., and Tarapchak, S.J. 1976. A chlorophyll *a* model and its relationship to phosphorus loading plots for lakes. *Water Res* Res. 12:1260-1264.

Cooke, G.D., Heath, R.T., Kennedy, R.H., and McComas, M.R. 1978. *The Effect of Sewage Diversion and Aluminum Sulfate Application on Two Eutrophic Lakes*, EPA-600/3-78-033.

Cooke, G.D., McComas, M.R., Waller, D.W., and Kennedy, R.H. 1977. The occurrence of internal phosphorus loading in two small, eutrophic, glacial lakes in northeastern Ohio. *Hydrobiologia* 56:129-135.

Edmondson, W.T. 1970. Phosphorus, nitrogen, and algae in Lake Washington after diversion of sewage. *Science* 169:690-691.

Edmondson, W.T. 1978. *Trophic Equilibrium of Lake Washington*. EPA-600/3-77-087.

Edmondson, W.T., and Lehman, J.R. 1981. The effect of changes in the nutrient income on the condition of Lake Washington. *Limnol Oceanogr* 26:1-29.

Edmondson, W.T., and Litt, A.H. 1982. Daphnia in Lake Washington. *Limnol Oceanogr* 27:272-293.

Fee, E.J. 1979. A relation between lake morphometry and primary productivity and its use in interpreting whole-lake eutrophication experiments. *Limnol Oceanogr* 24:401-416.

Fitzgerald, G.P. 1964. The biotic relationships within water blooms. In D.F. Jackson, ed., *Algae and Man*, New York: Plenum, 300–306.

Jacoby, J.M., Lynch, D.D., Welch, E.B. and Perkins, M.A. 1982. Internal Phosphorus loading in a shallow eutrophic lake. *Wat Res* 16:911–919.

Jones, J.R., and Bachmann, R.W. 1976. Prediction of phosphorus and chlorophyll levels in lakes. *J Water Pollut Control Fed* 48:2176–2182.

Kamp-Nielsen, L. 1975. Seasonal variation in sediment-water exchange of nutrient ions in Lake Esrom. *Verh Int Ver Limnol* 19:1057–2182.

Kamp-Nielsen, L. 1980. The influence of sediments on changed phosphorus loading to hypertrophic L. Glumsø. In J. Barica and L.R. Mur, eds. *Developments in Hydrobiology*, Vol. 2, 29–36.

Larsen, D.P., Malueg, K.W., Schultz, D.W., and Brice, R.M. 1975. Response of eutrophic Shagawa Lake, Minnesota U.S.A., to point-source phosphorus reduction. *Verh Int Ver Limnol* 19:884–892.

Larsen, D.P., Van Sickle, J., Malueg, K.W., and Smith, P.D. 1979. The effect of wastewater phosphorus removal on Shagawa Lake, Minnesota: Phosphorus supplies, lake phosphorus and chlorophyll a. *Water Res* 13:1259–1272.

Larsen, D.P., Schultz, D.W., and Malueg, K.W. 1981. Summer internal phosphorus supplies in Shagawa Lake, Minnesota. *Limnol Oceanogr* 26:740–753.

Lorenzen, M.W. 1973. Predicting the effects of nutrient diversion on lake recovery. In E.J. Middlebrooks, D.H. Falkenborg, and T.E. Maloney, eds., *Modelling the Eutrophication Process*, Logan, UT: Utah State Univ., 205–210.

Rast, W. Jones, R.A., and Lee, G.F. 1983. Predictive capability of U.S. OECD phosphorus loading-eutrophication response models. *J Water Pollut Control Fed* 55:990–1003.

Reckhow, K.H., and Chapra, S.C. 1983. *Engineering Approaches for Lake Management, Volume I: Data Analysis and Empirical Modeling*, Boston: Butterworth.

Ryding, S.O. 1980. *Monitoring of Inland Waters*. OECD Eutrophication Programme, The Nordic Project, Nordic Cooperative Organization for Applied Research.

Ryding, S.O., and Forsberg, C. 1980. Short-term load-response relationships in shallow, polluted lakes. In J. Barcia and L.R. Mur, eds., *Hypereutrophic Ecosystems*, The Hague: Dr. W. Junk, 95–103.

Shantz, F. 1977. Effects of wastewater treatment on Lake Zürich. In W.K. Downey and G. Ni Vid, eds. *Lake Pollution Prevention by Eutrophication Control*, Proceedings of a seminar, Killarney, Ireland, May 1977, 131–139.

Shantz, F., and Thomas, E.A. 1980. Die Durchsichtigkeit des Zurichseewassers von 1897 bis 1980. *Viertel Naturforsch Gesell Zürich* 125:239–248.

Stauffer, R.E., and Lee, G.F. 1973. The role of thermocline migration in regulating algal blooms. In E.J. Middlebrooks, D.H. Falkenborg, and T.E. Maloney, eds., *Modelling the Eutrophication Process*, Logan, UT: Utah State University, 73–82.

Stewart, K.M. 1976. Oxygen deficits, clarity and eutrophication in some Madison lakes. *Int Rev Ges Hydrobiol* 61:563–579.

Stumm, W., and Leckie, J.O. 1971. Phosphate exchange with sediments: Its role in the productivity of surface waters. *Proc 5th Int Conf Water Pollut Res* III-26:1–16.

USDL. 1983. *Handbook of Basic Economic Statistics*. U.S. Dept. of Labor, Bureau of Labor Statistics, 37, 1, p. 99.

Uttormark, P.D., and Hutchins, M.L. 1980. Input/output models as decision aids for lake restoration. *Water Res Bull* 16:494–500.

Vanderboom, S.A., Pastika, J.D., Sheehy, J.W., and Evans, F.L. 1976. *Tertiary Treatment for Phosphorus Removal at Ely, Minnesota AWT Plant, April, 1973 through March, 1974*, EPA-600/2-76-082.

Vollenweider, R.D. 1976. Advances in defining critical loading levels for phosphorus in lake eutrophication. *Mem Inst Ital Idrobiol* 33:53-83.

Welch, E.B. 1977. *Nutrient Diversion: Resulting Lake Trophic State and Phosphorus Dynamics*, EPA-600/3-88-003.

Welch, E.B., Perkins, M.A., Lynch, D.D. and Hufschmidt, P. 1979. Internal phosphorus related to rooted macrophytes in a shallow lake. In J.E. Breck, R.T. Prentke and O.L. Loucks, eds., *Aquatic Plants, Lake Management and Ecosystem Consequences of Lake Harvesting*, Proceedings, Madison, Wisc., 81-99.

Welch, E.B., Rock, C.A., Howe, R.C., and Perkins, M.A. 1980. Lake Sammamish response to wastewater diversion and increasing urban runoff. *Water Res* 14:821-828.

4

Hypolimnetic Withdrawal

The principal purpose of this technique is to change the depth at which water leaves the lake, from the surface to within the hypolimnion, so that nutrient-rich instead of nutrient-poor water is discharged from the lake. Coincidentally, the hypolimnion detention time is shortened, the chance for anaerobic conditions to develop is decreased, and the availability of nutrients to the epilimnion through entrainment is lessened. This technique is applicable to stratified lakes and small reservoirs in which anaerobic hypolimnia restrict the habitat for fish and promote the release of phosphorus and toxic metals from sediments.

The technique is accomplished by installing a pipe along the lake bottom from near the deepest point to the outlet, and possibly beyond. The device, which acts as a siphon, is referred to as an "Olszewski tube" after its originator (Olszewski, 1961). While stratification may be weakened somewhat because epilimnetic water will be drawn downward, destratification will probably not occur. Destratification could increase the transport of hypolimnetic nutrients and water without dissolved oxygen to the epilimnion. To further lessen the chances of destratification, it may be possible to direct inlet water to the metalimnion or hypolimnion. This modification has been installed in Lake Ballinger near Seattle, Washington (Fig. 4-1). While the lake has remained stratified with the inflow water directed to depth, the system has not been tested with inflow water entering the epilimnion but undirected to depth.

Preferentially removing hypolimnetic water, and therefore decreasing the residence time of the hypolimnion, should increase the oxygen content at the sediment-water interface and decrease the internal loading of P. This is an obvious alternative to accelerate recovery in stratified lakes where little improvement has followed wastewater diversion or AWT because of high internal loading.

The process works effectively in reservoirs where hypolimnetic waters are discharged for power generation. The procedure has not been considered a benefit to water quality in the reservoir itself, however, because the oxygen content of discharged water is usually low and adversely affects downstream uses. Multiple and shallower outlets have been incorporated into reservoir design to counteract such problems.

CASE STUDIES

Through the middle 1970s, nine Olszewski-type tubes were installed in European lakes (Björk, 1974). The four reported on here are Mauensee in Switzerland (Gächter,

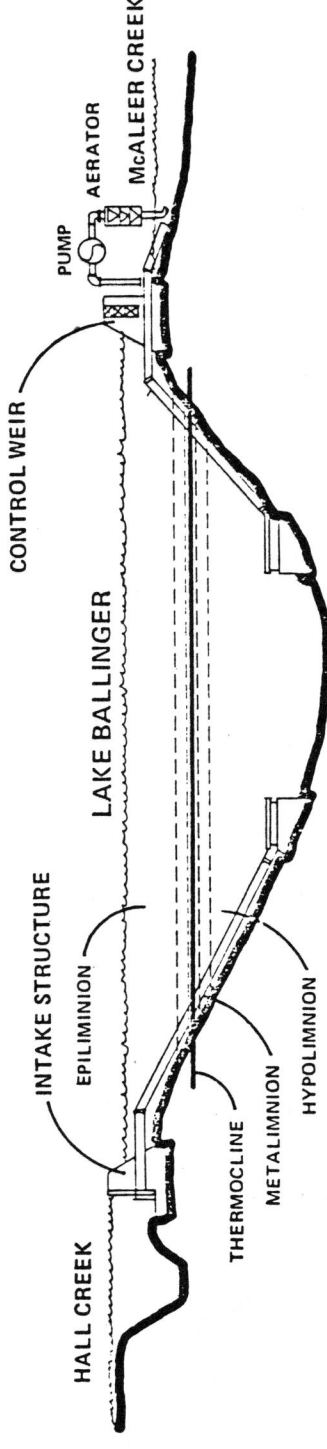

Figure 4-1 Inlet and outlet structures designed for hypolimnetic withdrawal in Lake Ballinger (KCM, 1981).

and Piburger See, Reither See, and Hechtsee in Austria (Pechlaner, 1978). There are three recorded installations in the U.S.; two are in Connecticut in Lake Wononscopomuc and Lake Waramaug (Kortmann et al., 1983 and E.R. Davis, pers. comm.) and the third is Lake Ballinger in Washington state (KCM, 1981).

Mauensee

Gächter (1976) has described the recovery of Mauensee, a 51-ha Swiss lake with a maximum depth of 6.8 m. An Olszewski tube was installed in this lake in 1968 to discharge hypolimnetic water at 67 l s^{-1}, providing a water residence time in the hypolimnion (> 4 m) of 0.2 years. The lake showed marked improvement following installation of the tube. Phosphorus and nitrogen decreased substantially, while oxygen and Secchi visibility increased. *Oscillatoria rubescens* decreased from a summer maximum of 152 g m^{-2} to 41 g m^{-2} seven years after installation. Before installation the internal loading of P from lake sediments during June and July was more than 200 times greater than the external loading; after installation it progressively decreased to only four times the external loading. There was a steady decline in P from the sediment and associated improvement in lake quality during the six years of observation following tube installation. During that time P export exceeded the external supply (360 kg yr^{-1}) by a total of 3700 kg, which resulted in a reduction in P content at the sediment surface.

Austrian Lakes

Pechlaner (1978) has reported on the response of three Austrian lakes (Piburger See, Reither See, and Hechtsee) following installation of Olszewski tubes. All three lakes are relatively small but important to local populations and tourists for recreation, especially swimming. Sewage effluent was diverted from the three lakes and Olszewski tubes were installed to accelerate the restoration process.

The tube in Piburger See draws water from a depth of 23 m, which is nearly the maximum depth of the lake (24.7 m). The total length of the tube is 639 m, with a diameter of 10 cm, and hypolimnetic water from the lake is discharged at eleven l s^{-1} at a point 13.5 m below the lake level. In 1970, the same year as the tube was installed, the oxygen content at the time of ice cover had increased by 63 percent over pretube (1969) conditions. Oxygen continued to be at least as high for the next seven years, but no recognizable oligotrophication has occurred in the 13.4-ha lake.

Piburger See tends to be morphometrically meromictic. Only twice during the nine years of observation did the lake completely mix, even though the lake's monimolimnion was effectively replaced about three times per year by means of the Olszewski tube. Increased circulation of deep water across the bottom sediments, as well as the removal of phosphorus-rich water overlying the sediments, has resulted in increased internal loading that has tended to compensate for the increased losses of P via the tube. P losses through the tube over three years have been 79 to 192 percent more with the tube than would have been the case without it. Unfortunately,

P concentration in the lake has not declined and therefore the lake's trophic state, except for oxygen, has not changed (Pechlaner, 1979).

In contrast to Piburger See, Reither See showed a marked improvement following installation of the Olszewski tube (Pechlaner, 1978). The tube was placed near the lake's maximum depth (8.2 m) in 1972. The lake area is only 1.5 ha and the 10-cm-diameter tube discharges hypolimnetic water at 4 l s^{-1}.

Total P decreased from pretreatment annual means of 38 and 43 $\mu\text{g l}^{-1}$ in 1974 and 1975 to a posttreatment mean of 21 $\mu\text{g l}^{-1}$ in 1977. Water transparency has about doubled over the four-year period following installation. Although there is some doubt about changes in phytoplankton biomass, because of interference with detritus, less blue-green algae were present after installation.

A larger tube (18 cm diameter) was placed in the Hechtsee in 1973. The depth of placement, however, was not near the maximum depth as in the other lakes. Because the lake was meromictic, the odor of the anaerobic monimolimnion was quite strong. The tube was placed at 23 m, considerably less than the 56.5-m maximum depth, to protect the recreational environment around the lake from nuisance odors. Hechtsee is the largest of the three lakes, with an area of 21.5 ha; discharge through the tube varies between 20 and 30 l s^{-1}.

Because the unmixed monimolimnion was not withdrawn, oxygen content in Hechtsee remained at zero from 25 to 55 m. Oxygen content increased significantly above 25 m between 1972 and 1977 (following installation of the tube). Even though the tube was not placed at the maximum depth, P transport was substantial. During the four years following tube installation, P output (203 kg) exceeded input (93 kg) by 110 kg, which was the actual decrease in lake content. P content above 25 m declined by 70 to 80 percent from 1973 to 1977, while as expected the content from 25 to 55 m changed little and actually showed some increase (Pechlaner, 1978).

USA Lakes

A hypolimnetic withdrawal unit was installed in Lake Ballinger, north of Seattle, Washington, during 1982. The lake is 41 ha in area with a mean depth of 4.5 m. The device allows the lake's inlet to be directed to the hypolimnion, as shown in Figure 4-1. The option also exists to allow all or some fraction of inflow to enter the epilimnion. A control weir exists on the outlet to adjust the fraction of hypolimnetic and epilimnetic water discharged from the lake. A maximum flow of 57 l s^{-1} can enter and be discharged from the hypolimnion through 40-cm-diameter pipes. The replacement time for hypolimnetic water at that rate is about three months. The cost of the device was estimated at about $145,000 ($236,000 in 1983 dollars) (KCM, 1977).

Preliminary results following the first year of treatment indicate that oxygen content of the hypoliminion has increased substantially and the anaerobic period was two weeks at most. As a result, O_2 was not a significant problem in the lake discharge. Soluble phosphorus content of the lake decreased markedly, but total P showed little or no change (KCM, personal communication). The biomass of attached algae apparently did increase in the discharge stream. Mean annual total P content was expected to decline from 48 to about 30 $\mu g\ l^{-1}$ following treatment, but it was really too early to evaluate the relative success of the treatment at decreasing internal P loading.

Withdrawal systems were installed in the shallower of two basins of Lake Wononscopomuc in 1980. The lake is 141 ha in area and has a maximum depth of 37 m. Hypolimnetic water was pumped at a flow rate of about $17\ l\ s^{-1}$ from a depth of 15 m during the stratified period, which was sufficient to replace the hypolimnetic water in 5.6 months (Kortmann et al., 1983).

The oxygen content of the hypolimnion was greatly improved by this treatment. Generally, the depth of low oxygen content increased following hypolimnetic withdrawal, but no consistent pattern of hypolimnetic oxygen depletion is apparent in the data. This treatment has helped eliminate a metalimnetic blooming of *Oscillatoria* (Kortmann et al., 1983). The authors further contend that internal P loading was reduced, although the magnitude of any reduction is unclear. Initial cost for pipes and pumping was about $35,000, with about $10,700/year additional for operation and maintenance (E.R. Davis, pers. comm.).

Two systems were installed in Lake Waramaug, Connecticut (maximum depth, 12 meters). One system was the normal hypolimnetic withdrawal to remove water from the 8-9 m depth via the lake outlet. The other system was in the upper lake to remove water from the same depth and return it, after areation, to the hypolimnion. Together the pumping systems removed hypolimnetic water at a rate slightly more than $100\ l\ s^{-1}$. The first year of operation showed little change in phosphorus content or dissolved oxygen in the deep (9-10 m) water, but the depth with dissolved oxygen and low P content had increased by 2 m. The authors suggest that this improvement was responsible for decreased P transport to the epilimnion following storms and improved transparency (Connors and Kortmann, 1984). Total installation cost for this project was about $131,000 in 1983 (E.R. Davis, pers. comm.).

CONSIDERATIONS

An important advantage of this technique is the continual dependability of the siphon and low operational costs. Because it operates by gravity, there are no moving parts and no worry about mechanical failure or power shortage, though an auxilliary pump may be installed to increase the discharge if sufficient head cannot be attained with the tube length dictated by other circumstances. Of course that increases operational expenses, and some measure of reliability is lost.

The effect of hypolimnetic withdrawal on internal loading of P is not entirely clear. Increased internal loading occurred in Piburger See and compensated for increased P export, but this apparently did not happen in the other three lakes, where P declined substantially following tube installation. Even if sediment P release is enhanced initially, the sediment P pool should eventually be depleted if output continues to exceed input. The morphometry of some lakes (e.g., Piburger See) may be inappropriate for hypolimnetic (or monimolimnetic) withdrawal if the water exchange rate is too low. To be successful the exchange rate should be severalfold per stratified period, the desired magnitude of which can be judged by the O_2 deficit rate.

For example, the average O_2 demand for the 0.47×10^6 m^3 hypolimnion in Lake Ballinger is 0.05 g m^{-3} d^{-1}, or 23.6 kg O_2 d^{-1}. To supply inflow water and oxygen at roughly double that rate (48 kg O_2 d^{-1}) required an inflow of 0.057 m^3 s^{-1} if DO in the inflow was 10 mg l^{-1}. The pipe was designed to provide such a flow (KCM, 1977). The mean flow of the inlet to Lake Ballinger prior to installation was 0.170 m^3 s^{-1}, which was more than adequate to supply the needed flow and DO to the hypolimnion. An inflow of that rate results in an exchange of the hypolimnetic water about every three months, which is nearly as frequent as the exchange time in Mauensee (2.4 months), the successfully restored Swiss lake. A hypolimnion exchange time of two to three months may, therefore, be considered a minimum for observable quality improvements in most lakes.

The potential negative impact of this technique is, of course, the effect of high nutrient and low oxygen content on downstream waters. If the outflow stream contains an important fishery and is otherwise used for recreation or water supply, then special precautions may be necessary to maintain water quality. The extent to which oxygen in the outflow water will be reduced can be estimated by comparing the existing O_2 deficit rate in the lake with the input load of oxygen (Pechlaner, 1979). If low DO is expected in the outlet, then aeration equipment can be installed. Whether the elevated P content will cause a nuisance attached algal and subsequent secondary BOD problem downstream will depend upon whether periphyton growth is limited by nutrients, light, or velocity and scouring. If the outflow rate can result in a short enough renewal time for the hypolimnetic volume, then O_2 may not be depleted to a level that would cause a problem downstream; presumably a content of 3 mg l^{-1} or more would be acceptable, considering the effect of natural aeration. On the other hand, if anaerobic conditions are completely prevented the potential for depleting sediments of P is diminished.

REFERENCES

Björk, S. 1974. *European Lake Rehabilitation Activities*, Inst. Limnology report, Univ. of Lund, Sweden.

Gächter, R. 1976. Die Tiefenwasserableitung, ein Weg zur Sanierung von Seen. Schweiz Z Hydrol 38: 1–28.

Conners, M.E. and Kortmann, R.W. 1984. Hypolimnetic withdrawal and withdrawal-treatment-rejection: Management of two deep Connecticut lakes. Ecosystem Consulting Services, Inc., Coventry, CT. Unpub. Ms.

Davis, E.R. The Hotchkiss School, Lakeville, CT.

Kortmann, R.W., Davis, E., Frink, C.R., and Henry, D.D. 1983. Hypolimnetic Withdrawal: Restoration of Lake Wonoscopomuc, Connecticut. In *Lake Restoration, Protection and Management,* EPA-440/5-83-001, R 46-55.

KCM 1981. *Lake Ballinger Restoration Project Interim Monitoring Study Report*, Kramer, Chin & Mayo, Inc., Seattle, WA.

KCM 1977. *Lake Ballinger Rehabilitation Study*, Kramer, Chin & Mayo, Inc., Seattle, WA.

Olszewski, P. 1961. Versuch einer Ableitung des hypolimnische Wassers ans einem See. Ergebnisse des ersten Versuchsjahres. *Verh Int Ver Limnol* 18:1792-1797.

Pechlaner, R. 1978. Erfahrungen mit Restaurierungsmassnahmen an eutrophierten Badeseen Tirols. *Oester Wasserwertsch* 30:112-119.

Pechlaner, R. 1979. Response of a eutrophied lake (Piburger See) to reduced nutrient load and selective water renewal. Lake Metabolism and Management. *Arch Hydrobiol Beih Ergeb Limnol* 13:293-305.

5

Dilution and Flushing

The technique of dilution and flushing can achieve improvement in eutrophic lakes in two ways: The concentration of limiting nutrient can be reduced (dilution) while at the same time the water exchange (flushing) rate is increased. Both processes can reduce the biomass of plankton algae. The addition of low-nutrient dilution water reduces the limiting nutrient concentration upon which the maximum potential biomass is dependent. By increasing the water input the flushing rate is increased, which in turn increases the loss rate of plankton algae from the lake. Dilution can be effective even when the increase in flushing rate is insufficient to cause a significant loss of algae. On the other hand, flushing rate increase can effect a significant loss without achieving a reduction in the limiting nutrient concentration. Other effects of dilution are also possible, such as increased vertical mixing and a decrease in the concentration of algal excretory products, which can influence the kinds and abundance of algae.

The technique is most important where large quantities of low-nutrient water are available for transport to the affected lake. Treatment effectiveness is greater the lower the concentration of limiting nutrient in the dilution water relative to that in the lake and its natural inflow. In some instances improvements can be achieved by adding water with moderate to high nutrient content but these results are less certain than with low-nutrient water.

Dilution and flushing have worked successfully in several lakes. Green Lake in Seattle, Washington, was improved markedly in the 1960s by adding city water (Oglesby, 1969) and planning for more varied and reliable sources of dilution water has recently been completed (URS, 1983). Moses Lake in eastern Washington has received Columbia River dilution water on a regular basis since 1977, resulting in substantial improvement in quality (Welch, 1979; Welch and Patmont, 1980; Welch and Tomasek, 1981; Welch et al., 1983). Lake Bled, Yugoslavia, was flushed intentionally with water from the River Radovna (Sketelj and Rejic, 1966). Dilution has been proposed or instituted in three other lakes in Washington State and was proposed for Clear Lake, California (Goldman, 1968). Maintenance of low phytoplankton concentrations by high natural rates of dilution and flushing is a commonly observed phenomenon (Dillon, 1975; Dickman, 1969; Welch, 1969).

In what must be one of the world's first lake flushing experiments, water was diverted from Switzerland's Ruess River to Rotsee in 1921–1922 to alleviate the lake's eutrophic condition (Stadelman, 1980). The flushing rate of this 460-ha lake was increased from 0.33 yr^{-1} to 2.5 yr^{-1} (or about 0.1% to 0.7% day^{-1}) following construction of a canal between the Ruess River and the lake. The lake's state did not improve because the increased flushing rate was not sufficient for a significant biomass washout, and because of high concentrations of nutrients in the river water used to flush the lake. The nutrients originated in sewage effluent from the upstream city of Luzern. There was still no improvement following diversion of direct inputs of sewage effluent to the lake in 1933. There was considerable improvement, however, after nutrient removal from Luzern's wastewater, in the 1970s, which resulted in the Ruess River inflow water being reduced more than ten times in P concentration.

The theoretical basis for the dilution and flushing technique will be discussed, followed by detailed reviews of the Moses Lake and Green Lake cases. General guidelines for application of the technique, including quantity and quality of water, frequency of application, and project and operating costs, will be given.

THEORY AND PREDICTIONS

The mechanisms involved in the dilution and flushing technique for the control of algal biomass in lakes are in many ways analogous to those in continuous culture systems. When low-nutrient dilution water is added to a laboratory culture system, the inflow concentration of limiting nutrient is reduced, the maximum biomass concentration possible in the reactor vessel is likewise reduced, and at the same time nutrients and algal biomass are more rapidly washed from the reactor vessel since the water exchange rate is increased. Concentration of limiting nutrient is the critical parameter that determines algal biomass in many lakes, as well as in continuous culture systems, although in lakes the nutrient concentration can be greatly affected by internal sources.

Another factor that greatly influences the lake nutrient concentration and that is not considered in an analysis of continuous cultures is nutrient loss to the sediment. At very high rates of water exchange loss to the sediment can decrease and may result in higher lake concentrations than at moderate exchange rates, where sedimentation loss is greater, even if inflow nutrient content is similar. This is described by the empirical relationship between the retention coefficient for phosphorus (R) and the flushing rate (ρ) derived by Larsen and Mercier (1976):

$$R = \frac{1}{1 + \sqrt{\rho}} \qquad (5.1)$$

Short-Term Predictions

The transient reduction in lake concentration of a nutrient that can be brought about by adding dilution water in rather large quantities is reasonably predictable in the short term, by a simple continuity equation:

$$C_t = C_{in} + (C_0 - C_{in}) e^{-\rho t}, \qquad (5.2)$$

where C_t is the concentration at time t; C_{in} is the concentration in the inflow water; C_0 is the initial lake concentration, and ρ is the water exchange or flushing rate. This equation assumes that the lake is well mixed, that no other sources of nutrients exist, and that the limiting nutrient or "percent lake water" can be treated as conservative. Because this equation does not include a sedimentation term it is useful only in the short term as a tracer for nutrient behavior and under conditions of rather large water exchange rates (several percent per day or more). It allows one to estimate the potential for reducing average lake concentrations with a given source of water, and the time necessary for that reduction. Predictions can be compared with the observed distribution of dilution water—as percent lake water—as indicated by a conservative variable such as Na or specific conductance.

Long-Term Predictions

A term for sedimentation should be included to predict long-term changes in the concentration of limiting nutrient by adding rather small quantities of low-nutrient dilution water. That requirement is best approximated by the following equation for steady-state phosphorus content (Vollenweider, 1976):

$$P = \frac{L}{\bar{Z}\rho} \cdot \frac{1}{1 + 1/\sqrt{\rho}} \qquad (5.3)$$

where L is the areal loading rate for P, $\bar{Z}$ is the mean depth, and ρ is the flushing rate. The term $1 + 1/\sqrt{\rho}$ is a correction for sedimentation loss, and $L/\bar{Z}\rho$ is the inflow concentration. Estimates of error and confidence intervals for predictions using this equation have been provided by Chapra and Reckhow (1979).

A principal assumption of this model is the existence of a steady-state condition, usually approximated by annual means. This may be a satisfactory assumption if seasonal internal loading from sediments is insignificant. If internal loading is significant, the transient form of the model may be advisable so that the effect on peak P concentrations during the summer recreational period can be evaluated; Larsen et al. (1979) have used this approach.

Uttormark and Hutchins (1980) have evaluated the use of this and similar equations for estimating the long-term effect of adding dilution water. They noted that adding more water with lower nutrient content also increases nutrient loading L, while the resulting increased flushing rate ρ can decrease nutrient loss through sedimentation. The processes could be counteracting in some instances, because, as the authors stated, "a reduction in the influent concentration tends to reduce in-lake concentrations, but a reduction in phosphorus retention tends to increase in-lake concentration." They showed that a large increase in the combined flushing rate obtained by adding low-nutrient water (40% of the normal inflow nutrient content) could theoretically increase the lake nutrient concentration if the original flushing rate ρ is low enough, e.g., 0.1 yr^{-1}. If the flushing rate is relatively large

86 *Physical and Chemical Methods*

($\geqslant 1.0$ yr^{-1}) initially, the effect of reduced sedimentation rate is minimized and a reduction in lake concentration will result, but large quantities of water are necessary. Of course, the amount of water needed to achieve a given reduction in inflow concentration is a function of the concentration difference between normal inflow and the dilution-water source.

CASE STUDIES

Two lakes where the dilution and flushing technique has been implemented can be used as guides for application elsewhere. Moses Lake lies in eastern Washington and has a surface area of 2,753 ha and a mean depth of 5.6 m. Dilution water has been added to one arm of Moses Lake during spring and early summer since 1977. Transport to previously undiluted portions of the lake, by pumping, was begun in 1982. Green Lake, in Seattle, Washington, has an area of 104 ha and a mean depth of 3.8 m. It has received dilution water from the city domestic supply since 1962, but inputs declined in recent years. The cost and effectiveness of other, more reliable dilution-water sources have been evaluated recently (URS, 1983). The suitability of the dilution-water sources for the treatment of these lakes is apparent from the large ratios of nutrient content in the lakes relative to the dilution inflows, which range from 5:1 to 10:1.

Moses Lake

Dilution water from the Columbia River has been added to Moses Lake via Parker Horn, the U.S. Bureau of Reclamation's East Low Canal, and Rocky Coulee Wasteway (Fig. 5-1). The addition of dilution water from Parker Horn to the previously undiluted Pelican Horn began in 1982, using a pumping facility designed by Brown and Caldwell Engineers (Fig. 5-1).

The patterns of dilution water addition have been varied but not systematic or conducive to determining an optimum quantity and seasonal distribution. Prior to the Mount St. Helen's ashfall in 1980 three periods of dilution were provided in 1977 and 1979, but only one in 1978. The total days of dilution per year ranged from 60 to 138 for those years. The lowest inputs occurred during the two post-ashfall years, 1980 and 1981 (34 and 69 $\times$ 10^6 m^3), with inputs ranging from 117 to 258 $\times$ 10^6 m^3 for the other four years. These dilution-water input quantities represented average flow rates over the dilution periods of about 10 to 20 m^3 s^{-1}, or overall exchange rates for Parker Horn of 0.1 to 0.16 day^{-1} and 0.01 to 0.02 day^{-1} for the whole lake volume.

The suitability of Columbia River water for dilution is shown in Table 5-1. Because the P and N concentrations in Crab Creek (Parker Horn natural inflow) are so high, apparently due to irrigation and fertilization practices in the watershed, relatively large quantities of Columbia River water are needed to significantly lower the composite inflow concentration, which is necessary to, in turn, lower the in-lake

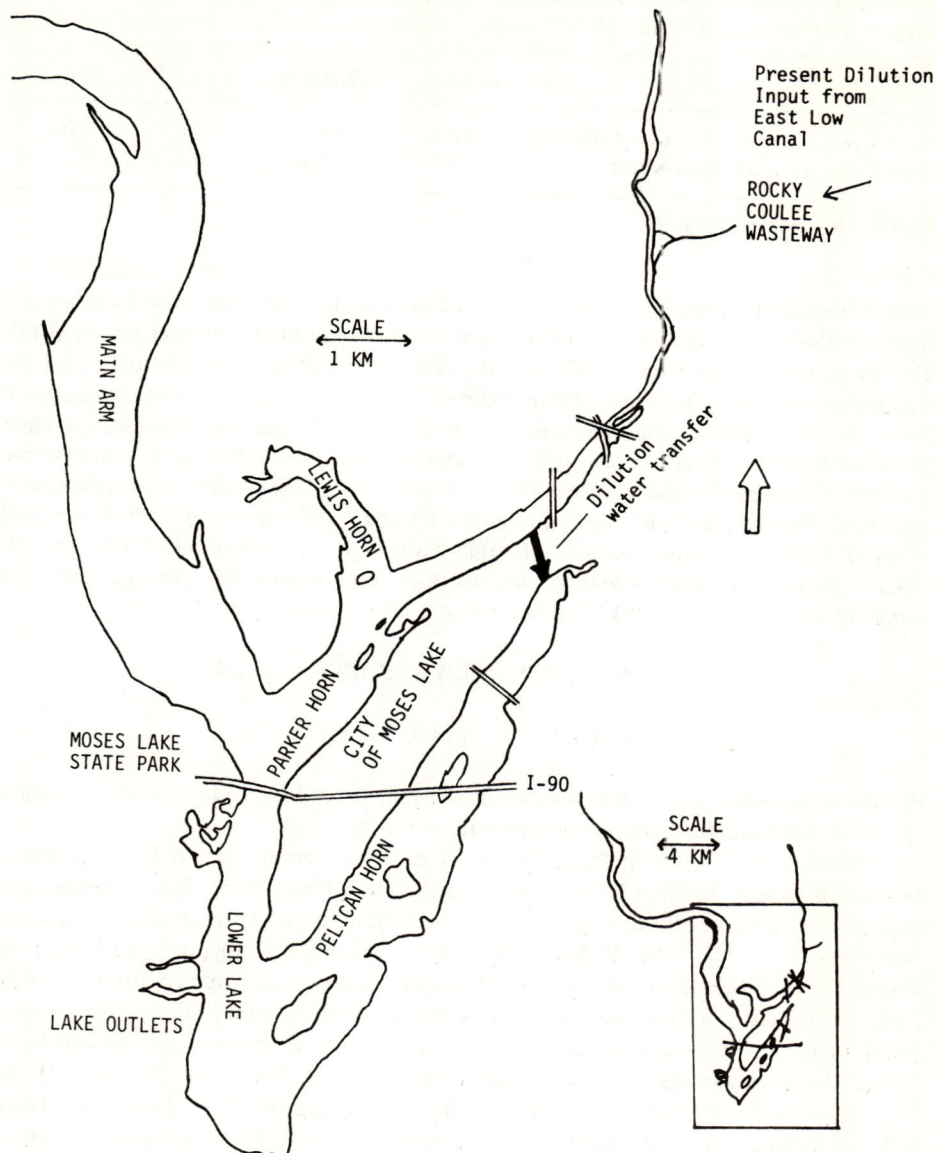

Figure 5-1 Moses Lake, Washington. Note source of dilution water from East Low Canal via Rocky Coulee waterway and point of pumped transfer of dilution water from Parker Horn to Pelican Horn.

Table 5-1 Nutrient Concentrations ($\mu g\ l^{-1}$) in Inflow Water to Parker Horn from May to September, 1977 and 1978.

	Total P	Total N	SRP*	NO_3-N
Crab Creek inflow without dilution	148	1331	90	1096
East Low Canal dilution water	25	308	8	19

*Soluble Reactive Phosphorus

concentration. This results in larger exchange rates than would otherwise be necessary without the Crab Creek inflow. Unfortunately, diversion of the normal inflow (Crab Creek) is economically infeasible in this case, but such a manipulation could be considered for other lakes to obtain more efficiency from dilution water quantities.

As a short-term phenomenon, the addition of dilution water to Moses Lake predictably and rapidly replaced lake water, as judged by tracing a conservative parameter, specific conductance. Values for percent lake water were calculated, assuming that 100 percent was represented by the conductance of Crab Creek and 0 percent by the conductance of Columbia River water. For example, if Crab Creek (CCW) specific conductance was 460 umhos cm^{-1}, lake water (LW) 250, and East Low Canal dilution water (ECDW) 120, then percent lake water is

$$100(LW - ECDW)/(CCW - ECDW) = \%\ LW$$
$$100(250 - 120)/(460 - 120) = 38$$
(5.3)

Percent lake water, which decreases with dilution, is used instead of percent dilution water, to represent the behavior of nutrients in the lake.

Percent lake water in Parker Horn, where water enters, was reduced to values in the 20s, much less than in other parts of the lake (Figure 5-2). This was expected, because the average dilution rate during the period from April to June, described here, was 0.15 day^{-1} for Parker Horn, which is a small (8%) portion of the lake volume. The dilution rate decreased, of course, as the water moved through other parts of the lake. As the dilution water input declined in June, the fraction of lake water in Parker Horn quickly rose to between 50 and 60 percent (see Figure 5-2). Part of that increase was no doubt caused by wind pushing lake water into Parker Horn.

Because Moses Lake is rather irregular in shape and most of the lake's volume (63% in the main arm) appeared to lie out of a direct path from the inflow, dilution water was expected to have little effect in areas other than Parker Horn and the lower lake, which together represent 29 percent of lake volume. The lake water residual decreased, however, similarly to that for the lower lake, even if the "whole lake" volume was used to calculate water exchange rate. Lake-water residuals reached levels between 50 and 60 percent in late May and early June and then began a more gradual return to normal as dilution input declined. In fact, there was little difference between actual and predicted removal of lake water in the whole lake compared to the lower lake (see Figure 5-2).

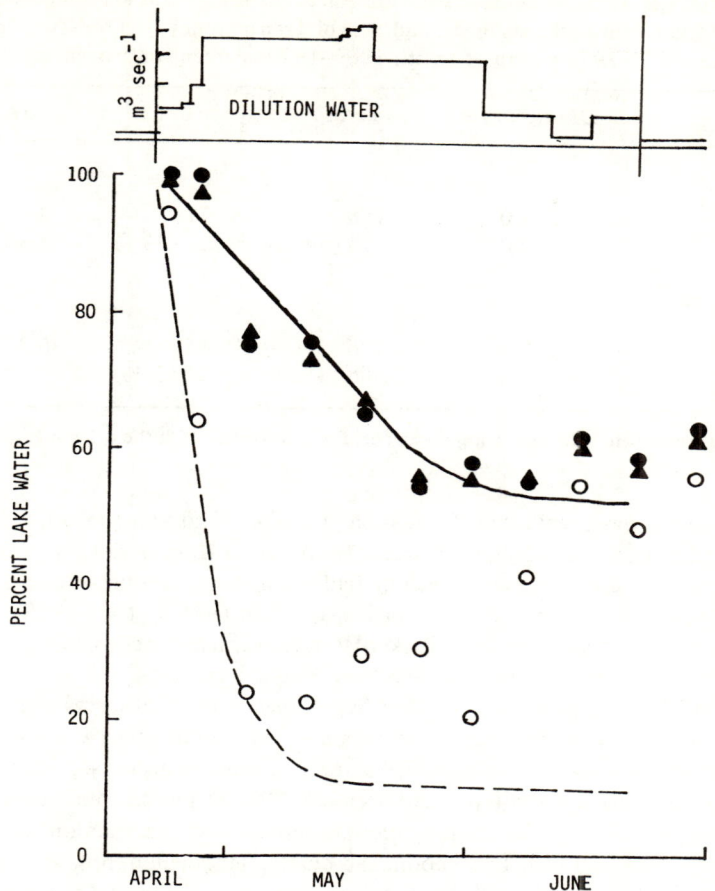

Figure 5-2 Residual lake water, in percent, remaining in Parker Horn (open circles), the lower lake (closed circles), and the whole lake (triangles) compared to that predicted (based on average inflow from mid-April to mid-June) for the whole lake (solid line) and Parker Horn (dashed line) in response to dilution water input in 1978. Parker Horn, the lower lake, and the whole lake represent, respectively, 8, 21, and 100 percent of the lake volume. Reprinted with permission from *Water Research*, vol. 14, Welch and Patmont. Lake restoration by dilution: Moses Lake, Washington. Copyright 1980, Pergamon Press Ltd.

Improvement of lake quality from 1977 to 1979, compared to 1969-1970, was near or in excess of 50 percent for P, chl *a*, and Secchi visibility, not only for Parker Horn but for most of the lake volume (Table 5-2). Total N decreased by about the same magnitude, although predilution data for N are not as complete as for P.

Improvement was greatest in Parker Horn, where the fraction of dilution water was highest, but most of the lake responded nearly as well. As shown by Figure 5-2,

Table 5-2 Average April-September Dilution Rates (% day^{-1}) and May-September chl a, Total P (μg l^{-1} near the surface), and Secchi Transparency (m) for Two Portions of Moses Lake, 1977-1979 Compared with 1969-1970 (Percent improvement).

Years	Dilution rate	Total P	Chl a	Secchi
Parker Horn				
1969-70	1.0	158	71	0.6
1977-79	10.0	71 (54%)	26 (63%)	1.3 (54%)
58% Lake volume				
1969-70	0.1	156	45	0.9
1977-79	1.4	86 (45%)	21 (53%)	1.5 (40%)

*Reciprocals were used to calculate % improvement, for comparison with the other measures.

the dilution water was distributed throughout the lake. Wind was probably the main force causing the transport of dilution water from Parker Horn up into the main arm, with the fraction of dilution water reaching half the distance through the arm, dependent largely upon the fraction existing in Parker Horn (Welch et al., 1982). Part of the improvement in areas away from Parker Horn (lower lake) is considered to be due to the natural depletion of usable fractions of nutrients over time.

Presentation of means for the May-September periods obscures the extreme conditions, such as a maximum Secchi transparency of 3 m in June (4 m was reached in 1982) throughout most of the lake. Chl a reached peaks near 50 μg l^{-1} in late July and August after dilution water input had been curtailed for two to four weeks. Unless dilution water was added continually, blooms returned as the fraction of dilution water left in the lake declined. This "boom and bust" phenomenon, promoted by large inputs followed by no input at all, has prompted the proposition of a continual input at low rates throughout the summer, employing similar total amounts of water. The large quantities added over a short period of time—exchanging water in Parker Horn at the rate of about 20 percent per day and in most of the lake at 2-3 percent per day—are probably unnecessary, considering the general response of the phytoplankton, particularly the blue-greens, to dilution water addition.

Determining the optimum quantity of dilution water and its distribution over time requires defining the cause(s) for the quality improvement (algal reduction). Several possibilities have been discussed (Welch and Patmont, 1979; 1980; Welch and Tomasek, 1981; Welch et al., 1982), but the reduction of N concentration is the most probable cause for the effect of dilution on algal biomass in this lake. Nitrate, rather than SRP (soluble reactive phosphorus) has been the nutrient that most frequently limited algal growth rate (Welch et al., 1972). NO_3 in the lake was not appreciably reduced by dilution, because it was limiting and remained rather low in the lake water during summer both before and after dilution. Although control of biomass was discernable at total N concentrations in lake water below about 600 μg l^{-1} (Welch and Tomasek, 1981), the best relationship exists between flow-weighted NO_3 concentration in the inflow (Crab Creek) and the average chl a concentration in Parker Horn and the lower lake (r = 0.97; Welch et al., 1984).

Prior to implementation of this project, it was thought that total P would be the most important nutrient to control in order to reduce algal biomass. Although NO_3 was known to limit growth rate during summer, fixation of atmospheric N by blue-green algae should supply enough N to complement the available P supply. N fixation is not a rapid process, however; maximum rates of cell N replacement and growth of about 5 percent and 10 percent day^{-1} have been reported (Horne and Goldman, 1972; Horne and Viner, 1971). A growth rate by N fixation of only 2.4 ± 1.8% day^{-1} was determined for Moses Lake algae (Brenner, 1983). With such slow growth, increasing the flushing rate by about tenfold may well prevent N uptake by fixation from fully utilizing the available P. In any event, there is a close relationship between inflow NO_3 and chl a, which has allowed estimates of optimum dilution water input.

Goldman (1968) had suggested that reduction of the N content in Clear Lake, California (a hypereutrophic, shallow lake in an arid region, like Moses) by adding Eel River water should reduce algal content, although there was uncertainty about the buffering effect of increased release of N from sediments once the lake-water N content decreased and a larger gradient between sediment interstitial N and the overlying water occurred. There has been no indication that the release of P from Moses Lake sediments increased following dilution and a 50 percent decrease in lake-water concentration. In fact, mass balance analysis indicated that less P was released from sediments after dilution (Patmont, 1980).

Other factors probably contributed to the decrease in algal biomass. The physical loss of algal cells by washout affected biomass in Parker Horn, where high rates of exchange (0.2-0.25 day^{-1}) existed for short periods. Biomass in upper Pelican Horn decreased from about 20 mm^3 l^{-1} to about 10 mm^3 l^{-1} in one month after water exchange was increased by pumping Parker Horn water (Carlson and Welch, 1983; Welch et al., 1984). The greatest decrease occurred with an exchange rate of 0.09 day^{-1}, although subsequent rates were higher (0.19 day^{-1}). Persson (1981) observed that the biomass of *Oscillatoria* was greatly influenced by flushing rate during the period (about a month) of maximum growth in a hypereutrophic brackish-water bay. At a flushing rate of 8.1 and 9.4% day^{-1} the average biomass was about one half the level at 4.7% day^{-1}. At 20.7% day^{-1} biomass was only about one-third the level at 4.7% day^{-1}.

For Parker Horn, where the mean flushing rate was 10% day^{-1} from May to September (Table 5.2), washout of cells was probably a significant control on biomass. In the remainder of the lake, where flushing averaged only 1.4% day^{-1}, washout was of less significance when compared to a 50% day^{-1} maximum growth rate, which has been observed for *Aphanizomenon* in the lake. If the N source is fixation, however, even relatively low flushing rates may effectively reduce biomass through washout.

Instability of the water column, as indicated by decreased vertical density gradient, probably contributed to the crash or prevention of blue-green algal blooms in Parker Horn, as well as elsewhere in Moses Lake (Welch and Tomasek, 1981). Because the flotation capability of blue-greens provides them with advantages over greens and diatoms when mixing is poor, decreased stability may hinder dominance by blue-greens (Knoechel and Kalff, 1975; Paerl and Ustach, 1982). Wind is considered to have more effect on water-column stability in Moses Lake than dilution water input.

There are yet other factors resulting from dilution that may have contributed to reduced biomass of algae and reduced contribution by blue-greens. Some of those considered are limitation of the N fixation process by iron, an observed increase in free CO_2, and dilution of blue-green excretory products. These are discussed by Patmont (1980) and by Welch and Patmont (1980).

The optimum use of dilution water for Moses Lake would be a moderate but continuous input from May through August. Water added too early (February-March) would be largely replaced by high-nutrient Crab Creek water by June, when algal blooms begin. Replacement with Crab Creek water is also a problem if dilution water is stopped in June or early July. Unfortunately, the lack of irrigation demand and greater irrigation return flows to the downstream impoundment reduces space available for dilution water routed through Moses Lake during late summer. Based on the relationship between May-August average inflow concentration of NO_3 and June-August average chl a, however, a dilution-water volume of about 100×10^6 m^3 from May through August should control chl a to an average of about 20 μg l^{-1}, whether that quantity comes as 10 m^3 s^{-1} for the whole period or is divided into 25 m^3 s^{-1} for May and 5 m^3 s^{-1} for June through August. If water input is not maintained through August, blooms with very high chl a levels will result.

Although the input of dilution water has not been distributed over the summer as desired, the primary cost of the project has been the pumping facility, at $324,000 (1983 dollars) plus planning, administrative, and monitoring and research costs. The water has cost the Moses Lake Irrigation and Rehabilitation District nothing, because water diverted through Moses Lake is used for downstream irrigation. To ensure a flow of 5 m^3 s^{-1} during July and August, however, it would be necessary for the District to buy water. Although this is not planned and costs and liabilities have not been considered, it is instructive to assume a cost m^{-3} (1983 dollars) similar to that encountered for diluting Green Lake ($0.074; see following section). For two months the volume would be about 26×10^6 m^3, with a cost of nearly 2×10^6. The cost for dilution water at Moses Lake would probably not be as high as that for the domestic water used in Green Lake. However, it becomes clear that to buy dilution water may be a practical restoration alternative for relatively small lakes only, which require much lower rates of dilution water input, such as Green Lake. Green Lake is not fed naturally by a large input of high-nutrient water that itself must be diluted in order to dilute the lake nutrient concentration. In that respect, Moses Lake may be a unique case of a large lake existing at a location where large amounts of low-nutrient water are available at no cost.

Green Lake

The dilution of Green Lake, beginning in 1962, represents another example of the benefits of this technique for restoring lakes. The setting is the Seattle metropolitan area; 47,000 people live within 1.6 km of the lake, and as many as 1,162 people per hour use the 4.5-km path around the lake. Perhaps Green Lake represents a more typical case where dilution would be a practical restoration alternative than does Moses Lake.

Dilution was proposed as the primary treatment for Green Lake in 1960 (Sylvester and Anderson, 1964) and was instituted in 1962. In contrast to Moses Lake, the dilution of Green Lake was a long-term process with a relatively low rate of water input. The average combined water exchange rate was greatly increased from an estimated 0.83 yr^{-1} to 2.3 yr^{-1} as a result of adding low-nutrient water from the Seattle domestic supply, which comes from diversions near the source of two Cascade mountain streams. The addition of dilution water to Green Lake from 1965 to 1978 produced a flushing rate, based on dilution water only, that ranged from 0.88 to 2.4 yr^{-1}.

A marked improvement in chl a, P concentration, and Secchi disc transparency was noted during the first few years after dilution was initiated. Only one predilution year's data existed for comparison with the three years of postdilution monitoring, which began in 1965, but water visibility during the summer increased nearly fourfold, to an average depth of 4 m (because the mean depth is 3.8 m, most of the lake bottom was visible), and chl a decreased more than 90 percent, from 45 to 3 $\mu g\ l^{-1}$. The summer average for total P decreased from 65 to 20 $\mu g\ l^{-1}$. A substantial decrease in the fraction of blue-green algae was observed, particularly during spring and early summer.

Regular monitoring was terminated in 1968, but the lake has recently been studied intensively for the purpose of proposing a new restoration scheme (Perkins, 1983; URS, 1983). Chl a and total P averaged 38 and 55 $\mu g\ l^{-1}$, respectively, during the summer of 1981; the lake quality had degraded markedly during the late 1970s, primarily as a result of declining dilution water inputs. No water was added in 1982, and massive blooms of blue-green algae resulted. Dilution water has again been added in 1984 and 1985 to avoid deterioration in lake quality.

The percent decrease in P concentration following initiation of dilution is about what would be expected using equation 5.2. The expected P concentration in Green Lake prior to dilution, calculated from estimated external loading, should have been about 80 $\mu g\ l^{-1}$, but in fact the P content was about 65 $\mu g\ l^{-1}$. Following dilution, the steady-state concentration should have been about 35 $\mu g\ l^{-1}$; it actually declined to a much lower level, 20 $\mu g\ l^{-1}$ by 1967. Note, however, that the decrease was the same (45 $\mu g\ l^{-1}$) for the expected and the actual values. The discrepancy is most likely due to overestimating external loading.

Recent work on Green Lake has shown that much of its problem is due to internal loading of P, in spite of nearly all of its area being unstratified and apparently aerobic during summer (Perkins, 1983; URS, 1983). Internal loading accounted for 21 percent of the total annual loading of P during 1981 (determined by difference in the annual mass balance of P). During the three summer months, however, when chl a averaged 38, with a maximum of 60 $\mu g\ l^{-1}$, the internal source accounted for 88 percent of the total. Three storm drains contributed 41 percent of the annual external loading to the lake.

Dilution remains the principal technique to improve Green Lake and maintain high water quality, the goal being a mean annual P concentration of 20 $\mu g\ l^{-1}$. Water from the city supply is most desirable, because the delivery facility is in place, and because the P concentration of the city water is only 10 μg l^{-1}, but the city water is expensive, with an assumed cost of \$0.074 m^{-3} (URS, 1983). Other sources of water were therefore considered, even though construction of delivery systems would be required. As Figure 5-3 shows, the quantity of city water needed on an annual basis to lower the average lake concentration to 20 μg l^{-1} is 5 mgd (19.2 $\times$ 10^3 m^3 day^{-1} or 7 $\times$ 10^6 m^3 yr^{-1}), whereas the quantity required if other sources were used would be greater. As can be seen, however, Lake Washington (15 μg l^{-1} P) is an entirely adequate supply, at no direct cost because the water would be returned.

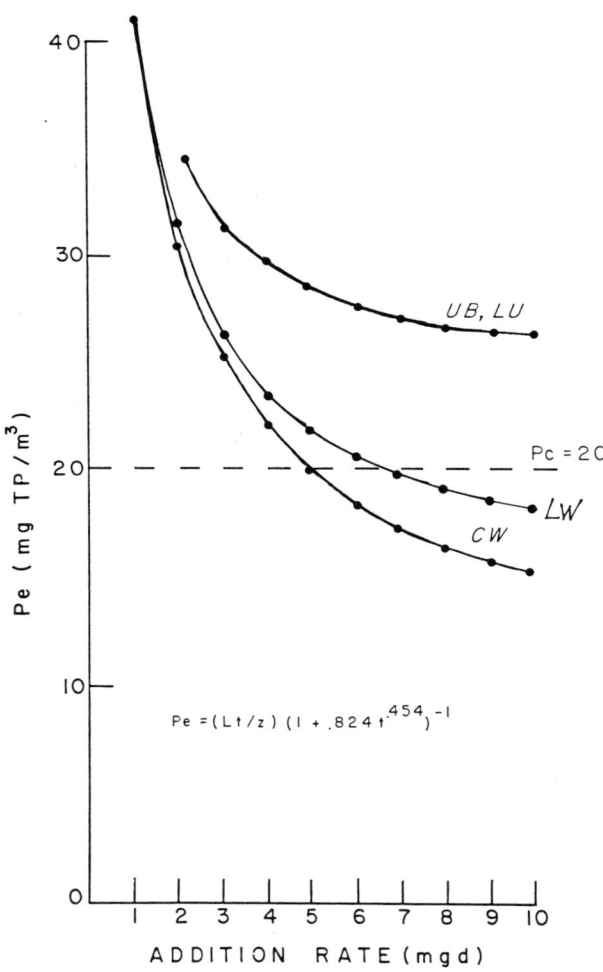

Figure 5-3 Effects of dilution alternatives on equilibrium phosphorus content of Green Lake (from Perkins, 1983).

The quantity of 7×10^6 m^3 yr^{-1} was arrived at by calculating the steady-state P concentration with incremental additions of city water, using the equation of Walker (1977) as illustrated in Figure 5-3. Equation 5-3 can be used similarly to illustrate the procedure. For that quantity of dilution water, corrected for evaporation, the flushing rate would increase from 0.92 (for 1981) to 2.38 yr^{-1}. The P loading would increase from 203 mg m^{-2} yr^{-1} (160 external and 43 internal) to 270, because dilution water contains 10 mg m^{-3} P. The sedimentation correction term would increase from 0.49 to 0.61, according to

$$\frac{1}{1 + 1/\sqrt{\rho}} = \frac{1}{1 + 1/\sqrt{2.38}} = 0.61,$$

which illustrates the expected reduction in sedimentation loss as flushing rate increases. Equation 5.3 predicts that with 7×10^6 m^3 yr^{-1} of city water the steady-state (annual mean) P concentration will be 18 mg m^{-3}:

$$P = L / (Z\rho) \cdot 1 + 1/\sqrt{\rho}$$

$$18 = 270 / (3.8 \cdot 2.38) \, 0.61.$$

The prediction is very similar to that using the Walker equation (Figure 5.3). The principal assumption is that Green Lake behaves similarly to the lakes from which equation 5.3 was derived. That is probably not the case, so there is a confidence band around the prediction, which can be expected to be quite large (Chapra and Reckhow, 1979). In fact, the model does not predict the lake concentration closely, using the 1981 data, even when internal loading is included. The predicted concentration is 29 µg l^{-1} and the observed concentration was 42 µg l^{-1}. The geometric mean is 32 µg l^{-1}, which statistically may be a better estimate of the steady-state concentration and is closer to the estimated value.

An alternative to adding dilution water to Green Lake at a constant rate throughout the year is to dilute only during the summer, when the internal loading is greatest and the problem water-quality conditions exist. A mass balance analysis for P during the three-month summer period showed that to achieve the 20 µg l^{-1} goal by the end of the period, the amount of dilution water required would be impractically high. It is more practical to add water at a low but constant rate throughout the year. Only half as much water would be required, and the P concentration would be lower during the summer.

The amount of water needed for the three summer months (July–September) to reduce the lake concentration to 20 µg l^{-1} can be estimated by the mass balance approach. The summer mean total P was 54.7 µg l^{-1}. The net decrease required is 34.7 µg l^{-1} (54.7 − 20) or 34.7×10^{-6} kg m^{-3} for the three months. That much P must be removed during the summer. The mass balance equation would be

$$dC/dt = 1/V \, [TP + (Q_d \cdot 10 \times 10^{-6} \text{ kg/m}^3) - ((Q_{out} + Q_d) \cdot (20 \times 10^{-6} \text{ kg/m}^3))], \quad (5.5)$$

where TP = total kg P input without dilution (internal loading = 259.5 kg and external loading = 36.8 kg)
Q_d = volume of dilution water required, m³
Q_d · 10 × 10⁻⁶ kg m⁻³ = kg P input with dilution water
$Q_{out} + Q_d$ = total outflow
20 × 10⁻⁶ kg m⁻³ = desired lake concentration.

Solving for Q_d,

$$(V \cdot dC/dt) - TP + Q_{out}(20 \times 10^{-6}) = Q_d(-10 \times 10^{-6})$$
$$(4.15 \times 10^6 \cdot 34.7 \times 10^{-6}) - 296.3 +$$
$$(1.12 \times 10^6 \cdot 20 \times 10^{-6}) = Q_d(-10 \times 10^{-6})$$
$$-129.9/-10 \times 10^{-6} = 13 \times 10^6 \text{ m}^3 = Q_d^-$$

The estimated present-worth cost for using Lake Washington water, which would be the capital and operating cost for the piping system, is $10 × 10⁶ for an annual flow rate of 7.2 mgd (27.6 × 10³ m³) over a 20 year period. The cost for city water at 5 mgd (19.2 × 10³ m³) is $13.7 × 10⁶ (1983 dollars) over 20 years. Other techniques, such as stormwater diversion and artificial circulation, were projected to lower the total project cost. The benefits from diversion and artificial circulation, however, are less certain than those from dilution. The inputs from stormwater occur largely in the winter and would not greatly affect the summer P concentration, which comes largely from internal sources. Because the mechanism that accounts for the internal loading of P is unknown quantitatively, the effect of artificial circulation is uncertain. A seasonal P modelling approach is planned to assist the decision process.

GENERAL APPLICATIONS AND PRECAUTIONS

Dilution is frequently used synonymously with flushing as a restoration technique. In fact, the effect of dilution includes both a reduction in the concentration of nutrients and a washout of algal cells, while flushing may only cause the latter. For dilution, or a reduction in nutrient concentration, to occur, the inflow water must be lower in concentration than that of the lake. Effectiveness will, of course, increase as the difference between inflow and lake concentrations becomes greater. For washout of algal cells to be an effective control on algal biomass, the flushing rate must be a sizeable fraction of or preferably approach the algal growth rate.

The ideal dilution scheme would be one to attain a long-term reduction of the limiting nutrient content through a low-rate input of low-nutrient water. Where there is an existing high-nutrient input, it should be diverted, if possible, for the low dilution rate to be most effective. This scheme would provide for reduction in biomass primarily through nutrient limitation. If diversion is not possible, one is faced with high-rate inputs over the short term to sufficiently reduce the inflow nutrient concentration. If only moderate- to high-nutrient water is available, short-term flushing may work well if the loss rate of cells is sufficiently great relative to the growth rate, and if washout becomes significant. Flushing rates on the order of 10% to 15% day^{-1} should afford some control through washout. Results from Moses Lake suggest that, until flushing rate reached 20% day^{-1} or more, the zooplankton crop was not adversely reduced and growth rate may even increase.

Costs are highly variable, depending upon the presence of facilities to deliver the water and the quantity and proximity of available water. If the lake is in an urban setting and domestic water is available, then improvement may be possible for less than $100,000 for construction, water costs, and the first year of maintenance and operation. For Wapato Lake in Tacoma, Washington, the cost was less than maintenance and operation of the city's swimming pools, and the lake had more swimmers (Entranco Engineers, personal communication). If the lake is near a free-flowing river and diversion of a portion of the river flow through the lake during summer is feasible, then the costs involve facilities, pumps and pipes, operation, and prevention of side effects (entraining fish).

The advantages of using dilution water include (1) relatively low cost if water is available, (2) an immediate and proven effectiveness if the limiting nutrient can be decreased, and (3) some success even if only moderate- to high-nutrient water is available, through physical limitations to large algal concentrations. The principal limitation for use of this technique is the availability of low-nutrient dilution water.

SUMMARY

The two cases described for the use of dilution to restore and maintain the quality of lakes are in Washington State: Green Lake in Seattle and Moses Lake in the eastern part of the State. Green Lake receives nutrients from urban runoff and internal loading from sediments. Low-phosphorus water from the city supply was added to the lake, beginning in 1962. The amount added raised the water exchange rate to an annual average of 2.4 yr^{-1} from 0.83 yr^{-1}. After five years of treatment, the summer Secchi transparency had improved fourfold, chl a decreased by over 90 percent, and total P declined by over 50 percent. Dilution-water inputs had declined by the late 1970s and early 1980s, and the lake was reinvestigated to plan a comprehensive restoration procedure. While city water to dilute the lake is still the option with the most promise, the cost is great, so other sources of dilution water have been considered, along with stormwater diversion, lake circulation, alum, and biological manipulation.

Moses Lake has received low-nutrient dilution water from the Columbia River, via an irrigation system during spring and summer periods since 1977. Although the lake is irregular in shape, dilution water entering one arm distributed throughout most of the lake, primarily as a result of wind. Prior to dilution the lake maintained very high nutrient and algal content and very low transparency; total P and chl a normally averaged about 150 and 45 μg l^{-1}, respectively. After dilution the respective average spring and summer averages have been 86 and 21 μg l^{-1}, while average Secchi transparency improved from 0.9 to 1.5 m, with maximums of 4 m. The biomass of algae was reduced in direct relation to the reduction of NO_3 in the inflow water, although other factors, such as iron, CO_2, and cell washout could have contributed to the effect. Facilitating the distribution of dilution water to another, previously undiluted arm of the lake demonstrated that algal biomass could be controlled by washout with moderate water exchange rates.

Dilution and flushing can be considered as effective techniques for restoring, or even temporarily improving the quality of eutrophic lakes. The techniques are especially effective and appealing if a supply of low-nutrient water exists. The costs involved are facilities to deliver the water and their maintenance and operation. In the two lakes mentioned here the facilities were largely in existence. While irrigation water had a cost, the Bureau of Reclamation was able to deliver the water to users via the treated lake. For Green Lake, a domestic supply was used with little operation cost. Maintenance of water quality in both systems has suffered, however, during late summer in Moses Lake and in Green Lake during some years when short supplies were encountered. A plan has been developed to either buy city water or build facilities to provide an adequate supply of no-cost water for Green Lake. For the most part, the use of dilution and flushing will not be limited as much by costs for facilities or water as by proximity to a supply of low-nutrient water.

REFERENCES

Brenner, M.V. 1983. The cause for the effect of dilution water in Moses Lake, M.S. Thesis, University of Washington, Seattle.
Carlson, K.L., and Welch, E.B. 1983. *Evaluation of Moses Lake Dilution: Phase II*, Water Res Tech Rep 80, Dept. Civil Engr., Univ. WA, Seattle.
Chapra, S.C., and Reckhow, K.H. 1979. Expressing the phosphorus loading concept in probabilistic terms. *J Fish Res Bd Can* 36:225-229.
Dickman, M. 1969. Some effects of lake renewal on phytoplankton productivity and species composition. *Limnol Oceanogr* 14:660-666.
Dillon, P.J. 1975. The phosphorus budget of Cameron Lake, Ontario: The importance of flushing rate relative to the degree of eutrophy of a lake. *Limnol Oceanogr* 29:28-39.
Goldman, C.R. 1968. *Limnological Aspects of Clear Lake, California with Special Reference to the Proposed Diversion of Eel River Water Through the Lake*, Report to Fed Water Pollut Control Admin.
Horne, A.J., and Viner, A.D. 1971. Nitrogen fixation and its significance in tropical Lake George, Uganda. *Nature* 232:417-418.
Horne, A.J., and Goldman, C.R. 1972. Nitrogen fixation in Clear Lake, California, I. Seasonal variation and the role of heterocysts. *Limnol Oceanogr* 17:678-692.
Knoechel, R., and Kalff, J. 1975. Algal sedimentation: the cause of a diatom-bluegreen succession. *Verh Int Ver Limnol* 19:745-754.
Larsen, D.P., and Mercier, H.T. 1976. Phosphorus retention capacity of lakes. *J Fish Res Board Can* 33:1742-1750.
Larsen, D.P., Van Sickle, J. Malueg, K.W., and Smith P.D. 1979. The effect of wastewater phosphorus removal on Shagawa Lake, Minnesota: Phosphorus supplies, lake phosphorus and chlorophyll A. *Wat Res* 13:1259-1272.

Oglesby, T.T. 1969. Effects of controlled nutrient dilution on the eutrophication of a lake. In *Eutrophication: Causes, Consequences and Correctives*, Washington, DC: National Academy of Science, 483-493.
Paerl, H.W., and Ustach, J.F. 1982. Blue-green algal scums: an explanation for their occurrence during freshwater blooms. *Limnol Oceanogr* 27:212-217.
Patmont, C.R. 1980. Phytoplankton and nutrient responses to dilution in Moses Lake. M.S. Thesis, University of Washington, Seattle.
Perkins, M.A. 1983. *Limnological Characteristics of Green Lake: Phase I Restoration Analysis*, Dept. Civil Engr., Univ. WA, Seattle.
Persson, P.E. 1981. Growth of *Oscillatoria agardhii* in a hypertrophic brackish-water bay. *Am Bot Finnici* vol. 18.
Sketelj, J., and Rejic, M. 1966. Pollutional phases of Lake Bled. In *Advances in Water Pollution Research*, Proc. 2nd Intl. Conf. Water Pollut. Res., London: Pergamon, 345-362.
Stadelman, P. 1980. Der Zustand des Rotsees bei Luzern, Kantonales amt fur Gewasserschutz, Luzern.
Sylvester, R.O., and Anderson, G.C. 1964. A lake's response to its environment. *J San Eng Div ASCE* 90: 1-22.
URS, 1983. *Green Lake Restoration Diagnostic Feasibility Study*. Seattle: URS Engineers.
Uttormark, P.D., and Hutchins, M.L. 1980. Input-output models as decision aids for lake restoration. *Wat Res Bull* 16:494-500.
Vollenweider, R.A. 1976. Advances in defining critical loading levels for phosphorus in lake eutrophication. *Mem Ist Ital Idrobiol* 33:53-83.
Walker, W.W. Jr. 1977. Some analytical methods applied to lake water quality problems. Ph.D. Dissertation, Harvard Univ.
Welch, E.B. 1969. Factors initiating phytoplankton blooms and resulting effects on dissolved oxygen in Duwamish River estuary, Seattle, Washington. U.S. Geol. Surv. Wat. Supp. Pap. 1873-A. 62 pp.
Welch E.B., Buckley, J.A., and Bush, R.M. 1972. Dilution as an algal bloom control. *J Water Pollut Control Fed* 44:2245-2265.
Welch, E.B., and Patmont, C.R. 1979. Dilution effects in Moses Lake. In *Limnological and Socioeconomic Evaluation of Lake Restoration Projects*, EPA-600/3-79-005, 187-212.
Welch, E.B. 1979. Lake restoration by dilution. In *Lake Restoration*, EPA-400/5-79-001, 133-139.
Welch, E.B., and Patmont, C.R. 1980. Lake restoration by dilution: Moses Lake, Washington. *Water Res* 14:1317-1325.
Welch, E.B., and Tomasek, M.D. 1981. The continuing dilution of Moses Lake, Washington. In *Restoration of Lakes and Inland Waters*, EPA-440/5-81-010, 238-244.
Welch, E.B., Carlson, K.L., Nece, R.E., and Brenner, M.V. 1982. *Evaluation of Moses Lake Dilution*, Water Res Tech Rep 77, Dept. Civil Engr., Univ. WA, Seattle.
Welch, E.B., Carlson, K.L., and Brenner, M.V. 1983. Control of algal biomass by inflow nitrogen. In *Lake and Reservoir Management*, EPA-440/5-84-001, 493-497.

6

Phosphorus Precipitation and Inactivation

Nuisance algal blooms can be reduced if the concentration of an essential nutrient in the lake is lowered to growth-limiting levels through diversion or pretreatment of income, through dilution by incoming water, or through a combination of methods. In cases where the amount of diversion is substantial, where the lake flushes rapidly, and where recycling from sediments is small, the lake's trophic state can be significantly improved. The case of Lake Washington (Edmondson, 1970) illustrates this type of response.

In many cases, even with diversion of a significant fraction of nutrient income, slow lake flushing and nutrient release from sediments and other internal sources may prolong the lake's enriched state and support continued algal blooms. Biological activities such as recycling by browsing fish (LaMarra, 1975) or sloughing and decomposition of macrophyte tissue (Barko and Smart, 1980) can be important. Or, as demonstrated by Stauffer and Lee (1973) for Lake Mendota, vertical entrainment of nutrient-rich hypolimnetic waters can be sufficient to account for all of the phosphorus required for the algal blooms of an entire summer. Cooke et al. (1977) reported that 65 to 100 percent of the summer increase in phosphorus content of two eutrophic Ohio lakes was from such internal loading. Lakes that experience significant internal loading of phosphorus to their water columns may be the rule rather than the exception, particularly lakes with extensive littoral areas or with close proximity between the epilimnion and anoxic sediments. In this lake type, additional in-lake steps may be necessary following diversion to prevent a prolonged eutrophic state. Shagawa Lake, Minnesota, is a case in point (Larsen et al., 1975; 1979; 1981).

Phosphorus precipitation and inactivation are lake-improvement techniques. Their purpose is to lower the lake's phosphorus (P) content by removal of P from the water column (P precipitation) and by retarding P release from lake sediments (P inactivation). An aluminum salt, usually aluminum sulfate (alum), sodium aluminate, or both, is added to the water column to form, depending upon water pH and alkalinity, a precipitate of aluminum phosphate or colloidal aluminum hydroxide, to which certain P fractions are tightly bound.

THE REACTIONS OF ALUMINUM AND PHOSPHORUS IN LAKE WATER

An understanding of how this technique works, how to treat a lake, and the reasons for previous successes and failures requires some knowledge of the chemistry of aluminum. Aluminum is the element most often chosen for P precipitation and inactivation applications because its complexes and polymers are apparently inert to redox changes such as occur in an anoxic hypolimnion. Phosphorus bound to these molecules should remain so, in contrast to iron which releases P as the redox potential falls during the development of anoxia. Also, these complexes are efficient in entrapment and removal of particulate and inorganic P and are apparently nontoxic to aquatic biota at the pH and dosage required to effect the desired change in lake P concentration. The chemistry of aluminum in water has been reviewed by Burrows (1977) and Hayden and Rubin (1974).

When aluminum sulfate [$Al_2(SO_4)_3 \cdot 18\,H_2O$] and sodium aluminate are added to water, the solution pH dictates the form of hydrolyzed aluminum species, with settleable, polymerized aluminum hydroxide predominant between pH 6 and 8, aluminate predominant above this range, and Al (III) below it. When $Al(OH)_3$ is formed by the addition of aluminum sulfate to water with carbonate alkalinity, pH and total alkalinity fall at a rate related to the initial alkalinity of the water. These facts are important to the determination of dose, and to the possible toxicity of aluminum applications to lake biota. Low alkalinity or an excessive dose, either of which would permit pH to fall well below 6.0, would bring about a sharp increase in potentially toxic dissolved Al (III) and a decrease in the amount of $Al(OH)_3$ floc available to remove P from the water or retard its release from lake sediments.

When either alum or sodium aluminate is added to lake water, P is removed by an $AlPO_4$ precipitate, by sorption of P on the surface of $Al(OH)_3$ polymer or floc, and by entrapment and sedimentation of P-containing particulate matter in the $Al(OH)_3$ floc (Cooke and Kennedy, 1981a). At the P concentration and pH found in the lake water, $AlPO_4$ does not form readily. Removal of inorganic P as the precipitate is maximized at high Al:P ratios (over 500 in Lake Mendota, Wisconsin, water; Eisenreich et al., 1977). Inorganic P is most effectively removed by sorption to the $Al(OH)_3$ floc. Removal of particulate P is most effective in the pH range of 6-8 when maximum floc [$Al(OH)_3$] occurs (Cooke and Kennedy, 1981a).

A potentially important P fraction of lake water is ineffectively removed by complexes of aluminum. Francko and Heath (1981) found that the dissolved organic P fractions were removed with less effectiveness than total P in West Twin Lake, Ohio. Kennedy and Cooke (1982) reported removal of dissolved organic P ranging from 1 percent to 36 percent in laboratory studies and 65 percent at Dollar Lake, Ohio. This lower removal percentage of organic P fractions could be of critical importance under P-limiting conditions because some blue-green algae adaptively form alkaline phosphatase under these conditions in quantities sufficient to obtain P from certain phosphomonoesters. This adaptive response can be a source of phosphorus sufficient to develop and maintain an algal bloom (Heath and Cooke, 1975).

Variability in P removal, depending upon P concentration and form and upon pH, thus suggests that lake treatments for P removal must be timed to a season when most P is in an inorganic fraction. This would normally correspond to spring turnover, a period when weather conditions might make application difficult. Because of the possible variability of P removal and because a P removal dose of aluminum (see section on dose) may be too small to give long-term control of P release from lake sediments, P removal has been recommended only for special cases (Cooke and Kennedy, 1981a). An example of an appropriate use of P removal is the interception of P released during macrophyte decay, a procedure used at Liberty Lake, Washington, by Funk et al. (1982).

If sufficient aluminum is added, the $Al(OH)_3$ floc that falls to the sediment continues to sorb P from interstitial water or to form $AlPO_4$, depending upon pH, P concentration, and amount of aluminum present (Kennedy, 1978). This approach, in which removal is accomplished in the process of applying a large dose to the sediments, is called phosphorus inactivation. Its purpose is to keep concentration in overlying waters low over very long periods by retarding P release from sediments during anoxic conditions, from groundwater seepage into the lake, or from sediment disturbance. As shown in the case studies, this approach can significantly lower the P content of a lake, maintain that low level for many years, and bring about a measurable and lasting improvement in trophic state.

DOSE DETERMINATION AND APPLICATION TECHNIQUES

Beginning with the experiments of Jernelöv in 1968 (Jernelöv, 1970), aluminum treatment of lakes has been used to control P concentration (Cooke and Kennedy, 1981a). Many of these treatments had little, if any, basis for determining the amount of aluminum to add. Presently there are two distinctly different approaches to calculating the aluminum dose. These differ in treatment objective, amount of aluminum, and sometimes in depth of application.

In the first approach, phosphorus removal or precipitation is achieved by adding just enough aluminum to the lake surface to remove the P in the water column at the moment chosen for application. Nearly all of the earlier lake treatments were of this type (see Cooke and Kennedy, 1981a for a tabular summary of each case history). The dose is determined by adding increments of aluminum sulfate to lake water samples until the desired removal of P is achieved. This dose is then used to calculate the amount needed to remove P from the entire lake. This procedure is an extension of the one used in wastewater facilities. A variant to this approach is to control pH and add alum to constantly mixed lakewater samples until the desired P removal is achieved. The Al:P ratio is then used, along with data on lake P concentration, to calculate the dose for P removal.

Small amounts of aluminum are usually needed to bring about P removal, but long-term control of P release from sediments may not be achieved. Also, some P fractions, particularly the dissolved organic fraction, may be incompletely removed, leaving a substrate for algal assimilation and growth. This approach to phosphorus limitation of algae is not recommended, for these reasons. As will be seen in the case studies, however, P removal has brought about lake improvement and can be effectively used in special cases, such as the removal of P from incoming waters.

A second method of determining aluminum dose is to apply as much aluminum as possible to lake sediments, consistent with environmental safety, with the objective being long-term control of P release from these sediments. Usually application is made to the lake's hypolimnion to avoid exposure of littoral biota to the heavy dose and because the target is the anoxic sediment, but surface treatment could be satisfactory, and P removal would thereby be accomplished as well (Kennedy and Cooke, 1980).

The relationships between aluminum dose, hydrogen ion concentration, carbonate alkalinity, and possible aluminum toxicity are important for determining the maximum amount of aluminum salt that could be added to a particular lake. In lake waters, pH decreases and alkalinity is reduced as aluminum sulfate is added. As pH falls through the 6-8 range, insoluble $Al(OH)_3$ is formed and dissolved aluminum Al (III) concentration is very small. With the addition of more aluminum sulfate (below pH 5.5), dissolved aluminum concentration increases rapidly (Figure 6-1). The "maximum" or optimum dose to a lake is pH-alkalinity dependent, therefore, and varies from lake to lake. In lakes with high carbonate alkalinity (above 100 mg $CaCO_3$-1), large quantities of aluminum sulfate can be added before pH falls sufficiently to bring about an appearance of dissolved aluminum (Kennedy and Cooke, 1982).

How is a maximum dose for P inactivation determined? As described in later paragraphs, there are few toxicity studies that take into account the relationship between aluminum dose, pH, alkalinity, and dissolved aluminum. Kennedy (1978) and later Cooke et al. (1978) were the first to suggest that the maximum dose could be based on these relationships. They adopted a standard of 50 μg Al l^{-1} (dissolved) as the safe maximum dose, based upon a study by Freeman and Everhart (1971), which reported that this level of dissolved aluminum had no obvious short-term or long-term effects on rainbow trout. Dissolved aluminum remains below this concentration in the pH range 5.5-9.0, and any dose producing posttreatment pH of about 6.0 would be considered environmentally safe from aluminum toxicity, according to this criterion (Kennedy and Cooke, 1980; 1982).

Kennedy and Cooke's (1980; 1982) reasoning is useful in that it provides a toxicological basis for adding enough aluminum sulfate to a lake to provide long-term control of sediment P release without significant damage to fish. Several problems are inherent in their approach, however. First, no toxicity studies (see later paragraphs) are available regarding the possible impacts of aluminum on lake community processes and organization. The use of data from a cool-water fish species as a basis for treating a lake may be too restrictive. There are also indications of potential problems of chronic toxicity and bioaccumulation (see Figure 6-4). Second, low pH itself can be detrimental. Third, some lakes have low carbonate alkalinity and only small amounts of aluminum sulfate could be added before pH 6.0 is reached. As noted later, this last problem is overcome by the use of sodium aluminate.

A step-by-step procedure to determine the maximum dose of aluminum sulfate for a lake was developed by Kennedy (1978). This method was used by Cooke et al. (1978), and the Appendix of their report provides the actual calculation for their

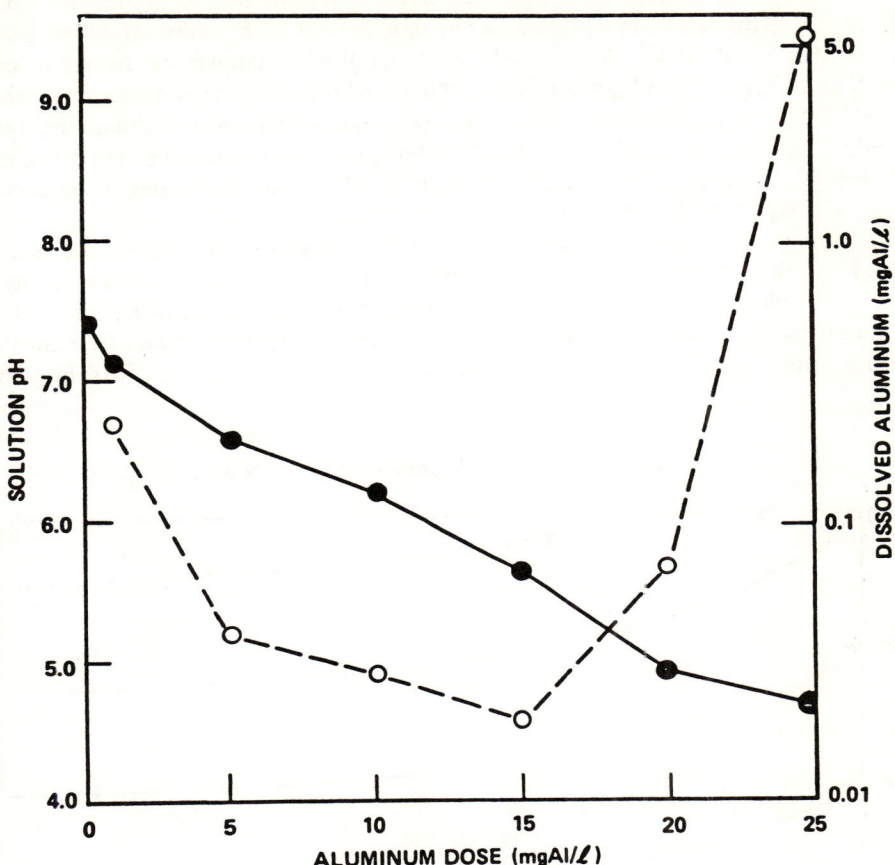

Figure 6-1 Changes in pH (closed circles) and posttreatment dissolved aluminum concentration (open circles) following additions of aluminum sulfate to lake water (initial total alkalinity of 98 mg $CaCO_3$ l^{-1} pH 7.3) (from Kennedy and Cooke, 1982; Kennedy, 1978).

treatment. Further guidelines for dose calculation were provided by Cooke and Kennedy (1981a) and Kennedy and Cooke (1980; 1982). The following steps for obtaining the correct dose are from these reports; the reader is urged to consult them before proceeding to treat a lake.

Aluminate Sulfate Dose Determination

1. Obtain water samples over the entire range of lakewater alkalinities. Normally this means a series of samples from surface to bottom. Determine alkalinity to a pH 4.5 endpoint.

2. An approximation of the dose for each stratum can be determined from Figure 6-2. This graph uses pH 6.0 as the endpoint of alum additions to the lake, rather than 50 μg Al l^{-1}. At usual lake dose, dissolved aluminum will remain below this limit as long as pH is between 5.5 and 9.0, so pH 6.0 was selected by Kennedy and Cooke (1982) as providing an adequate safety margin with respect to aluminum, thus avoiding problems to biota due to excessive hydrogen ion concentration, and probably allowing the addition of sufficient aluminum to lakes with alkalinities above about 50 mg $CaCO_3$ l^{-1} to give long-term control of P release.

A more accurate determination of the dose is made by titrating the water samples with a stock solution of aluminum sulfate of a known aluminum concentration (a solution that contains 1.25 mg Al ml^{-1} is made by dissolving 15.4211 g technical grade $Al_2(SO_4)_3 \cdot 18\ H_2O$ in distilled water and diluting to 1.0 liter). Adding 1.0 ml of this stock to a 500-ml water sample is a dose of 2.5 mg Al l^{-1}. As the alum

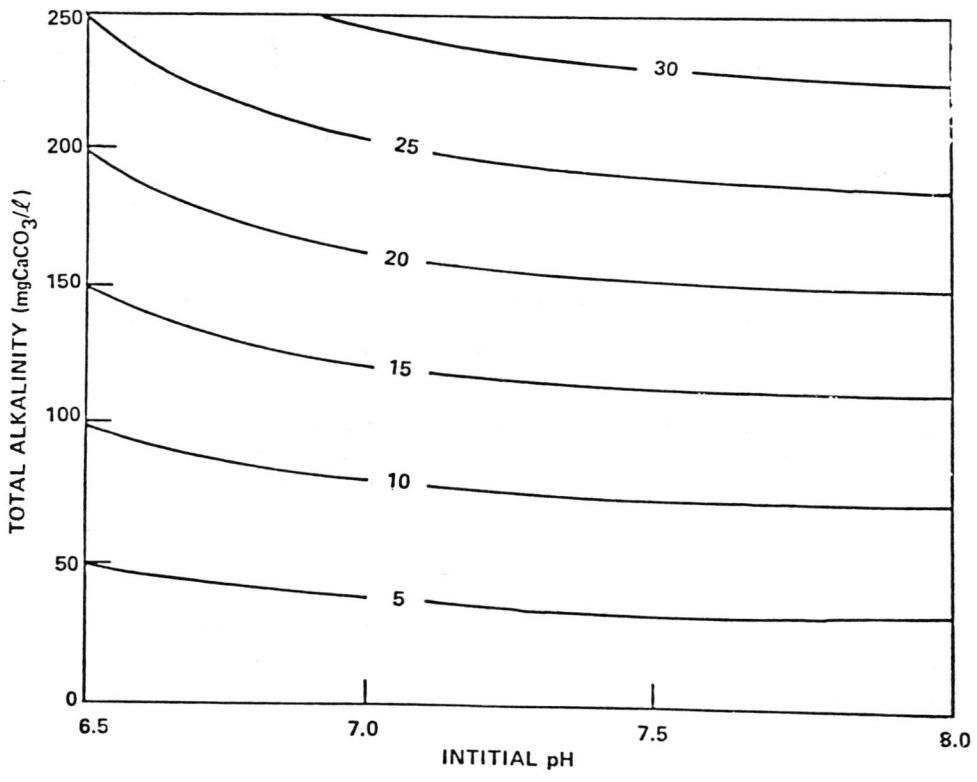

Figure 6-2 Estimated aluminum sulfate dose (mg Al l^{-1}) required to obtain pH 6 in treated water of varying initial alkalinity and pH (from Kennedy and Cooke, 1982).

is added, the samples are mixed with an overhead stirrer and pH changes are constantly monitored. Optimum dose for each sample is the amount that produces a stable pH of 6.0.

Linear regression is used to determine the relationship between dose and alkalinity. The resulting equation can then be used to obtain dose for any alkalinity for this particular lake, or reservoir, over the alkalinity range tested.

The maximum dose for each depth interval from which the alkalinities were obtained is calculated by converting the dose in mg Al l^{-1} to pounds (dry) alum m^{-3}, using a formula weight of 666.19 [$Al_2(SO_4)_3 \cdot 18 H_2O$] and a conversion factor of 0.02723 to change mg Al l^{-1} to pounds (dry) alum m^{-3}. Nonmetric units must be used because alum is sold by the pound or gallon. A conversion factor of 0.02428 is used for $Al_2(SO_4)_3 \cdot 14 H_2O$.

3. If liquid rather than granular alum is to be used, as is the usual case, further calculations are necessary to express the dose in gallons of alum m^{-3}. Alum ranges from 8.0 to 8.5 percent Al_2O_3, which is equivalent to 5.16 to 5.57 pounds dry alum per gallon at 60°F. It is shipped by tank truck at about 100°F and will thus have lower density. The percent Al_2O_3 at 60°F will be stated by the shipper. Convert this to density, expressed as degrees Baumé, using Figure 6-3. Then obtain the shipment temperature and use Figure 6-4 to adjust the 60° Baumé number. This is done by subtracting the adjustment value from the 60° Baumé number. Pounds per gallon is then obtained from Figure 6-5, using the adjusted Baumé number.

Maximum dose for each depth interval sampled for alkalinity was calculated earlier as pounds dry alum m^{-3}. This is converted to gallons m^{-3} by dividing pounds (dry) m^{-3} by the value in pounds per gallon obtained from Figure 6-5. Total dose to the lake is then the sum of each interval's dose.

4. Accuracy in treating the lake is obtained by dividing the lake into treatment areas, marked by buoys. The volume and alkalinity in each area is measured and the

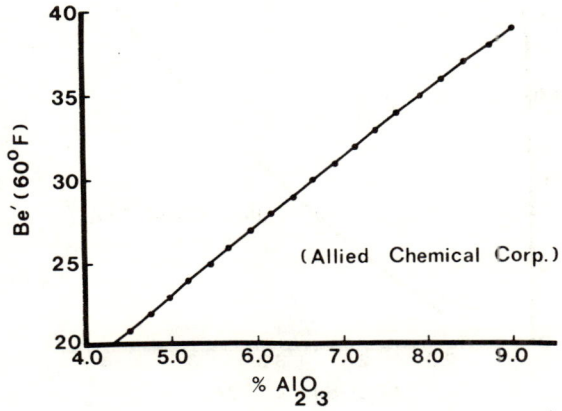

Figure 6-3 Relationship of Baumé (60° F) and percent Al_2O_3 (from Cooke et al., 1978).

108 *Physical and Chemical Methods*

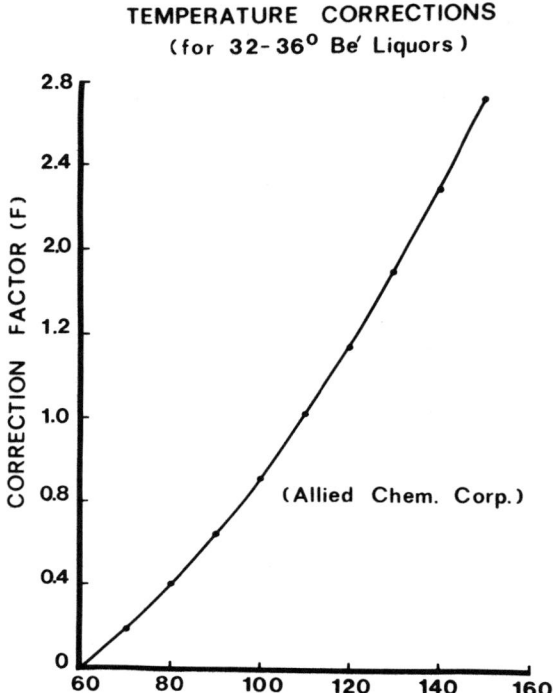

Figure 6-4 Temperature correction factors for 32-36° Baumé liquors (from Cooke et al., 1978).

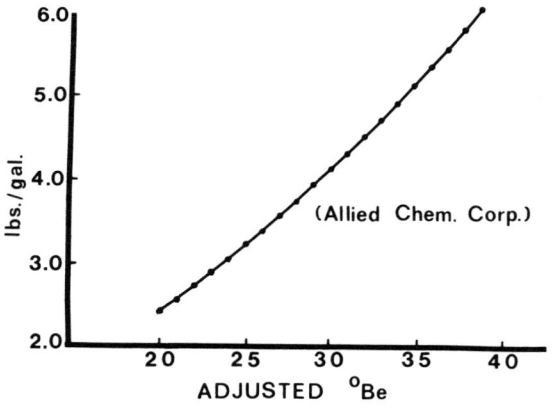

Figure 6-5 Curve to determine pounds of alum gallon^{-1}, based on adjusted Baumé (from Cooke et al., 1978).

gallons per treatment area determined. This approach will prevent underdosing deep areas and overdosing shallow ones, as would occur if an even application were made to the whole lake or reservoir area.

In soft-water lakes only small amounts of aluminum sulfate can be added before the pH falls below 6.0. A.R. Gahler and C.F. Powers (pers. comm., 1970) of the Corvallis Environmental Research Laboratory (USEPA, Corvallis, Oregon) were perhaps the first to suggest that the use of sodium aluminate, which increases the pH of an aqueous solution, could be used with aluminum sulfate to maintain a pH between 6.0 and 8.0. Dominie (1978) was apparently the first to successfully use this dose approach on a large scale, when Annabessacook Lake, Maine (alkalinity 20 mg $CaCO_3 l^{-1}$), was treated with this mixture in an empirically determined ratio of 1:1.6 alum to sodium aluminate.

A basis for aluminum dose now exists. The Kennedy procedure is based on the assumption that as much aluminum sulfate as possible should be added, up to the pH where significant amounts of dissolved aluminum appear, since the goal is long-term lake improvement. Currently it is not known what relationship exists between dose to the sediments and length of time that P release is controlled. It is likely to vary based on P concentration of sediment, groundwater inflow concentration, and sediment chemistry. It is important to examine this question, because it is possible to add to lake sediments an amount of aluminum hydroxide limited only by funds, using the buffering action of a mixture of alum and sodium aluminate. The use of alum-sodium aluminate may be a better way to treat lakes because it could provide for enough P sorption to keep lake P content at low levels for an indefinite period. More research on this subject is needed.

Application systems for adding aluminum to lakes have been based upon the system developed for the treatment of Horseshoe Lake, Wisconsin, in 1970 (Peterson et al., 1973). They mixed dry alum with lake water in a slurry tank on board a barge. The mixture was then pumped into a manifold trailing the barge and and spread and mixed just below the lake surface by the mixing action of the manifold and the propeller of the barge's motor. Nearly all subsequent lake treatments have used liquid alum, delivered to the site in chemical tank trucks, along with the Peterson barge system.

Figure 6-6 is an illustration of the basic application equipment used at Dollar Lake and West Twin Lake, Ohio. This system delivered 140 m^3 (39,000 gal) aluminum sulfate, mixed 50-50 with lake water, to the hypolimnion of West Twin Lake over a three-day period. Details of the complete system are in Kennedy (1978) and Cooke et al. (1978). This design is effective in providing even coverage of the target sediments, but is fairly labor intensive. The West Twin Lake treatment required 4.5 man-days (1 person working 8 hours = 1 man-day) for each of 16 hectares of hypolimnion. While this cost for controlling P release is less than the cost of removing those nutrient-rich sediments through dredging, a faster application procedure must be sought to make this lake-improvement technique easier to use and less expensive. Dominie (pers. comm.) has suggested the use of helicopters.

Pond applications can be simpler, but published experiences are limited. Alum could be added to small ponds by a pump and hose from the shore, or, as apparently

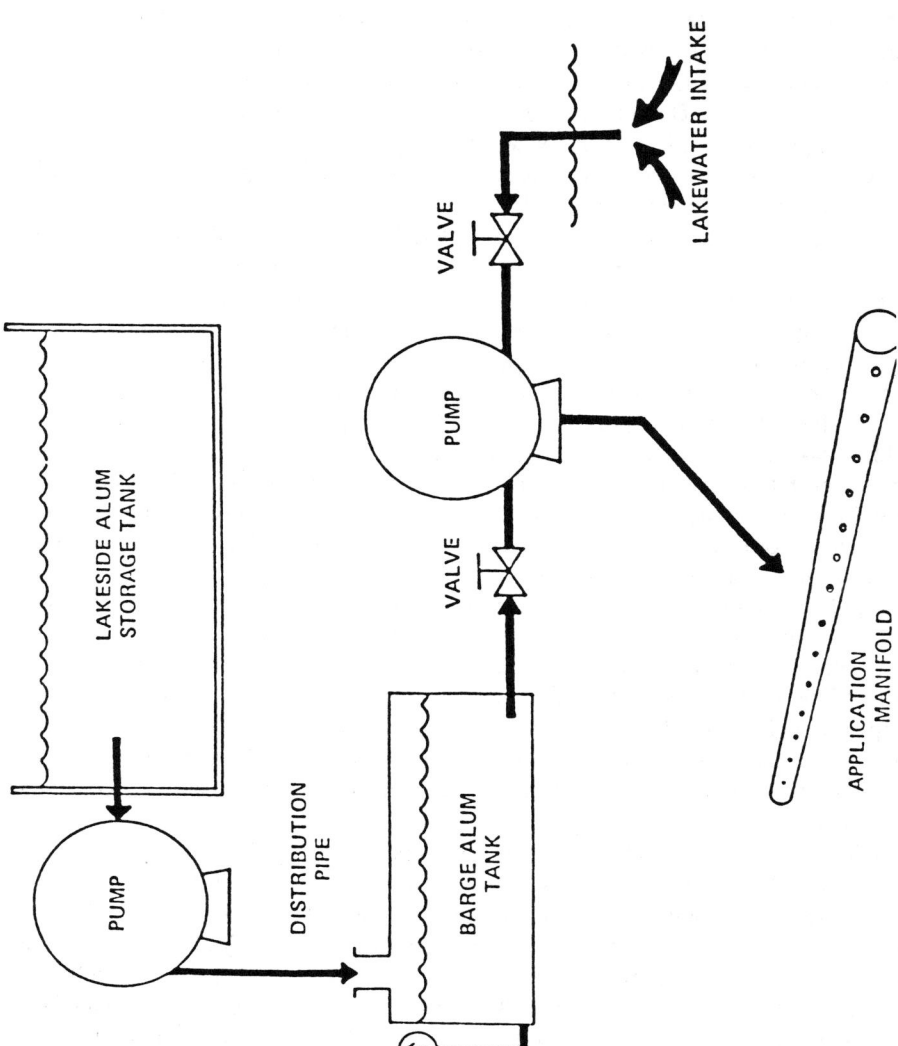

Figure 6-6 Basic components of a lake application system (from Cooke and Kennedy, 1981a; Kennedy and Cooke, 1982).

first developed by May (1974), blocks of ferric alum can be suspended at middepth in the pond, allowed to dissolve, and replaced as needed (see Section 6.3.6 for more on ponds).

CASE STUDIES

Nearly every aluminum treatment has been successful in bringing about a reduction in phosphorus concentration and an improvement in trophic state. Treatment areas up to 121 hectares (Annabessacook Lake, Maine; Dominie, 1980), doses up to 936 metric tons (Medical Lake, Washington; Gasperino et al., 1980), and treatment effectiveness up to 12 years (Garrison and Knauer, 1984) have been reported. There have also been some limited successes and some clear failures. Problems have been caused by insufficient dose, lake mixing (polymixes), or insufficient diversion of nutrient income. The following case studies are intended to cover a range of lake types, lake problems, and techniques of treating them with aluminum. A very wide range of experiences exists; these are reviewed by Cooke and Kennedy (1981a).

Horseshoe Lake and Snake Lake, Wisconsin

The use of aluminum sulfate to improve lakes was pioneered in the United States by the Wisconsin Department of Natural Resources. The first U.S. lake to be treated was Horseshoe Lake, a dimictic, eutrophic, hard-water lake (8.9 ha, Z_{max} = 6.7 m, $\bar{Z}$ = 4 m), which received a surface application of 2.6 g Al m^{-3} in May, 1970. The lake receives agricultural drainage, and until 1965 also received effluent from a cheese and butter factory. Winter fish kills and summer blooms of blue-green algae were common (Peterson et al., 1973).

The P content of Horseshoe Lake in 1971 and 1972 was lower than in 1966, particularly in the hypolimnion. Concentrations climbed somewhat in 1973 and then leveled (Peterson et al., 1973). In 1982, 12 years after treatment, hypolimnetic and whole-lake phosphorus concentrations remained low (about 0.30 mg P l^{-1} and 0.05 mg P l^{-1}, respectively) compared to pretreatment years, when hypolimnetic concentrations were as high as 1.5 mg P l^{-1} at the height of summer stagnation and whole lake P reached 0.3 mg P l^{-1}. Plankton blooms have declined and transparency appears to have increased. The lake continues to receive agricultural runoff, and the aluminum hydroxide layer appears to have been buried, so that P is reentering the water column from recent sediments, while P in deeper sediment layers continues to be controlled by the aluminum hydroxide layer (Garrison and Knauer, 1984).

Snake Lake is also dimictic, but it is smaller than Horseshoe Lake (5 ha, Z_{max} = 5.5 m, $\bar{Z}$ = 2 m) and has very soft water (alkalinity 13 mg CaCO$_3$ l^{-1}). Snake Lake received sewage-treatment-plant effluent from 1942 to 1965 and developed symptoms of severe eutrophication. An attempt to lower nutrient levels through dilutional pumping was made in 1970 (Born et al., 1973). The surface waters were treated in May 1972 with a mixture of aluminum sulfate and sodium aluminate, and an aluminum concentration of 12 mg Al l^{-1} was achieved over 80 percent of the lake's volume. The treatment was effective in lowering P concentrations, and water transparency improved to the levels found before discharge of sewage in 1942. In 1982, ten years after treatment, water transparency had again declined during blue-green algal blooms, but P concentrations were very similar to those in the first post-treatment year. The floc appears to have continued sorbing P, but storm-sewer runoff triggers algal blooms (Garrison and Knauer, 1984).

These early lake treatments were highly successful and have demonstrated the longevity of an aluminum treatment, since their treated sediments continue to sorb P 10 and 12 years after the application. It should be mentioned that neither lake received a maximum dose, yet the treatment has been lasting and effective. This work also points out the need for more complete diversion of nutrient income. There have been no indications of toxicity to benthic invertebrates (Narf, 1978). Studies of Horseshoe Lake and Snake Lake have been summarized by Garrison and Knauer (1984).

Medical Lake, Washington

Medical Lake ($\bar{Z}$ = 10 m; Z_{max} = 18 m) had become eutrophic and supported intense blue-green algal blooms. Internal P loading from the sediments was high (Gasperino et al., 1980; Soltero et al., 1981) and phosphorus removal was chosen to control it. Because lake alkalinity was high (750 mg $CaCO_3$ l^{-1}), large amounts of alum could be added. Jar tests were conducted to determine the dose to remove 87 percent of the ortho-P in the water column, 12.2 gm Al m^{-3}. In August–September 1977, 936 metric tons of alum as a liquid slurry were applied over a five-week period. Seven subsurface (4 m) and four surface applications were made, using a marker system to divide the lake into treatment areas.

The treatment produced a greatly improved trophic state over the three years of monitoring thus far reported (Gasperino et al., 1980; Soltero et al., 1981). The treatment had its greatest effect on phosphorus concentrations after fall overturn, not immediately after the application in August. The reasons for this observation, which was also made in Lake Annabessacook, Maine (Dominie, 1978), are not known. Mean ortho-P was reduced 90 percent and total phosphorus 30 percent; these lowered concentrations persisted through 1980 (Soltero et al., 1981). The overall mean concentration of chlorophyll *a* was reduced by 87 percent; growing-season chlorophyll *a* was reduced from a 1977 mean of 16.7 to posttreatment means near 2.5 mg m^{-3}. The phytoplankton community shifted from one dominated by blue-green algae to one dominated by Chlorophyceae and Cryptophyceae. The limiting nutrient, as tested with algal assays, shifted from nitrogen before treatment to phosphorus. Oxygen levels did not improve immediately; this was attributed to the rich organic matter deposited with the floc.

Unfortunately, the Medical Lake restoration may be shortened by a subsequent lake manipulation. After the treatment, the zooplankton community shifted from one dominated by Rotifera to dominance by Crustacea (Mires et al., 1981), a shift consistent with observations by others regarding zooplankton community changes and trophic state (e.g., Gannon and Stemberger, 1978). Mires et al. (1981) note that the dominance of rotifers prior to treatment was probably due to the absence of fish predation [the only pretreatment fish were tench (*Tinca tinca*) and carp (*Cyprinis carpio*), according to Gasperino et al. (1980)] and to intense predation on microcrustacea by *Chaoborus* and *Cyclops bicuspidatus thomasi*.

The State of Washington Department of Game chose to stock Medical Lake with a total of 31,000 rainbow trout (*Salmo gairdneri*) fingerlings in 1978, 1979, and 1980 to take advantage of the greatly improved water quality, which could now support a recreational fishery. Unfortunately, this action may have negated some or all of the benefits achieved through the alum treatment, because the fish are intense grazers upon zooplankton. *Daphnia pulex*, a large-bodied herbivorous microcrustacean that can contribute significantly to water clarity through its grazing activities on phytoplankton (Shapiro et al., 1982), had become abundant in the water column of Medical Lake after the alum treatment and may have contributed to its low post-treatment chlorophyll. An investigation of the diet of trout in Medical Lake revealed that they were eating *Chaoborus* and *D. pulex*, and that fish predation was enhanced by the clarity of the water and the improved hypolimnetic oxygen.

Knapp and Soltero (1983) predict that trout predation will alter the plankton community in two ways: Small-bodied zooplankton will have enhanced survival, because they are not preyed upon and because they will not have to compete with the more efficient grazer, *D. pulex*; and phytoplankton biomass will increase as a result of the decreased grazing by *D. pulex*. Ultimately, Knapp and Soltero believe the trout fishery may decline if the main food item, *D. pulex*, is eliminated. A successful lake restoration project however may be compromised, along with the recreational fishery. Knapp and Soltero (1983) correctly point out that lake manipulations that are ecologically unsound, as it appears this attempt to develop a trout fishery in Medical Lake may be, can negate an otherwise effective prior lake treatment. Further discussions of the effect of zooplanktivory by fish such as trout on phytoplankton biomass are presented in Chapter 13 on Biological Controls.

Annabessacook Lake, Maine

Annabessacook Lake, Maine (574 ha), is the largest lake to be treated with aluminum. The lake had experienced blue-green algal blooms over the past 40 years, particularly at the end of summer stratification. Sewage diversion and land management controlled significant P sources, but blooms continued, supported by internal P loading. Due to low alkalinity (20 mg $CaCO_3$ l^{-1}), the lake posed special dose problems. A dose of aluminum sulfate sufficient to give long-term control of P release from sediments would increase hydrogen ion and dissolved aluminum concentrations sufficiently to create toxic conditions. A 1:1.6 mixture of aluminum sulfate to sodium aluminate, determined empirically in the laboratory as a ratio that would keep pH in the 6-7 range, was applied to the top of the hypolimnion (121 ha) over an 18-day period in the summer of 1978 (Dominie, 1980). A concentration of 25 g Al m^{-3} was applied to the 8-10 m contour, and 35 g Al m^{-3} was applied to the 10-m contour and deeper.

A significant lake improvement followed. In the summer before treatment, 1800 kg P were added to the water column through internal loading. In the summer of 1979, one year after treatment, only 625 kg P (a 65% reduction) were added, and there was no autumn blue-green bloom. No adverse effects were observed. This treatment demonstrates that an adequate dose can be determined for soft-water lakes.

Dollar Lake and West Twin Lake, Ohio

Dollar Lake and West Twin Lake (Cooke et al., 1978; 1982; Cooke and Kennedy, 1978; 1981a, b; Kennedy, 1978) were treated with "maximum" hypolimnetic doses (as defined earlier) of aluminum sulfate, in 1974 and 1975 respectively, to control P release from anoxic hypolimnetic sediments. Dollar Lake received 21 g Al m^{-3}, West Twin Lake 26 g Al m^{-3}. These lakes are small (3 and 35 ha), shallow ($\bar{Z}$ = 3.2 and 4.3 m), dimictic, and have extensive internal P loading (Cooke et al., 1977). Septic tank diversion was accomplished in 1972, but high P concentrations and algal blooms persisted. After alum treatment, trophic state in both lakes improved from eutrophic to mesotrophic (Carlson, 1978, total phosphorus TSI index number of about 60 to an index number of about 48). Algal blooms were less intense, transparency increased, and lake P content decreased sharply and has remained low (Figs. 6-7 and 6-8). In West Twin Lake, for example, volume-weighted mean total P concentration for the entire water column was 79 μg P l^{-1} in 1974 but was 32 μg P l^{-1} in 1976, one year after the treatment.

Volume-weighted epilimnetic mean total P declined from a 1974 value of 47 μg P l^{-1} to a 1976 value of 28 (Cooke et al., 1982). Dollar Lake remained in an improved condition for six years, after which monitoring was discontinued. Kraus (unpublished) has demonstrated that the sediments of West Twin Lake in 1982 were still significantly more sorptive of P than the downstream control lake, East Twin, and that hypolimnetic concentrations were still low; thus the treatment was effective for at least seven years.

Internal P release in West Twin Lake during the summer was only partially controlled, despite laboratory and field evidence showing good control of P release from anoxic sediments after alum application. This suggests that the littoral zone (30% of the lake's area) represents a significant P source to the open water, and will help maintain the lake in a permanent mesotrophic state.

Shallow Lakes: Pickerel Lake, Wisconsin, Long Lake, Washington

Successful treatments of shallow, polymictic lakes have been thought to be unlikely, since it was believed that the floc would be readily dispersed and relocated during severe weather. This belief is supported by the P removal project at Pickerel Lake, Wisconsin, a shallow, hard-water lake ($\bar{Z}$ = 2.4 m; alkalinity = 110 mg CaCO$_3$ l^{-1}) with a low water residence time (0.63 years). A dose of 7.3 gm Al m^{-3} was applied to the surface as liquid aluminum sulfate in April 1973. After holomixes in August,

Phosphorus Precipitation and Inactivation 115

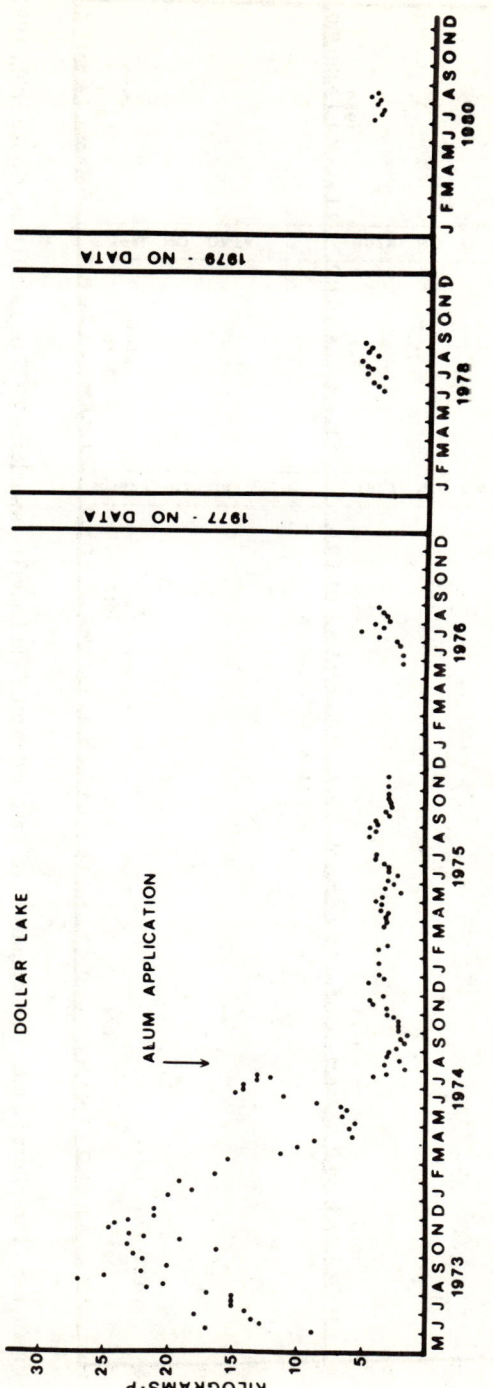

Figure 6-7 Phosphorus content (kilograms P) of Dollar Lake (no samples in 1977 or 1979) (from Cooke et al., 1982).

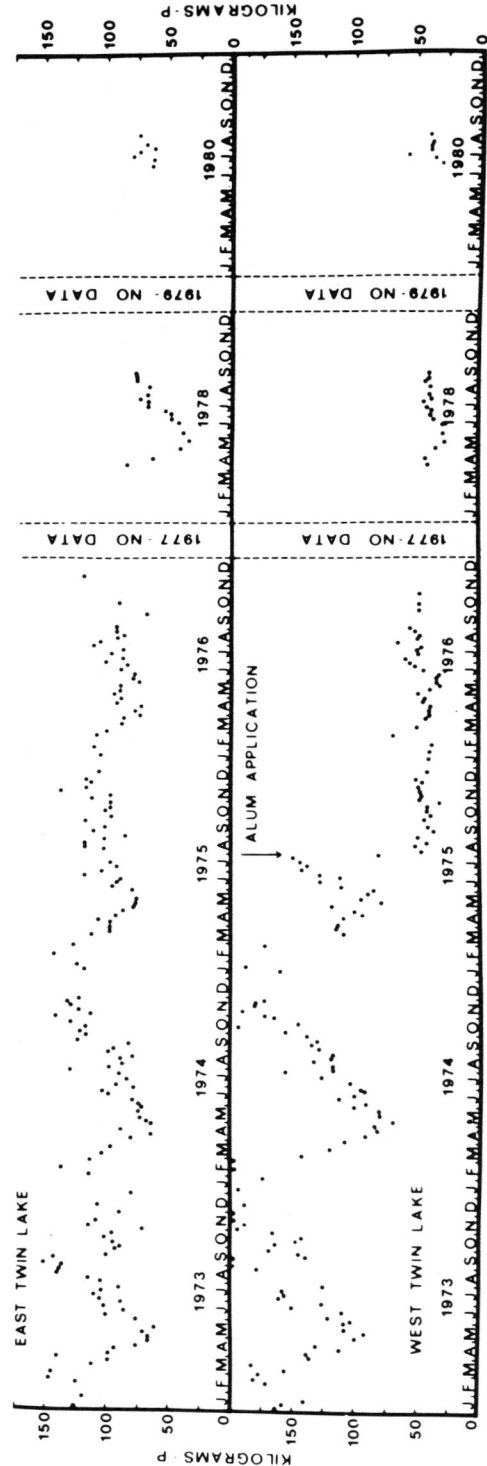

Figure 6-8 Phosphorus content (kilograms P) of East and West Twin Lakes (no samples in 1977 or 1979) (from Cooke et al., 1982).

total P was sharply reduced, but returned to pretreatment levels by March 1974. Analysis of aluminum in sediment cores showed that the floc had been redistributed to the lake's center following a series of holomictic occurrences, which left areas of the sediment free to release P to the water (Knauer and Garrison, 1980).

Experience was the opposite at Long Lake, Washington, which is a similarly shallow ($\bar{Z}$ = 2.0 m), highly flushed lake (t_w = 0.13-0.28 yr^{-1}), but with lower alkalinity (10-40 mg CaCO$_3$ l^{-1}) (Jacoby et al., 1983; Welch et al., 1982). The lake was drawn down in 1979 in a successful attempt to control *Elodea densa*. This plant was believed to be responsible for the high internal P loading through its decomposition and subsequent regeneration of P from the sediments, especially at high pH. In 1980, 5.5 gm Al m^{-3}, as liquid aluminum sulfate, was applied over an 11-day period (Welch et al., 1982). This was a maximum dose under the guidelines of Kennedy and Cooke (1981, 1982). Total P concentration fell and remained low for three years (Jacoby et al., 1983; Welch and Spyridakis, 1984). Associated with this was lowered phytoplankton biomass, chlorophyll, and pH, and increased transparency. The macrophytes recovered in 1981 to levels similar to before drawdown, perhaps due to increased light penetration after the alum treatment. Jacoby et al. believe that the alum treatment was successful, in that internal P loading did not resume after the macrophyte growth-death-decay cycle began again. The floc was apparently not redistributed despite a winter of high winds and rapid flushing.

Only further studies of shallow lakes will reveal whether lake-bottom slope, application procedure or dose, sediment chemistry, or other factors were responsible for the differences in these two treatments. Welch et al. (1982) point out that alum may not be appropriate for shallow lakes anyhow, especially those with macrophyte problems. Alum makes the water clear, by flocculation and by the control of phytoplankton biomass through control of internal P release. This increases the quantity of light and the outer depth of light penetration, and thus enhances macrophytes. Dredging, drawdown, or harvesting may be more effective controllers of internal P loading and macrophytes in shallow lakes. Alternatively, as pointed out by Funk and Gibbons (1977) and Funk et al. (1978; 1980; 1982) from experiments at Liberty Lake, Washington, alum could be added annually at a time appropriate to intercept the maximum P release from the annual die-back and decomposition of macrophytes.

Ponds

The number of ponds and small swimming lakes that have problems with nuisance blooms of algae is very large. The most common treatment is the temporary and costly expedient of adding copper sulfate or some other algicide. While these treatments can be briefly effective, May (1974) and May and Baker (1978) have suggested the alternative of adding alum or blocks of ferric alum [Fe$_2$(SO$_4$)$_3$ · 24 H$_2$O].

Their object was to reduce total P concentrations below 50 μg P l^{-1}. They found that a dose of 50 g alum m^{-3}, as blocks of ferric alum, would accomplish this. Blooms of toxic blue-green algae (*Anacystis cyanea* and *Anabaena circinalis*) were suppressed in most trials in small farm ponds in New South Wales, Australia.

May and Baker (1978) now recommend a single treatment of 100 g alum m^{-3} as ferric alum applied just before the onset of warm weather. The blocks are placed in cloth bags and suspended from an anchored float. For larger ponds, several floats are used. The blocks are allowed to dissolve over the season and do not need replacement for 12 months (May and Baker, 1978).

This technique could be an extremely satisfactory replacement for algicides. As before, however, it must be noted that alum lowers pH and alkalinity. Unless care is taken, one problem (expensive, repeated applications of a toxin) may be replaced by another (low pH). Much additional work on this new idea is needed, including detailed studies on dose and effectiveness.

LAKE AND RESERVOIR PRETREATMENT

The treatment of incoming streams to remove phosphorus and other nutrients is an unexplored means of improving lakes and reservoirs. Drainage basins are often very large, and the principal stream may receive nutrients and silt from many nonpoint sources, making sufficient diversion to significantly reduce loading nearly impossible, or at the least, very expensive. Alternatives are treating the incoming water with chemical precipitants such as aluminum sulfate or iron, constructing sedimentation basins, and diverting the stream through a marsh. Such treatments would be possible or economically feasible only in cases where just one or two streams formed the main sources of water.

Fiala and Vasata (1982) provide an example of the use of a prereservoir detention basin. The basin was placed above Jesenice Reservoir, Czechoslovakia, to eliminate up to 70 percent of the total P entering the reservoir. The authors point out that the success of such an operation depends upon the detention time of the basin, with P elimination increasing exponentially with increase in detention. The retention of P by a basin such as this could fall markedly if the water became anaerobic. Dissolved oxygen should therefore be monitored and an aeration system installed if necessary to maintain the iron III system.

Bernhardt (1981) describes the phosphorus elimination process used at Wahnbach Reservoir, Federal Republic of Germany. The reservoir had become very eutrophic, and treatment for drinking water was thus costly and increasingly difficult. A prereservoir, containing 500,000 m^3, was constructed as a sedimentation basin, and a phosphorus elimination plant capable of handling 5 m^3 sec^{-1} of water, a flow five times greater than the average flow of incoming water, was constructed. Water is pumped from the prereservoir sedimentation basin into the plant. Iron III is used to precipitate orthophosphate, and is followed by a cationic polyelectrolyte to form large floc. Then the water is filtered through a layered system of activated carbon, hydro anthracite, and quartz sand. About 95 to 99 percent of phosphorus-containing compounds were eliminated, and the output P concentration into the reservoir averaged 4 μg P l^{-1} over two years. Very high (95-99%) removal of coliform bacteria, chlorophyll, and turbidity was also achieved, with lesser removal of COD (77%) and DOC (58%). The reservoir quality has greatly improved with regard to the production of finished drinking water.

Another means of improving river water before it enters the main basin is to add P precipitants directly to the inflowing water. There have been very few published descriptions of this approach. Bannink and van der Vlugt (1977) reported on the use of iron, added as 10 mg $FeSO_4$ l^{-1}, to the inflowing water of a reservoir on the Rhine River in the Netherlands for P removal. Concentrations near 600 μg P l^{-1} in the river were lowered to about 100 μg P l^{-1} in the reservoir, and chlorophyll levels also declined. Hayes et al. (1984) describe the use of ferric sulfate as a treatment of the river water entering Foxcote Reservoir, England. Dose was obtained through jar tests to determine the amount of precipitant needed to maintain soluble phosphorus below 50 μg P l^{-1}. Doses ranged from 3.0 to 5.4 mg Fe l^{-1}. In practice, the system kept soluble phosphorus nearer 10 μg P l^{-1}, and the reservoir had greatly reduced chlorophyll. The costs of the iron treatment were offset by lower costs to treat the reservoir water for domestic drinking purposes. Cooke and Carlson (1985, unpub. ms., Kent State University) have initiated an examination of the feasibility and effectiveness of adding alum to incoming river water to remove soluble reactive phosphorus and thereby control the development of algal blooms in a reservoir's upper basin.

Harper et al. (1983) may have been the first to attempt to use $Al(OH)_3$ sludge, formed during the flocculation-clarification process of drinking water treatment, to attempt P removal from water entering a lake. Lake Eola, Florida, had become eutrophic from stormwater inflows. An underground filtration system was built, and stormwaters were collected in a basin, filtered through a 50-50 mixture of sludge and coarse sand, and then discharged to the lake. The filtration system removed about 99 percent of orthophosphate and 80 percent of total P, as well as large amounts (> 70%) of suspended solids and organic N. $Al(OH)_3$ sludge was added directly to the sediments of Lake Eola, as a P inactivation agent. Laboratory studies revealed that doses of five to ten metric tons of wet sludge per hectare of bottom sediments would inhibit orthophosphate release. No results of the treatment of eight hectares of Lake Eola, a procedure that took 25 days, are presently available.

Caution should be observed in use of $Al(OH)_3$ sludge from drinking water treatment plants. Raw water from eutrophic reservoirs, treated with aluminum sulfate, produces an aluminum hydroxide sludge that is very high in total phosphorus and organic matter, with little ability to retain further phosphorus. The sludge produced by the treatment of Lake Rockwell Reservoir, Ohio, water for drinking purposes contains 0.172 g P l^{-1} (sample taken on 2 August 1983; Cooke and Carlson, unpublished). This means that treatment plant sludges may be saturated and unable to take up significant additional amounts of phosphorus. Addition of such sludge to a lake or reservoir may constitute addition of excessive amounts of BOD and phosphorus, thereby defeating the purpose of treatment.

NEGATIVE ASPECTS

The acceptability of any lake restoration method hinges on cost, effectiveness, and the absence of significant or permanent negative impacts to the lake. The P precipitation and inactivation technique is the only lake restoration technique discussed in this book in which a potentially toxic substance is introduced to the lake.

Our knowledge of aluminum toxicity to aquatic organisms and more generally to aquatic communities is very limited. In part this is due to a failure in early studies to recognize the relationships between pH, alkalinity, and dissolved aluminum concentration (described earlier, in Reactions of Aluminum and Phosphorus in Lake Water). In many of these studies, aluminum was added as a salt, such as aluminum chloride, to test waters of unspecified pH and alkalinity. Dissolved aluminum was usually not measured but was assumed to equal the amount added in the salt. We now know that the appearance of dissolved aluminum is pH dependent, and studies that related concentration as calculated to observed mortality are not useful in estimating the impact of this lake treatment.

Burrows (1977) has reviewed the chemistry and toxicity of aluminum. There appear to be only a few laboratory studies of relevance to this lake restoration technique (e.g., Biesinger and Christensen, 1972; Everhart and Freeman, 1973; Lamb and Bailey, 1981; 1983; S.A. Peterson et al., 1974; 1976). Collectively these studies indicate that a dissolved aluminum concentration below 50 μg Al l^{-1} will have no harmful effects on *Daphnia magna*, rainbow trout (*Salmo gairdneri*), or chironomid larvae (*Tanytarsus dissimilis*). Field studies on the impact of aluminum on aquatic communities and species are equally sparse, although this might change rapidly with increased concern over acid precipitation and the effect of reduced pH on solubility of metals.

Mortality to fish in acidified environments appears to be a combination of effects due to aluminum and lowered pH. Aluminum is most toxic to fish at a pH of 5.0 (see Haines, 1981, for a review). Buergel and Soltero (1983) examined bioaccumulation of aluminum by rainbow trout (*Salmo gairdneri*) in alum-treated Medical Lake, Washington (see case study for details of the treatment). No mortality, physiological stress, gill hyperplasia or necrosis, or retardation of growth was observed. Gill tissues contained aluminum (6.25, 5.88, and 6.08 μg Al gm^{-1} in 0+, 1+, and 2+ aged fish), and there were increases up to sixfold in liver and kidney tissues between ages 0+ and 2+. Increases in stomach muscle and heart were much smaller, but reproductive tissues showed a 20-fold increase between 1+ and 2+ fish. Interestingly, gill tissues from fish sampled in a nearby hatchery and untreated lake were higher in aluminum than Medical Lake trout. Aluminum levels in composite plankton samples from Medical Lake were ten times greater than in trout tissues. No comparisons were made to plankton of untreated lakes. Lamb and Bailey (1983) also found no acute toxicity or sublethal effects to rainbow trout from the Liberty Lake, Washington, treatment (Funk et al., 1982) when caged 2.5-cm trout were exposed to either an alum application or to the floc.

Lake treatment with alum may be a procedure with significant potential hazards, particularly if bioaccumulation of aluminum occurs. This may be especially true in poorly buffered waters where concentration of dissolved aluminum could become high. More studies on the bioaccumulation of aluminum in various lake organisms are needed.

Lamb and Bailey (1981) examined the acute and chronic effects of alum on *Tanytarsus dissimilis* (Insecta, Chironomidae) in laboratory systems in which pH was held constant at 7.8 in acute and at 6.8 in chronic test solutions. The acute tests demonstrated no apparent effect on second or third instar larvae at doses ranging from 6.5 to 77.8 mg Al l^{-1}. Dissolved Al remained below 0.1 μg Al l^{-1} and the larvae used the floc to build tubes. In the chronic tests, using a dose range from 0.8 to 77.8 mg Al l^{-1}, mortality occurred at all concentrations. At a dose of 77.8 mg Al l^{-1}, 50 percent mortality occurred in 23 days. Dissolved aluminum remained below 0.1 μg Al l^{-1} at all doses except the dose of 19 mg Al l^{-1}. The 77.8 mg Al l^{-1} dose created a heavy floc that interfered with movements and feeding. No larvae pupated in the 55 days of study at any concentration, although the accepted life cycle to pupation time for this species is 14 days. Lamb and Bailey (1981) thus found that a common lake insect species exhibits chronic mortality over a dose range exceeded in 19 of the 28 case histories reviewed by Cooke and Kennedy (1981a) and in laboratory test solutions in which dissolved aluminum remained below 0.1 μg Al l^{-1}. As Lamb and Bailey point out, there are in-lake factors that may mitigate these chronic toxic effects in actual practice, including the predominant hardness-causing ions. The physical effect of the floc itself may be more damaging. Narf's (1978) report, in which no damage to invertebrate populations in four Wisconsin lakes could be seen during several years of monitoring, supports the view of either minimal damage or prompt recovery.

Moffett (1979) found a significant decrease in the species diversity of planktonic microcrustacea in West Twin Lake, Ohio, in two years of intense posttreatment sampling (1976 and 1978) of both the treated lake and a downstream reference lake. These samples were compared to each other and to a detailed series of samples taken in 1968-1969, before alum treatment (Heinz, 1971). This diversity change could not be attributed to dissolved aluminum in the water column, because it was always below detection limits (2 μg Al l^{-1}), except on treatment day, and because pH and alkalinity returned to normal levels (pH 7-9; alkalinity 100-150 mg $CaCO_3$ l^{-1}) after treatment. The decrease in diversity could be due to effects of the floc on resting stages in treated sediments, to the shift in phytoplankton from blue-greens to dinoflagellates, or to increased transparency and possible increased visual predation by fish. Gibbons et al. (1984), in contrast, did not find any lasting adverse impact to zooplankton by the aluminum sulfate treatment at Liberty Lake, Washington. While species diversity was not measured, the abundance of Cladocera, Copepoda, and Rotatoria was reduced only briefly after treatment and no unusual losses of species of any of these groups were reported.

The studies by Buergel and Soltero (1983), Lamb and Bailey (1981), and Moffett (1979), as well as those appearing in connection with acid precipitation investigations, suggest that lake managers must exercise caution in treating lakes with aluminum salts, and must maintain pH above 6.0 and below 9.0. Far more research is needed on the impact of aluminum salts on species populations and upon lake communities. We particularly need investigations, at the community level of organization, in which the potential mitigating influences of the environment as well as such potential negative features as bioaccumulation and magnification can be observed.

Cooke and Kennedy (1981a) report that increased transparency may bring about a lakeward extension of the macrophyte community. There is some evidence that this has occurred in West Twin Lake, Ohio, and it was believed to have occurred in Long Lake, Washington (Jacoby et al., 1983). This effect could be considered as a negative response to alum applications.

A special problem with aluminum sulfate or sodium aluminate treatments might occur in softwater lakes. Long-term exposure to acid rain and acid rain runoff eventually removes a substantial fraction of the buffering capacity of the water, bringing about a sharp decrease in pH and an increase in Al (III). This could prove to be highly deleterious to lake organisms. Lake managers and consultants should exercise great caution when recommending this treatment for an acid-rain-sensitive lake.

Finally, Bulson et al. (1984) found that alum application is extremely effective in removing bacteria, but *Escherichia coli* and presumably pathogenic bacteria survive for short periods in the floc. They suggest that the floc could be hazardous if ingested by swimmers or taken into a potable water supply plant. They suggest a ban on lake use for a period following application.

COSTS

Funk and Gibbons (1979) tabulated the costs for four lake treatments and concluded that costs are highly variable, depending upon local salaries, rentals, and the price of chemicals. A detailed comparison of costs and effectiveness for P precipitation and inactivation and for sediment removal, as procedures for controlling internal P loading, has been provided by S.A. Peterson (1981). He concluded (see Chapter 8, Sediment Removal) that P precipitation and inactivation was superior in overall economy and effectiveness, but that sediment removal has the benefit of removing the nutrient source from the lake. The effectiveness of the alum treatment over 10- and 12-year periods in Snake Lake and Horseshoe Lake, Wisconsin, respectively, (Garrison and Knauer, 1984) supports Peterson's conclusion of high benefit–cost ratio.

The principal cost for alum treatment of a lake is labor. Figure 6-9 illustrates the relationship between alum dose and man-days ha^{-1} (one man-day = one person working eight hours/day) for Horseshoe Lake, Wisconsin (J.O. Peterson et al., 1973), Welland Canal, New York (Shannon et al., 1974), Dollar Lake and West Twin Lake, Ohio (Cooke and Kennedy, 1981a), Medical Lake, Washington (Gasperino et al., 1980), and Liberty Lake, Washington (W. Funk, personal communication). Annabessacook Lake, Maine (Dominie, 1980) was omitted from the analysis due to the large amount of volunteer labor used there. The linear relationship (r^2 = 0.9411) from these six data points provides a guideline from which to estimate the major expense, labor, by obtaining the necessary data on alkalinity and volume and calculating the dose.

Equipment costs can be expected to be several thousand dollars. If properly designed, the equipment can be disassembled and used elsewhere for the cost of reassembly. To date, no commercial apparatus is available for alum application. If a weed harvester is available, it could easily be modified for alum application.

NEEDED RESEARCH

Lake treatments need to be monitored to determine their long-range effectiveness in controlling P release; too little is known about this. We do not know whether there is a direct relationship between the amount of aluminum hydroxide deposited

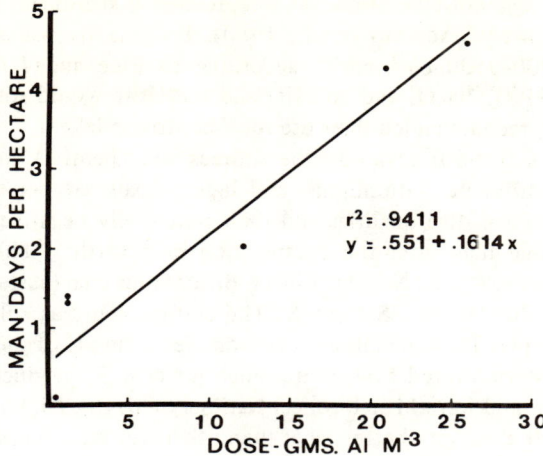

Figure 6-9 Relationship between dose of aluminum sulfate and labor (man-days ha^{-1}) for six North American lakes (see text for lakes and data sources).

on the sediments and the duration of effectiveness. Laboratory and field enclosure experiments could be used to test this question, and the several in-lake treatments should be followed up. Theoretically, as much aluminum hydroxide as affordable could be added, if both aluminum sulfate and sodium aluminate are used. Also, we do not know why treatments remain effective after organic matter containing P has settled on them. Large numbers of sites for P adsorption to the floc may remain.

The potential toxic effects of this treatment to aquatic communities or to lake species are virtually unknown. In lakes that are vulnerable to acidification from acid rain, aluminum toxicity could become a problem long after treatment for P inactivation.

Finally, cost of application could be lowered if a more rapid application technique could be developed. High velocity–high volume shore-based pumps and hoses should be tested.

USE OF FLY ASH FOR PHOSPHORUS INACTIVATION

An alternative to using aluminum salts to remove and inactivate phosphorus in lake or reservoir waters was apparently first suggested by Tenney and Echelberger (1970). They believed that fly ash, the airborne particles (0.5 to 100μ) trapped by electrostatic precipitators in the smokestacks of coal-fired power generating plants, could be applied to lake sediments to inhibit P release and to control macrophyte growth. Very large quantities of this material are produced in the United States, particularly since the oil embargo of the mid-1970s. In 1977, 56 million metric tons of fly ash were produced and only about 20 percent were used, for purposes such as cement manufacture. The rest must be stored and represents a significant disposal problem for the power industry (Adriano et al., 1980). Because fly ashes have very large sorptive areas (4500-5500 cm^2 cm^{-3}, according to Fine and Jensen, 1981), and are high in CaO, MgO, Na_2O, and Al (III) and therefore would sorb P, Tenney and Echelberger (1970) recommended their use for P control in lakes.

Adriano et al. (1980) reviewed the sources and chemical composition of fly ash. Ashes from anthracite, bituminous, and lignite coals vary in their composition and hence in their P-sorptive qualities and their potentially negative effects on lakes. Storage and leaching may alter the composition even further. The major elements in them are Si, Al, Fe, Ca, K, Na, Mg, and Ti. Bituminous coal (eastern United States) is high in Ce, Cr, Cu, Mo, As, Sr, and S. The sulfur occurs as sulfate, and aquatic solutions have low pH. These metals are very soluble at the pH of the typical solution. Lignitic coals (western United States) are much lower in S, produce high pH (above pH 12) in solution, and have higher concentrations of B, Cu, Ni, Pb, and Sr (Adriano et al., 1980). Lignite coals give better ashes than others for removal of P from solution, due to their higher CaO, MgO, Na_2O, and Al (III) concentrations.

Despite its favorable attributes for P removal and inactivation, fly ashes pose potentially severe hazards to lakes and reservoirs. Theis and De Pinto (1976) list five possible adverse effects: high pH of treated waters, dissolved oxygen depletion, biological reduction of sulfate to sulfide, heavy metal release, and physical crushing or clogging of biota or biotic activities. A series of experiments, culminating in the treatment of a portion of Lake Charles East, Indiana, was conducted by Theis and his associates to evaluate the efficacy of fly ash treatment of lakes. These experiments are briefly summarized here, as well as other experiments on release of heavy metals from fly ashes, to familiarize the reader with these data and with the reasons why fly ashes are not recommended for use in lakes at this time (Cooke, 1980).

Yaksich (1972) investigated several substances, including various fly ashes and clays, for their ability to retard P release, and adopted an optimum dose of 10 g ash l^{-1}. Hampton (1972) used several doses of an ash that increased pH, to test its effects on laboratory populations of *Daphnia magna* and *Lepomis macrochirus* (bluegill). Doses of 10 g ash l^{-1} and above were always fatal to the fish, due to clogging of gills or to high pH (which reached 10). These doses were also invariably fatal to *D. magna*. Hampton then treated a small pond (0.5 ha) with 10 g ash l^{-1}. pH increased from 8.9 to 10.9 in 24 hours and returned to 9.0 in five weeks. One day after treatment about 1500 fish were killed (99% of fish community) and zooplankton were absent from the pond for six weeks after fly ash application. Clogging of gills, high pH, or low dissolved oxygen might have caused these mortalities.

The toxic effects of fly ash to aquatic biota are well known. In one of the first reports, Cairns et al. (1972) observed an extensive fish and benthic macroinvertebrate kill over 160 km (100 miles) reach of the Clinch River, Virginia, following an accidental spill of 130 × 10^6 gallons of a fly ash slurry. A major concern about toxic effects involves the release of heavy metals to lake water following fly ash application. Guthrie et al. (1982), summarizing data from Oglesby (1975) and Natusch (1978), listed the concentrations of elements in coal and fly ash (Table 6-1). All elements except cadmium, chromium, mercury, and tin can be more concentrated in the fly

Table 6-1 Ranges of Elemental Concentration (ppm) in Coal and in Fly Ash [from Guthrie et al., 1982, based on data from Oglesby, 1975 (coal) and Natusch, 1978 (fly ash)].

Element	Coal			Fly Ash		
Antimony	0.05	–	8.9	1.0	–	15
Arsenic	0.5	–	93	2.0	–	500
Barium	20	–	1600	500	–	7000
Beryllium	0.2	–	4.0	1.0	–	10
Boron	1.0	–	216		—	
Cadmium	0.01	–	300	0.1	–	50
Chromium	22	–	400	500	–	300
Cobalt	1.0	–	34	5.0	–	100
Copper	3.0	–	180	50	–	650
Lead	1.0	–	218	5.0	–	1000
Manganese	5.0	–	240	50	–	500
Mercury	0.02	–	1.91	0.002	–	0.4
Molybdenum	1.0	–	30	5.0	–	40
Nickel	3.0	–	90	5.0	–	100
Selenium	0.04	–	7.7	1.0	–	20
Tellurium	0.02	–	3.0	2.0	–	30
Thallium	0.2	–	36	15	–	70
Tin	0.1	–	51	30	–	30
Uranium	0.09	–	3.7	5.0	–	20
Vandium	2.0	–	78	100	–	500
Zinc	1.0	–	1444	50	–	5000

ash. As Guthrie et al. (1982) point out, sulfur as sulfate will become concentrated in the ash from some coals, producing low pH and a favorable environment for solubilization of metals. Cherry et al. (1979) reported on the elemental concentrations of sediment, water, and benthic macroinvertebrates in a drainage system receiving effluent from an ash settling basin. Some bioaccumulation of heavy metals was found. Guthrie and Cherry (1979), studying the same ash disposal system, found significant bioaccumulation of heavy metals in organisms of several trophic levels. Others, including Joensuu (1971) and Gutenmann et al. (1976) provide additional data to indicate that mercury and selenium could accumulate in organisms exposed to fly ash.

Theis and Wirth (1977) and Phung et al. (1979) point out that the metal desorption from fly ashes is pH dependent, although the experiments of Theis and Wirth indicate that arsenic desorbs readily at high pH. Desorption at a pH of 7 or below is high, and eastern bituminous fly ashes, which have high sulfur content, would thus pose the greatest risk unless buffered.

Lake Charles East, Indiana, was the site of a partial lake treatment with fly ash in summer 1975 (Theis et al., 1979). Lake Charles is a small (8.7 ha), shallow (Z_{max} = 3 m, Z = 2 m), polymictic lake that had received sewage effluent until just prior to the treatment. The lake is eutrophic; the algal problem had been treated with copper sulfate and the weeds with herbicides, including arsenic. A section (3.0 ha) was isolated from the rest of the lake by a barrier attached to a highway bridge, and 1430 metric tons of fly ash (about 20 g l^{-1}) and 275 metric tons of CaO were added, with the objective of controlling P release from the sediments. The lime apparently was to buffer the fly ash. The results of the experiment are equivocal, due to collapse of the barrier during winter 1975 and to a paucity of chemical and biological data about the lake before treatment. The effect of sewage diversion, which was 95 percent completed in 1974, cannot be separated from the effect of the fly ash treatment on water-column P concentrations. Concentration in the treated section fell from 225 μg P l^{-1} to 150 μg P l^{-1} and remained at this level through 1977. Laboratory studies revealed that P release rates from treated sediments were about half those of untreated sediments, although rates in the experimental columns were high ($\bar{x}$ = 14 ± 12 mg P m^{-2} day^{-1}). Heavy metals, including Fe, Mn, Zn, and Ni, were elevated in lake waters after treatment, but their sources are unknown since pretreatment data are sparse. Neither pH nor alkalinity of the lake was reported.

The fly ash and lime application killed thousands of bluegills, but no bass. The oligochaete portion of the benthos was nearly eliminated, as were aestivating copepods. Cladocera and Rotifera populations declined sharply. Algal blooms were reduced and water transparency increased as the phytoplankton community shifted from blue-greens to flagellates. This shift could be due to a drop in pH according to Shapiro's (1983) theory. Spencer et al., (1983) have examined the plankton community of Lake Charles and have found the negative impact of the treatment to be short-lived, probably because organisms from the untreated two-thirds of the lake repopulated the treated section.

The evidence to this point strongly supports the view that application of fly ashes to lakes and reservoirs to control P release from sediments causes more problems than it solves. Until further laboratory and field studies are completed on the problems of pH, heavy metal release, bioaccumulation, and clogging of fish gill surfaces, this material should not be added to lake waters.

REFERENCES

Adriano, D.C., Page, A.L., Elseewi, A.A., Chang, A.C., and Straughan, I. 1980. Utilization and disposal of fly ash and other coal residues in terrestrial ecosystems: A review. *J Environ Qual* 9:333-344.

Bannink, B.A., and van der Vlugt, J.C. 1977. Hydrobiological and chemical response to the addition of iron and aluminum salts, studies in three Lund-type butyl-rubber reservoirs. *Verh Int Ver Limnol* 20:1816-1821.

Barko, J.W., and Smart, R.M. 1980. Mobilization of sediment phosphorus by submerged freshwater macrophytes. *Freshwater Biol* 10:229-238.

Bernhardt, H. 1981. Recent developments in the field of eutrophication prevention. *Z Wasser Abwasser Forsch* 13:14-26.

Biesinger, K.E., and Christensen, G.M. 1972. Effects of various metals on survival, growth, reproduction, and metabolism of *Daphnia magna*. *J Fish Res Board Can* 29:1691-1700.

Born, S.M., Wirth, T.L., Peterson, J.O., Wall, J.P., and Stephenson, O.A. 1973. *Dilutional Pumping at Snake Lake, Wisconsin. A Potential Technique for Small Eutrophic Lakes,* Tech. Bull. 66, Dept Natural Resources, Madison, WI.

Buergel, P.M., and Soltero, R.A. 1983. The distribution and accumulation of aluminum in rainbow trout following a whole-lake alum treatment. *J Freshwater Ecol* 2:37-44.

Bulson, P.C., D.L. Johnston, H.L. Gibbons and W.H. Funk. 1984. Removal and inactivation of bacteria during alum treatment of a lake. *Appl and Environ Microbiol* 48:425-430.

Burrows, H.D. 1977. Aquatic aluminum chemistry, toxicology, and environmental prevalence. *CRC Crit Environ Control* 7:167-216.

Cairns, J. Jr., Dickson, K.L., and Crossman, J.S. 1972. The response of aquatic communities to spills of hazardous materials. *Proc. National Conference Hazardous Material Spills,* 179-197.

Carlson, R.E. 1977. A trophic state index for lakes. *Limnol Oceanogr* 22:361-369.

Cherry, D.C., Guthrie, R.K., Sherburger, F.F., and Larrick, S.R. 1979. The influence of coal ash and thermal discharges upon the distribution and bioaccumulation of aquatic invertebrates. *Hydrobiologia* 62:257-267.

Cooke, G.D. 1980. Covering bottom sediments as a lake restoration technique. *Water Res Bull* 16:921-926.

Cooke, G.D., and Kennedy, R.H. 1978. The effects of a hypolimnetic application of aluminum sulfate to a eutrophic lake. *Verh Int Ver Limnol* 20:486-489.

Cooke, G.D., and Kennedy, R.H. 1981a. *Precipitation and Inactivation of Phosphorus as a Lake Restoration Technique,* EPA-600/3-81-012.

Cooke, G.D., and Kennedy, R.H. 1981b. State-of-the-art summary of phosphorus inactivation as a lake restoration technique. In *Proceedings of Workshop on Algal Management and Control,* Tech. Rept. E-81-7. U.S. Army Corps of Engineers, Vicksburg, MS, 32-56.

Cooke, G.D., Heath, R.T., Kennedy, R.H., and McComas, M.R. 1978. *Effects of Diversion and Alum Application on Two Eutrophic Lakes,* EPA-600/3-78-033.

Cooke, G.D., Heath, R.T., Kennedy, R.H., and McComas, M.R. 1982. Change in lake trophic state and internal phosphorus release after aluminum sulfate application. *Water Res Bull* 18:699-705.

Cooke, G.D., McComas, M.R., Waller, D.W., and Kennedy, R.H. 1977. The occurrence of internal phosphorus loading in two small, eutrophic glacial lakes in Northeastern Ohio. *Hydrobiologia* 56:129-135.

Dominie, D.R. 1978. *Cobbossee Watershed District. Lake Restoration Report. Progress Report #4,* East Winthrop, ME: Cobbossee Watershed District.

Dominie, D.R. II. 1980. Hypolimnetic aluminum treatment of softwater Annabessacook Lake. In *Restoration of Lakes and Inland Waters,* EPA-440/5-81-010, 417-423.

Edmondson, W.T. 1970. Phosphorus, nitrogen, and algae in Lake Washington after diversion of sewage. *Science* 169:690-691.

Eisenreich, S.J., Armstrong, D.E., and Harris, R.F. 1977. *A Chemical Investigation of Phosphorus Removal in Lakes by Aluminum Hydroxide,* Tech. Rept. 77-02, Water Resources Center, University of Wisconsin, Madison.

Everhart, W.H., and Freeman, R.A. 1973. *Effects of Chemical Variations in Aquatic Environments. Vol. II. Toxic Effects of Aqueous Aluminum to Rainbow Trout,* EPA-R3-73-011b.

Fiala, L., and Vasata, P. 1982. Phosphorus reduction in a man-made lake by means of a small reservoir on the inflow. *Arch Hydrobiol* 94:24-37.

Fine, L.O., and Jensen, W.P. 1981. Phosphate in waters: I. Reduction using northern lignite fly ash. *Water Res Bull* 17:895-897.

Francko, D.A., and Heath, R.T. 1981. Aluminum sulfate treatment—short-term effect on complex phosphorus compounds in a eutrophic lake. *Hydrobiologia* 78:125-128.

Freeman, R.A., and Everhart, W.H. 1971. Toxicity of aluminum hydroxide complexes in neutral and basic media to rainbow trout. *Trans Am Fish Soc* 100:644-658.

Funk, W.H., and Gibbons, H.R. 1977. *Effectiveness of an Alum Treatment at Liberty Lake, Washington*, Pullman, WA: Annual Pacific Northwest Pollution Control Association Conference.

Funk, W.H., and Gibbons, H.L. 1979. Lake restoration by nutrient inactivation. In *Lake Restoration*, EPA-400/5-79-001, 141-151.

Funk, W.H., Gibbons, H.L., and Bailey, G.C. 1978. *Effect of Restoration Procedures upon Liberty Lake. Third Status Report*, Pullman, WA: Dept Environmental Engineering, Washington State Univ.

Funk, W.H., Gibbons, H.L., and Bailey, G.C. 1980. *Lake Assessment in Preparation for a Multiphase Restoration Treatment*, Pullman, WA: Dept of Environmental Engineering, Washington State University.

Funk, W.H., Gibbons, H.L., Bailey, G.C., Moore, B., Woodwick, F., Mawson, S., Gibbons, M., Nelson, R., Bennett, P., Breithaupt, S., Bulson, P., Le Cain, G., Lamb, D., and Hein, J. 1982. *Preliminary Assessment of Multiphase Restoration Efforts at Liberty Lake, Washington. Report No. 43,* Pullman, WA: Water Research Center, State of Washington.

Gannon, J.E., and Stemberger, R.S. 1978. Zooplankton (especially crustaceans and rotifers) as indicators of water quality. *Trans Am Microsc Soc* 97:16-35.

Garrison, P.J., and Knauer, D.R. 1984. Long term evaluation of three alum treated lakes. In *Lake and Reservoir Management*, EPA 440/5-84-001, 513-517.

Gasperino, A.F., Beckwith, M.A., Keizur, G.R., Soltero, R.A., Nichols, D.G., and Mires, J.M. 1980. Medical Lake improvement project: Success story. In *Restoration of Lakes and Inland Waters*, EPA-440/5-81-010, 424-428.

Gibbons, M.V., Woodwick, F.D., Funk, W.H., and Gibbons, H.L. 1984. Effects of a multiphase restoration, particularly aluminum sulfate application, on the zooplankton community of a eutrophic lake in eastern Washington. *J Freshwater Ecol* 2:393-404.

Gutenmann, W.H., Bache, C.A., Youngs, W.D., and Lisk, D.J. 1976. Selenium in fly ash. *Science* 191:966-967.

Guthrie, R.K., and Cherry, D.S. 1979. Trophic level accumulation of heavy metals in a coal ash drainage basin. *Water Res Bull* 15:244-248.

Guthrie, R.K., Davis, E.M., Cherry, D.S., and Walton, J.R. 1982. Impact of coal ash from electric power production on changes in water quality. *Water Res Bull* 18:135-138.

Haines, T.A. 1981. Acidic precipitation and its consequences for aquatic ecosystems: A review. *Trans Am Fish Soc* 110:669-707.

Hampton, T.K. 1972. Evaluation of the effects on biotic communities of fly ash reclamation. M.S. Thesis, Notre Dame University, South Bend, IN.

Harper, H.H., Wanielista, M.P., and Yousef, Y.A. 1983. Restoration of Lake Eola. In *Lake Restoration, Protection, and Management*, EPA-440/5-83-001, 13-22.

Hayden, P.L., and Rubin, A.J. 1974. Systematic investigation of the hydrolysis and precipitation of aluminum (III). In A.J. Rubin, ed., *Aqueous-Environmental Chemistry of Metals*, Ann Arbor, MI: Ann Arbor Science, 317-381.
Hayes, C.R., Clark, R.G., Stent, R.F., and Redshaw, C.J. 1984. The control of algae by chemical treatment in a eutrophic water supply reservoir. *J Inst Water Eng Sci* 38:149-162.
Heath, R.T., and Cooke, G.D. 1975. The significance of alkaline phosphatase in a eutrophic lake. *Verh Int Ver Limnol* 19:959-965.
Heinz, M.H.E.F. 1971. A limnological study of the Twin Lakes, Portage County, Ohio; The annual variations of microcrustacea and physical, chemical, and biological parameters. M.S. Thesis, Kent State University, Kent, OH.
Jacoby, J.M., Welch, E.B., and Michaud, J.P. 1983. Control of internal phosphorus loading in a shallow lake by drawdown and alum. In *Lake Restoration, Protection, and Management*, EPA-440/5-83-001, 112-118.
Jernelöv, A. 1970. Phosphate reduction in lakes by precipitation with aluminum sulphate. 5th International Water Pollution Research Conference. Pergamon Press, New York.
Joensuu, O.I. 1971. Fossil fuels as a source of mercury pollution. *Science* 172: 1027-1028.
Kennedy, R.H. 1978. Nutrient inactivation with aluminum sulfate as a lake reclamation technique. Ph.D. Dissertation, Kent State University, Kent, OH.
Kennedy, R.H., and Cooke, G.D. 1980. Aluminum sulfate dose determination and application techniques. In *Restoration of Lakes and Inland Waters*, EPA-400/5-81-010, 405-411.
Kennedy, R.H., and Cooke, D.G. 1982. Control of lake phosphorus with aluminum sulfate. Dose determination and application techniques. *Water Res Bull* 18:389-395.
Knapp, S.M., and Soltero, R.A. 1983. Trout-zooplankton relationships in Medical Lake, WA, following restoration by aluminum sulfate treatment. *J Freshwater Ecol* 2:1-12.
La Marra, V.J. Jr. 1975. Digestive activities of carp as a major contributor to the nutrient loading of lakes. *Verh Int Ver Limnol* 19:2461-2468.
Lamb, D.S., and Bailey, G.C. 1981. Acute and chronic effects of alum to midge larva (Diptera: Chironomidae). *Bull Environ Contam Toxicol* 27:59-67.
Lamb, D.S., and Bailey, G.C. 1983. Effects of aluminum sulfate to midge larvae (Diptera: Chironomidae) and rainbow trout (*Salmo gairdneri*). In *Lake Restoration, Protection, and Management*, EPA-440/5-83-001, 307-312.
Larsen, D.P., Malueg, K.W., Schults, D.W., and Brice, R.M. 1975. Response of Shagawa Lake, Minnesota, U.S.A. to point-source phosphorus reduction. *Verh Int Ver Limnol* 19:884-892.
Larsen, D.P., Schults, D.W., and Malueg, K.W. 1981. Summer internal phosphorus supplies in Shagawa Lake, Minnesota. *Limnol Oceanogr* 26:740-753.
Larsen, D.P., Van Sickle, J., Malueg, K.U., and Smith, P.D. 1979. The effect of wastewater phosphorus removal on Shagawa Lake, Minnesota: Phosphorus supplies, lake phosphorus, and chlorophyll A. *Water Res.* 13:1259-1272.
May, V. 1974. Suppression of blue-green algal blooms in Braidwood Lagoons with alum. *J Aust Inst Agric Sci* 40:54-57.
May V., and Baker, H. 1978. *Reduction of Toxic Algae in Farm Dams by Ferric Alum, Tech. Bull. 19,* New South Wales: Dept. Agriculture.

Mires, J.M., Soltero, R.A., and Keizur, G.R. 1981. Changes in the zooplankton community of Medical Lake, WA, subsequent to its restoration by a whole-lake alum treatment and the establishment of a trout fishery. *J Freshwater Ecol* 1:167-178.

Moffett, M.R. 1979. Changes in the microcrustacean communities of East and West Twin Lakes, Ohio, following lake restoration. M.S. Thesis, Kent State University, Kent, OH.

Narf, R.P. 1978. *An Evaluation of Past Aluminum Sulfate Lake Treatments: Present Sediment Aluminum Concentrations and Benthic Insect Communities*, Madison, WI: Dept Natural Resources, State of Wisconsin.

Natusch, D.F.S. 1978. Potentially carcinogenic species emitted to the atmosphere by fossil-fuel power plants. *Environ Health Perspectives* 22:79-80.

Oglesby, S. Jr. 1975. *A Survey of Technical Information Related to Fine Particle Control*, Birmingham, AL: Southern Research Institute, Electric Power Research Institute, Pub. 259.

Peterson, J.O., Wall, J.T., Wirth, T.L., and Born, S.M. 1973. *Eutrophication Control: Nutrient Inactivation by Chemical Precipitation at Horseshoe Lake, Wisconsin*, Tech. Bull. 62, Madison WI: Dept Natural Resources, State of Wisconsin.

Peterson, S.A. 1981. *Sediment Removal as a Lake Restoration Technique*, EPA-600/13-81-013.

Peterson, S.A., Sanville, W.D., Stay, F.S., and Powers, C.F. 1974. Nutrient inactivation as a lake restoration procedure—laboratory investigation. EPA-660/3-74-032.

Peterson, S.A., Sanville, W.D., Stay, F.S., and Powers, C.F. 1976. Laboratory evaluation of nutrient inactivation compounds for lake restoration. *J Water Pollut Control Fed* 48:817-831.

Phung, H.T., Lung, L.J., Page, A.L., and Bradford, G.R. 1979. Trace elements in fly ash and their release in water and treated soils. *J Environ Qual* 8:171-175.

Shannon, E.E., Ludwig, F.J., Vachon, D.T., and Munawar, I.F. 1974. *The Welland Canal Water Quality Control Experiments (Phase II)*, EPS-4-WP-74-10 (Canada).

Shapiro, J., Forsberg, B., La Marra, V., Lindmark, G., Lynch, M., Smeltzer, E., and Zoto, G. 1982. *Experiments and Experiences in Biomanipulation*, Interim Rept. 19, Minneapolis, MN: Limnological Research Center, University Minnesota.

Shaprio, J. 1984. Blue-green dominance in lakes—the role and management significance of pH and CO_2. *Int Rev ges Hydrobiol* 69: 765-780.

Soltero, R.A., Nichols, D.G., Gasperino, A.F., and Beckwith, M.A. 1981. Lake restoration: Medical Lake, Washington. *J Freshwater Ecol* 1:155-165.

Spencer, D.F., Yeung, H.Y., and Green, R.W. 1983. Alterations in the zooplankton community of a fly ash treated lake. *Hydrobiologia* 107:123-130.

Stauffer, R.E., and Lee, G.F. 1973. The role of thermocline migration in regulating algal blooms. In E.J. Middlebrooks, D.H. Falkenborg, and T.E. Maloney, eds., *Modeling the Eutrophication Process*, Logan, UT: Utah State Univ. 73-82.

Tenney, M.W., and Echelberger, W.F. 1970. Fly ash utilization in the treatment of polluted waters. Bureau of Mines Arch. 8488, *Ash Utilization Proc.* 237-265.

Theis, T.L., and DePinto, J.V. 1976. *Studies on the Reclamation of Stone Lake, Michigan*, EPA-600/3-76-106.

Theis, T.L., and Wirth, J.L. 1977. Sorptive behavior of trace metals on fly ash in aqueous systems. *Environ Sci Technol* 11:1096-1100.

Theis, T.L., Greene, R.W., Sturm, T.W., Spencer, D.F., McCabe, P.J., Higgins, B.P., Yeung, H., and Irvine, R.L. 1979. *Treatment of Lake Charles East, Indiana Sediments with Fly Ash*, EPA-600/3-79-060.

Welch, E.B., Michaud, J.P., and Perkins, M.A. 1982. Alum control of internal phosphorus loading in a shallow lake. *Water Res Bull* 18:929-936.

Welch, E.B., and Spyridakis, D.E. 1984. Persistence of an alum treatment in a shallow lake. Unpublished ms. Dept Civil Engineering, University of Washington, Seattle, Washington.

Yaksich, S.M. 1972. The use of particulates to control phosphate release from eutrophic lake sediments. Ph.D. Dissertation. Notre Dame University, South Bend, IN.

7

Sediment Oxidation

A restoration technique to oxidize the top 15 to 20 cm of anaerobic lake sediment has been developed by Ripl (1976). This relatively new technique has been applied to lakes through equipment development and promotion fostered by the Atlas Copco Co. under the name of Riplox. The objective is reducing the internal phosphorus loading in lakes with anaerobic sediment and high interstitial P concentrations in which iron controls P interchange between sediment and the overlying water. By oxidizing the organic matter through increased denitrification, greater binding of interstitial P with ferric hydroxide complexes should occur, resulting in lower release rates for P (Ripl, 1976; Ripl and Lindmark, 1978).

Ripl and Lindmark (1978) argued that high interstitial P concentrations, which may lead to high internal P loading, are due largely to metabolic processes. To reduce the organic matter content in sediments, and thereby restore an oxidized state, a solution of $Ca(NO_3)_2$ is injected into the sediment to stimulate denitrification. Nitrate is preferred to oxygen added with air as the electron acceptor because, being a liquid solution, it penetrates into the sediment more readily and to a greater depth. Ferric chloride ($FeCl_3$) is added initially to remove hydrogen sulfide (H_2S) and form ferric hydroxide ($Fe(OH)_3$), which binds interstitial P. Lime ($Ca(OH)_2$) is added next, to raise the pH to an optimum level and encourage microbial denitrification.

Ferric chloride and lime additions may be unnecessary in some cases. The pH may be sufficiently high to promote denitrification, and the iron content of the sediments may be adequate (30-50 mg g^{-1}) for P binding. Willenbring et al. (1984) found that adding iron and lime was unnecessary to reduce the P release from sediment in Long Lake, Minnesota, so long as the calcium nitrate dose was sufficient. The cost savings were significant. The need for all three ingredients and the doses required should be optimized through laboratory experiments for each particular lake sediment. A range of calcium nitrate additions should be studied, with and without iron. The optimum dose of calcium nitrate for Long Lake, Minnesota, sediments was about the same as that recommended by Ripl (1976).

EQUIPMENT AND APPLICATION RATES

The chemical solutions are applied by direct injection into the sediment. Solutions are pumped from onshore storage reservoirs to a "harrow" device, which is about

6-10 m wide and is equipped with flexible tubes that penetrate the sediment (Figure 7-1). The vertical position of the harrow on the lake bottom is adjusted by regulating the injection of compressed air through the tubes. As the harrow is dragged along the lake bottom at a rate of 4-5 m/min by an airmotor-driven raft or onshore winch, sediment is disrupted to a depth of about 20 cm and the chemical solutions are injected into the sediment through tubes at the rear of the device. Control of the device's vertical position by air input, as well as other safety features, apparently helps prevent the device from being obstructed by debris on the lake bottom. The device was described by Ripl (1976) for use in Lake Lillesjön, a small (4.2 ha), shallow (2 m) lake in Sweden.

The dosages used in Lake Lillesjön were 13 tons of ferric chloride (146 g Fe m^{-2}), five tons of lime (180 g Ca m^{-2}), and 12 tons of calcium nitrate (141 g N m^{-2}). To treat Lake Trekanten below the 3-m contour (49 ha) with calcium nitrate only required 160 tons (56 g N m^{-2}) in a 50 percent solution. The iron naturally present in Lake Trekanten sediments was considered adequate (Atlas Copco Co.). The solution was further diluted to 10 percent with lake water, prior to injection into the sediments.

The iron content in Lake Lillesjön sediments prior to treatment was considerably less than that in other lakes in the surrounding area: 9-23 mg Fe g^{-1} dry weight, compared to more typical levels of 30-50 mg Fe g^{-1}. Sediment P content in Lillesjön was also quite low (1.4-3 mg g^{-1}) in spite of high interstitial concentrations (Ripl, 1976).

CASE STUDIES

Following the treatment of Lake Lillesjön sediments in 1975, interstitial P content dropped to 70 percent to 85 percent of 1974-1975 levels in the top 20 cm. The decreased levels persisted through at least 1977 (Figure 7.2). In spite of the high loading of NO_3 to the sediments, nitrogen was largely lost through evolution of N_2 gas, and ammonium actually decreased (Figure 7-2). Denitrification of the added NO_3 was complete after 1.5 months (Ripl, 1981). The oxygen demand of the sediment also decreased by about 30 percent. Recycling of P and N to the water overlying treated sediment was reduced to between 10 percent and 20 percent of the rate observed prior to restoration (Ripl and Lindmark, 1978; Ripl, 1981).

The interstitial P content of sediments in Lake Trekanten decreased from 2-4 mg l^{-1}, prior to the May 1980 treatment, to 0.01-0.3 mg l^{-1} by July (Atlas Copco Co.). There is some question regarding the overall success of this treatment; continued phosphorus input from external sources and from sediments by other controlling mechanisms than iron redox have been suggested as the reason lake P did not decrease substantially. (G. Lindmark and B. Bostrom, personal communication; Pettersson and Bostrom, 1981.).

COSTS

The total cost of the Lake Lillesjön sediment restoration was about $101,000 (1983 dollars). Of that amount, 44 percent was for development of the device and preliminary

Sediment Oxidation 135

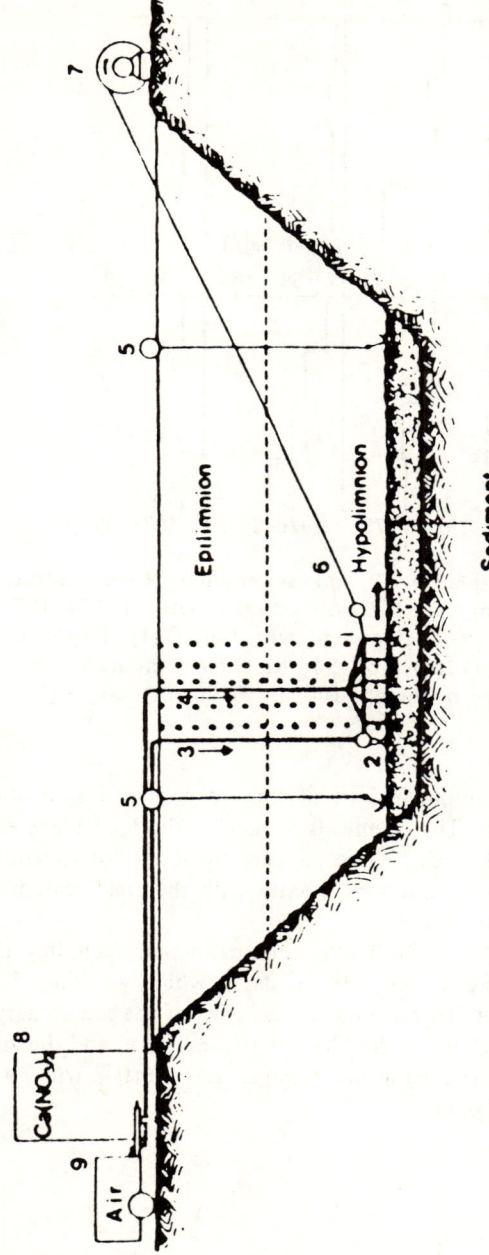

Figure 7-1 Riplox treatment system, showing (1) sediment "harrow," (2) injection point of chemicals, (3) supply tube for chemicals, (4) supply tube for compressed air, (5) buoys, (6) towing line, (7) cable winch, (8) tanks for chemicals, and (9) air compressor (from Ripl, 1981).

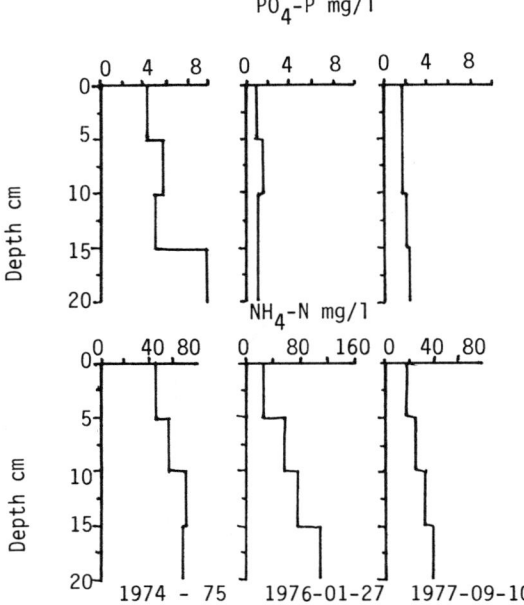

Figure 7-2 Phosphorate P and ammonium N concentration in interstitial water of Lake Lillesjön sediment before (1974–1975) and after (1976 and 1977) treatment with $Ca(NO_3)_2$. Reprinted with permission from *Vatten*, vol, 34, Ripl and Lindmark, Ecosystem control by nitrogen metabolism in sediment, copyright 1978, Bolms Boktr. AB.

lake investigations. The equipment part of this, of course, would not be necessary in subsequent lake projects. The chemicals applied to the 1.2-ha lake area represented only 6 percent of the total cost, while 28 percent of the total went for equipment installation. The latter would not vary greatly with the area treated. The remainder was for equipment rental and labor.

The cost for the 1980 Lake Trekanten treatments, excluding the preliminary investigation, was about $264,000 (1983 dollars), which was only 5 percent of the cost per area of the Lake Lillesjön project, exclusive of the preliminary investigation. The total area of Lake Trekanten (87 ha) was treated once, and the area greater than 3 m in depth (49 ha) was treated again. A larger device (10 m wide) was used for the treatment of this rather large lake.

RELATED EFFECTS

This technique appears to be a reasonable alternative to alum treatment to inactivate sediment P. Its greatest appeal may be that the chemicals added are normally found in high concentrations even in unpolluted sediments and are introduced directly to and confined largely in the sediments. In contrast to other popular P-inactivating chemicals such as aluminum, which is added to the surface of the water, toxicity to animals is probably not an issue in sediment oxidation. Furthermore, the effect of this technique may prove to be more permanent than that of alum, which only covers the sediment and can settle considerably below the surface of sediment with high water content, leaving the original surficial sediments in contact with overlying water.

There is, nevertheless, an important precaution in using the sediment oxidation technique. If the internal loading of P is controlled by iron redox reactions, then the P loading should decrease significantly following treatment. If, however, the lake is shallow and internal P loading is controlled to a large extent by high pH and temperature in the near-sediment water during summer, then sediment oxidation may not significantly decrease P loading (Pettersson and Boström, 1981).

Another application of this technique, the promotion of organic matter decomposition and increased sediment P binding through addition of NO_3, involves the use of effluent from advanced wastewater treatment (AWT) plants (Ripl et al., 1979). By directing AWT effluent, with its high NO_3 content, along the sediment-water interface of lakes with large internal P loading, such internal loading should decrease and the oxygen content should increase, according to Ripl et al., (1979). Furthermore, the nitrogen-fixing blue-green algae, normally favored by high PO_4 and low NO_3, should be discouraged in lakes receiving NO_3-rich waters.

These authors observed an indication of such changes in three small reservoirs. One received four treatments of NO_3, another received two treatments, and the third received none. A progressive improvement was observed in transparency, total P, and pH (decrease), and the presence of nonblue-green algae increased in response to NO_3 increase. Redirecting the effluent discharge in lakes being restored by AWT, but in which internal loading is retarding recovery, offers a complementary method to speed the recovery process.

The sediment oxidation (Riplox) technique represents an alternative to alum treatment and dredging as a long-term control of internal P loading from anaerobic sediment release. The technique should be considered in preliminary restoration studies. Sediment treatment experiments can be conducted in vitro (e.g., Ripl and Lindmark, 1978; Willenbring, et al., 1984) to test the method's potential prior to wholelake application. Care must be taken, however, to simulate conditions of pH and temperature that occur in the lake, in addition to anaerobic conditions. If these conditions are not duplicated, the potential of the technique may be overestimated. The technique would appear to be cost-effective at a level comparable with alum in the long run. Although the evidence of its success through actual test cases is limited at present, the principle is sound and it should be tried in appropriate situations.

REFERENCES

Atlas Copco Co., Wayne, New Jersey, and Wilrijk, Belgium.
Pettersson, K., and Boström, B. 1981. En kritisk gränskning av föreslagna metoder for nitratbehandling av sediment. *Vatten* 38:74-82.

Ripl, W. 1976. Biochemical oxidation of polluted lake sediment with nitrate—A new restoration method. *Ambio* 5:132-135.

Ripl, W. 1981. Lake restoration methods developed and used in Sweden. In *Restoration of Lakes and Inland Waters*, Proc. Inter. Sym. on Inland Waters and Lake Restoration, EPA-440/5-81-010, 495-500.

Ripl, W., Leonardson, L., Lindmark, G., Andersson, G., and Cronberg, G. 1979. Optimering av reningsverk/recipient-system. *Vatten* 35:96.

Ripl, W., and Lindmark, G. 1978. Ecosystem control by nitrogen metabolism in sediment. *Vatten* 34:135-144.

Willenbring, P.R., Miller, M.S., and Weidenbacher, W.D. 1984. Reducing sediment phosphorus release rates in Long Lake through the use of calcium nitrate. In *Lake and Reservoir Management*. EPA 440/5-84-001. p. 118-121.

8

Sediment Removal

When properly conducted, sediment removal is an effective lake management technique. This chapter describes the purposes of sediment removal, environmental concerns, the appropriate depth of sediment removal, sediment removal techniques, suitable lake conditions, dredge selection and disposal area design, exemplary case histories, and costs.

PURPOSES OF SEDIMENT REMOVAL

Deepening

When recreational activities are impaired due to shoaling, the only practical means of restoring use of a lake is to deepen it through sediment removal. The water volume of a lake must be sufficient to exceed water loss by seepage and evaporation, a depth, according to the U.S. Department of Agriculture (1971), anywhere from 1.5 m to 4.5 m, depending on the region of the country. Sufficient depth is also required to avoid winter kill of fish—in colder parts of the United States at least 4.5 m (Toubier and Westmacott, 1976). These and other factors, such as intended lake use, availability of an area for disposal of material, and available funds, must be considered in the design of a lake deepening project. The reasons for deepening and the means of measuring the success of such a project are the most straightforward of the sediment removal objectives. Since modern dredging equipment is reasonably efficient at moving large volumes of sediment, all lake dredging projects undertaken for the purpose of deepening are considered to be successful (Pierce, 1970).

Nutrient Control

Shallow eutrophic lakes frequently are susceptible to periodic nutrient gains from the sediment. These inputs of phosphorus and nitrogen most often occur when a passing summer cold weather front depresses the thermocline (Stauffer and Lee, 1973). Thus obnoxious algal blooms occur most frequently during peak recreation periods.

It has been estimated that sediment-regenerated phosphorus amounted to approximately 45 percent of the phosphorus loading to Linsley Pond, Connecticut

(Livingston and Boykin, 1962). Welch et al. (1979) estimated phosphorus inputs to Long Lake, Washington, were 200-400 kg yr^{-1}, or about 25-50 percent of the external loading. Shagawa Lake, Minnesota, experiences summer internal (from the sediment) phosphorus pulses of approximately 2000-3000 kg during June, July, and August. This compares to an annual phosphorus loading from the City of Ely, Minnesota of 5000-5500 kg before advanced waste treatment (AWT) and about 1000-1500 kg after AWT (Larsen et al., 1981). Before AWT, sediment phosphorus loading to Shagawa Lake was about 28-35 percent of the total loading. Although total phosphorus loading to the lake decreased following AWT, the sediment input portion of the total loading increased to about 66 percent (Peterson, 1981). Sediment-recycled phosphorus in Shagawa Lake has been sufficient to produce large summer algal blooms, thus slowing the lake's predicted rate of recovery (Larsen et al., 1981).

Phosphorus loading from sediment constitutes a significant portion of total phosphorus loading in some lakes. In cases where this can be documented, sediment removal might be expected to reduce the rate of internal nutrient recycling, thus improving overall water quality.

Removing Toxic Substances

Toxic substances are becoming a common concern among industrialized nations. Insufficient treatment and disposal practices of the past have materialized into problems for today. Large-scale surveys and improved analytical techniques are demonstrating that toxicants are more common to freshwater sediments than previously suspected (Bremer, 1979; Horn and Hetling, 1978; Matsubara, 1979). Many of the toxicants recycle from the sediment to the overlying water column, where they may be bioaccumulated by aquatic organisms. Perhaps the most infamous incident of this type was the mercury pollution of Minimata Bay, Japan, in 1956 (Fujiki and Tajima, 1973). More recent incidents, in the United States, have involved kepone contamination of the James River, Virginia (Mackenthun et al., 1979), and PCB contamination of Waukegan Harbor in the Great Lakes (Bremer, 1979).

The most obvious solution to the problem of contaminated sediment is removal, but removal of contaminated sediment is frequently complicated by secondary pollution of the overlying water column, through sediment agitation. Most conventional dredges can remove large volumes of sediment, but unfortunately also can cause resuspension of fine sediment (Suda, 1979; Barnard, 1978). Resuspension of sediment while dredging toxic substances must be minimized to prevent environmental damage. Proper selection of equipment therefore becomes important to the success of toxic sediment removal.

Rooted Macrophyte Removal

An overabundance of rooted aquatic plants may interfere with fishing, boating, and swimming activities and be aesthetically displeasing. Respiration by large plant masses in the littoral zone during hours of darkness may significantly reduce oxygen concentrations. In addition, there is an increasing literature on the effects of macrophytes on internal nutrient cycling. Their role in this process, with its effect on algal dynamics, may be an important reason for attempting to control macrophytes by removing them from a lake. Wetzel (1975) indicates that most of the organic matter found in small lakes may be derived from their littoral zones. Reimold (1972) showed that salt-marsh plants (*Spartina*) act as nutrient pumps, extracting sediment nutrients, translocating them to stems and leaves, and excreting them to the surrounding water on each successive tidal cycle. Other researchers have documented that various species of freshwater aquatic plants extract nutrients chiefly from the sediment and translocate them to the surrounding water (Carignan and Kalff, 1980; Schults and Malueg, 1971; Twilley et al., 1977).

There is evidence that healthy freshwater aquatic macrophytes do not excrete large quantities of nutrients to the surrounding water while in the active growth phase (Barko and Smart, 1980). They do tend, however, to concentrate sediment-supplied nutrients in their tissues. These nutrients are recycled to the lake when plants fruit and during the senescence, death, and decay stages (Barko and Smart, 1979; Lie, 1979; Welch et al., 1979). Barko and Smart (1979) estimated that in-lake mobilization of phosphorus by *Myriophyllum* in Lake Wingra, Wisconsin, might amount to 62 percent of the annual external phosphorus loading. Welch et al. (1979) indicated that much of the "sediment" phosphorus loading in Long Lake, Washington, was probably due to rapid plant die-off and decay. Current information indicates that any long-range lake restoration project concerned with in-lake nutrient controls will need to focus on both macrophytes and sediment (Barko, 1980; Carignan and Kalff, 1980). It should be noted that aquatic plant control, rather than eradication, is the most sound management approach.

ENVIRONMENTAL CONCERNS

Peterson (1979) pointed out a number of potential environmental problems associated with sediment removal. The problems generally can be classified as in-lake or disposal-area types.

In-Lake Problems

In-lake problems commonly center around the resuspension of sediment during its removal by dredging. One of the most common problems is liberation of nutrients. Phosphorus is of particular concern, because of its high concentration in the interstitial waters of eutrophic lakes. Dredge agitation and wind action tend to move nutrient-laden sediment into the euphotic zone of the lake, creating the potential for algal blooms. Churchill et al. (1975) reported increased phosphorus concentration in Lake Herman, South Dakota, coincident with cutter-head hydraulic dredging, but they noted no increased algal production. This lack of algal increase presumably was due to the high turbidity level. Dunst (1980), on the other hand, found increased algal production in Lilly Lake, Wisconsin, when hydraulic dredging began, but it was short-lived and never posed a nuisance-level problem. While nutrient enrichment due to dredging can become a problem, in most cases the effects are short-term and negligible relative to the long-term benefits. Hydraulic dredging might produce the additional problem of an unplanned drawdown if pumping rates are high.

Another, and potentially greater problem associated with resuspended sediments is the liberation of toxic substances. Although small-lake sediment removal projects for the purpose of removing toxic substances have been relatively uncommon, there is a growing concern (Bremer, 1979; Matsubara, 1979; Sakakibara and Hayashi, 1979; Spencer Engineering, 1981). Fine particles pose the major concern. Murakami and Takeishi (1977) have shown that up to 99.7 percent of the polychlorinated biphenyls (PCB) associated with sediment is attached to particles less than 74 μm in diameter. This could pose a particular problem for freshwater dredging projects, where particle settling times are significantly greater than for marine waters. Added precautions should therefore be taken where contaminated sediments are present. Such precautions might include special dredges (see Sediment Removal Techniques), special disposal and treatment techniques (Barnard and Hand, 1978; Matsubara, 1979), or both.

A relatively common concern about dredging, among fisheries managers, is destruction of the benthic fish-food organisms. If the lake basin is dredged completely, two to three years may be required to reestablish the benthic fauna (Carline and Brynildson, 1977). If portions of the bottom are left undredged, however, reestablishment may be almost immediate (Andersson et al., 1975; Collett et al., 1981) or within one to two years (Crumpton and Wilbur, 1974). In any case, the effect on benthic communities appears to be short-lived and generally tolerable relative to the longer-term benefits derived. The same appears to be true for game fish populations (Carline and Brynildson, 1977; Spitler, 1973).

These concerns are associated primarily with *dredging* as a sediment removal technique. Another technique for sediment removal involves lake drawdown (lowering the water level) to expose the littoral sediments, or in some cases (Born et al., 1973) the entire lake basin, followed by removal of sediment with earth-moving equipment. Drawdown accompanied by bulldozer operation probably is more destructive of the benthic community than dredging. It may also pose additional nuisance problems such as noise, dust, and truck traffic. The section on Sediment Removal Techniques will address dredging techniques that minimize many of these concerns.

The major nonlake impact of sediment removal concerns the area chosen for disposal of dredged material. The problem of finding disposal sites in urban areas has become more acute in the United States with the promulgation of Section 404 of Public Law 92-500; this law prohibits the filling of any wetland area exceeding 4.0 ha (10 acres) without a federal permit.

The flooding of wooded areas with dredged material should be avoided. Flooding kills trees, providing unsightly evidence of an improper disposal plan. Disposal areas may become attractive nuisances in the legal sense and can be extremely dangerous. It may be advisable to fence and post them for safety.

A disposal method used frequently in recent years employs diking in upland areas. A not uncommon problem with these sites is dike failure and flooding of adjacent areas (C.C. Calhoun, personal communication). Groundwater contamination may become a problem in upland disposal if sand or gravel underlies the disposal area. Water wells in these areas could become contaminated, potentially adding liability to the sediment removal project. Upland disposal areas for dredged material are commonly used for a variety of purposes once they have been closed and dewatered. However, dewatered and apparently dried disposal areas can be quite deceiving. Even those with strong surface crusting, deep cracking, and vegetation can swallow earth-moving equipment if excavation is attempted too early. Disposal areas covered to depths greater than one meter should be thoroughly tested for support capabilities before attempting to rework them with heavy equipment.

Another problem not uncommon to lake dredging is underdesign of disposal-area capacity. Unfortunately, these failings usually become apparent only after the project is fully operational. The problem may be associated with the slow settling rate of suspended sediment in fresh water (Wechler and Cogley, 1977) and reduced ponding depth as the project proceeds. This may result in failure to meet the requirements of suspended solids discharge permits. If that happens there are two choices: shut down until seepage and evaporation allows additional filling, or treat the discharge water. Either alternative adds additional cost to the project. Disposal area designs must be formulated for end-of-project efficiency, not average discharge requirements over the entire use period. Palermo et al. (1978) have summarized important technical information that will assist with the proper design, construction, and maintenance of disposal areas for dredged material. The subject is also addressed in a later section of this chapter. Barnard and Hand (1978) describe when and how to treat disposal area discharges if standards cannot be met. Other reports, by Brannon (1978), Chen et al. (1978), Gambrell et al. (1978), and Lunz et al. (1978), provide valuable information that will help minimize environmental problems at disposal sites.

SEDIMENT REMOVAL DEPTH

When restoring a lake for sailing, power boating, and associated activities, the deepening requirements are relatively straightforward. When deepening to control internal nutrient cycling and macrophyte growth, the criteria are less clearly defined.

Lake Trummen, Sweden, is perhaps the most thoroughly documented case of sediment removal to control internal nutrient cycling and macrophyte encroachment. Sediment removal depth in Lake Trummen was determined by mapping both the horizontal and the vertical distribution of nutrients in the sediment. Digerfeldt (1972), as cited by Bjork (1972), determined that approximately 40 cm of fine surface sediment accumulated from 1940 to 1965. Aerobic and anaerobic release rates of $PO_4^=$-P and NH_4^+-N from this surface layer were markedly greater than for the underlying sediment. Based on these differences a plan was developed to remove the upper 40 cm of sediment. Results of the project are presented as a case study later in this chapter.

Another approach to determining sediment removal depth has been proposed by Stefan and Hanson (1979) and by Stefan and Ford (1975). This approach is similar to that developed by Stauffer and Lee (1973), which described thermocline erosion by wind in northern temperate lakes. Stefan and Hanson (1979) have used their model to predict the depth to which Hall Lake, Minnesota, must be dredged to control adverse nutrient exchange from the sediment during the summer.

The Stefan and Hanson (1979) model assumes stable summer stratification is necessary to prevent enriched hypolimnetic waters from mixing into the epilimnion. Based on that assumption they calculated that Hall Lake (one of the Fairmont, Minnesota, lakes) would require dredging to a maximum depth of 8.0 m. Dredging volume to obtain the 8.0-m depth would be enormous, given Hall Lake's 2.25-km^2 surface area and 2.1-m mean depth.

There is little apparent chemical or physical distinction between shallow and deep sediments in Hall Lake. Phosphorus concentration is relatively uniform from the sediment surface to a depth of 8.5 m (737-1412 mg kg^{-1} for 37 samples, with a mean of 1097 mg kg^{-1}). It is possible, however, that the phosphorus release rates from deeper sediment could be less than those of surface sediments (they were not measured). Nutrient release from the deeper sediment could be enough to significantly reduce the adverse impact of nutrients on the overlying water, even though stratification might not be permanent (Bengtsson et al., 1975). If that is the case, surface sediment skimming might produce nearly the same result as deep dredging, and at a considerable saving. It therefore seems advisable to conduct incremental nutrient-release-rate experiments prior to adopting a lake-temperature-modeling approach to determine dredging depth for nutrient control.

Dredging will remove rooted macrophytes from the littoral zone of lakes, but there have been few detailed studies to determine the depths necessary to prevent regrowth of nuisance plants. Factors influencing the areas in which rooted macrophytes grow include temperature, sediment texture, sediment nutrient content, and light level.

Using field data developed by Belonger (1969) and Modlin (1970), the Wisconsin Department of Natural Resources developed a guide to prescribe dredging depths necessary to control the regrowth of macrophytes. The guide was developed by plotting a regression between the maximum depth of plant growth in several Wisconsin lakes and their average summer water clarity as measured by a Secchi disc. The relationship is described by the equation

$$Y = 0.83 + 1.22X, \qquad (8.1)$$

where Y = maximum depth of plant growth (meters)
X = average summer water clarity (meters).

According to Dunst (1980) the above relationship is used as a rough guide to develop dredging plans for macrophyte control. Dunst indicated, however, that dredging depths do not always need to exceed the predicted Y value to achieve control. He indicated further that slight deepening frequently changes plant speciation to less objectionable forms.

Work by Collett et al. (1981) attempted to establish the depth of dredging necessary to prevent plant regrowth in the usually turbid Tuggarah Lakes of New South Wales. The strategy was to bracket the light compensation depth by dredging three 30-m^2 test plots 1.0 m, 1.4 m, and 1.8 m deep in a 30-by-180-m rectangular area parallel to and about 300 m from the lake shore. Three control plots of the same size (30 m^2) were left undredged. Results indicated rapid recolonization (within four months) in the plot dredged to 1.0 m. One year after dredging, macrophyte biomass in the 1.0 m plot was about 60 percent of the predredging level, but macrophytes had not reestablished in the 1.4 m and 1.8 m test plots during the same year. Sediment nutrient levels were found to be similarly high in all test plots, so nutrient deficiency was ruled out as a probable cause of reduced growth. The authors speculated that reduced light penetration at the 1.4 m and 1.8 m depths may have limited regrowth, but also noted that deeper plots tended to fill with plant debris and lake detritus, altering the texture of the substrate. Unfortunately, no quantitative measurement of light level or sediment particle size was made.

That macrophytes ordinarily grow to depths up to 2 m (Higginson, 1970) in the Tuggarah Lakes seems to imply that light alone should not have prevented regrowth at 1.4 m and 1.8 m. The more flocculent sediment texture in deeper plots may have been a greater influence than was indicated by Collett et al. (1981). Their study did not answer conclusively the question of the influence of light on regrowth of plants. It may even raise some question about the rationale for using light level to determine dredging depth. This seems, however, to be a reasonable approach given what we know about macrophyte growth characteristics and in situ light requirements. The macrophyte depth-light level relationship is a very basic one. Hutchinson (1975), describing the work of Maristo (1941), indicated that the maximum depth at which autotrophic plants can grow is highly dependent upon the transparency of the water.

Canfield et al. (1985) have reevaluated the relationship between macrophyte maximum depth of colonization (MDC) and Secchi disc (SD) transparency. They found a significant positive relationship ($r = 0.70$ between MDC and SD transparency for 27 Finnish lakes, 55 Wisconsin lakes and 26 Florida lakes when analyzed as a group). Based on their analysis, they developed an empirical model to predict MDC values from SD data. The model (log MDC = 0.61 log SD + 0.26) was developed from half the data set and verified against the other half. The correlation between the measured and predicted MDC values was significant ($r = 0.66$), but weak.

Examination of Canfield et al. (1985) data shows there to be a wide range of maximum depths of aquatic macrophyte colonization for any given Secchi disc value. This means that factors other than Secchi disc values contribute to determining the maximum depth to which macrophytes grow. Canfield et al. (1985) state that among these factors is sampling error, MDC and SD values taken at different times, regional differences in the response of plants to changes in available light, and the type of plant itself. Nutrient availability was also mentioned as a contributing variable. Another factor that could be of considerable importance is condition of the substrate. Because of the wide range of variation between MDC and SD over broad geographic areas, it is suggested that at least minimum data be collected for the region of interest and that a regression equation be developed for that area. This procedure should increase greatly the confidence one has in using the equation to predict the MDC of macrophytes for other lakes in the immediate area. The work of Canfield et al. (1985) supports the hypothesis that SD can be used to predict the depth of macrophyte colonization, but points out that it is not an exacting science.

Macrophyte control, whatever the method, must consider the need for fish spawning, waterfowl feeding, and general wildlife habitat requirements. Macrophyte control should never be equated with eradication. In Lake Trummen, Sweden, for example, about 30 percent of the lake area (a bay) was left undredged, to provide a wildlife preserve.

SEDIMENT REMOVAL TECHNIQUES

There are two major variations on removal of sediment from freshwater lakes and reservoirs. The first one, lake drawdown followed by bulldozer and scraper excavation, has limited application. It has been used most successfully in small reservoirs (Born et al., 1973). The obvious limitation of this technique is that water must be drained or pumped from the basin. A second drawback is that the basin must be allowed to dewater sufficiently before earthmoving equipment can operate. In spite of these problems, plus the added concern of truck traffic to transport the removed sediment, this approach has been used with some success at Steinmetz Lake, New York (Snow et al., 1980).

The second and by far the most common sediment removal technique involves the use of dredges. Huston (1970) reviewed the many types of dredges that are in use today. This chapter will address only dredges commonly used in lake dredging and those with special features that afford them a means for minimizing many of the adverse effects of dredging mentioned above.

The most common means of classifying dredges is by dividing them into bucket (mechanical) and hydraulic types. A third category, "special purpose dredges," is included here to highlight a move toward low-turbidity systems for dredging finegrained and toxic sediments.

Grab Bucket Dredges

The only bucket dredges commonly used in lake restoration are the grab type illustrated in Figure 8-1. Figure 8-1A shows a clamshell bucket dredge in operation. Figure 8-1B depicts a typical Sauerman grab bucket setup. A common limitation of all grab bucket dredges is that they must discharge in the immediate vicinity of the sediment removal area or into barges or trucks for transportation to the disposal area. They normally are limited to a reach of no more than 30 to 40 meters. A major disadvantage of the grab bucket in lake restoration is the rough, uneven bottom contours that they create. Production rates are relatively slow, due to the time-consuming bucket swing, drop, close, retrieve, lift, and dump operating cycle. Grab dredges commonly create very

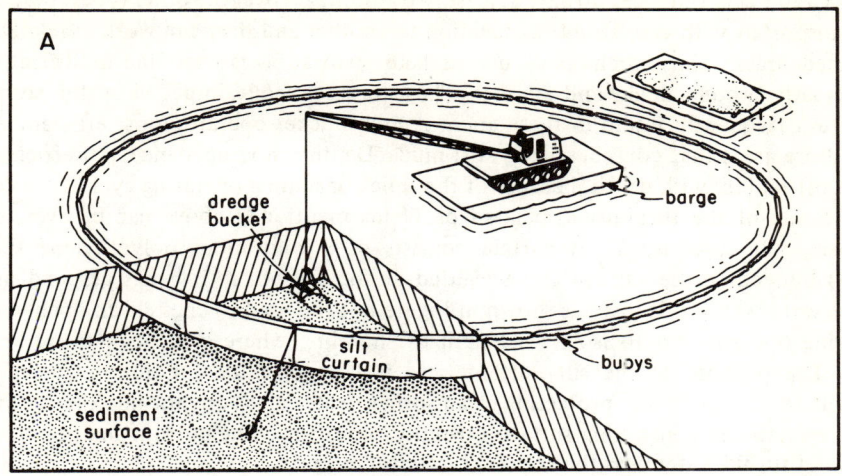

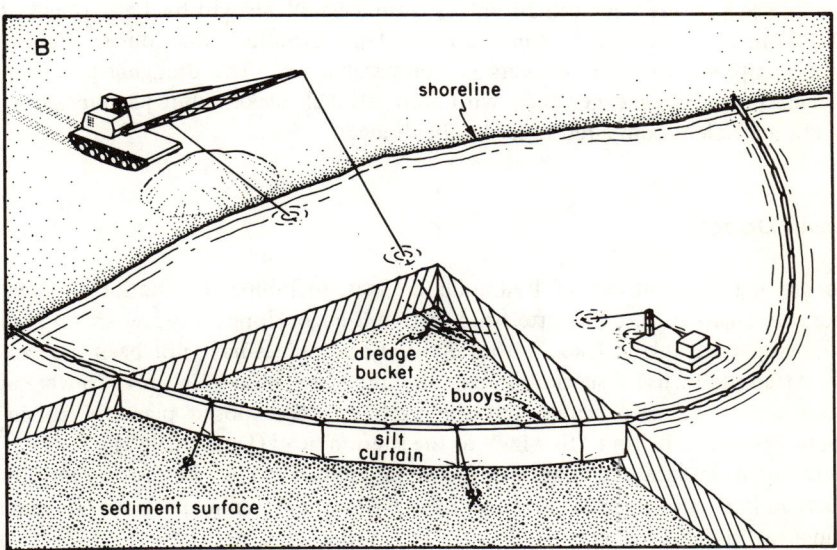

Figure 8-1 (A) Silt-curtain encirclement of an open-water grab dredge operation. (B) Shoreline isolation of a bucket dredge operation, using a silt curtain.

turbid water conditions, due to bucket drag on the bottom as it pulls free from the sediment, dragging an open bucket through the water column, bucket leakage once it clears the water surface, and the occasional intentional overflow of receiving barges, to increase their solids content. Yet another disadvantage is that many lake sediments are highly flocculent, which reduces the pickup capabilities of a grab bucket.

Grab bucket dredges do have at least two advantages over the other types: they can be transported with ease from one location to another and they can work in relatively confined areas. Thus, their chief use in lake restoration is shoreline modification, particularly around docks and marinas. They are not readily impeded by the stumps and trash frequently found in these areas. A grab bucket operates most efficiently in near-shore areas that contain soft to stiff muds. Depth is no impedance, but efficiency drops off rapidly with depth, because of the time-consuming operating cycle.

Some of the turbidity-associated problems mentioned above can be overcome by using silt curtains. A silt curtain consists of a continuous polyethylene sheet (skirt) buoyed at the surface and weighted at the bottom so it hangs perpendicular to the water surface. It may, as shown in Figure 8-1, be used to encircle an open-water dredging operation or to isolate a length of shoreline where grab buckets are being used. The purpose of the silt curtain is to isolate turbidity conditions within the immediate dredging area, protecting clean surface-water areas downwind. Silt curtains, while effective in controlling surface turbidity, are open at the bottom and permit the escape of turbid water near the sediment-water interface.

Another means of minimizing turbidity from grab-bucket-type dredging operations is to use a covered, watertight unit (Figure 8-2). Watertight buckets come in sizes ranging from 2 m^3 to 20 m^3. Manufacturers claim turbidity reductions from 30 to 70 percent compared to open buckets of comparable size. The dredging process with watertight buckets is cleaner than with conventional buckets, but production is still relatively inefficient compared to hydraulic dredges.

Hydraulic Dredges

There are many variations of hydraulic dredges, including the suction dredge, the hopper, the dustpan, and the cutterhead section dredge. Hopper dredges are impractical for dredging small inland lakes. Cutterless suction dredges have not been used extensively. Attempts to use a suction dredge at Lilly Lake, Wisconsin, in 1978 were abandoned when it was discovered that the partially decomposed plant matrix in the sediment prevented it from "flowing" to the suction head (Dunst, 1982).

Dustpan dredges per se are not commonly employed in lake restoration, though a "dustpan-like" dredge was used to remove flocculent sediment from Green Lake, Washington, in 1961-62 (Pierce, 1970). The device consisted of a 15.25-m suction manifold with slot openings. The total size of the inlet ports was designed to produce inlet velocities of at least 300 cm sec^{-1}. As sediment consistency increased with depth, some of the inlet ports were sealed to increase flow velocity in the ones remaining open. The dustpan-like suction head was barge mounted and designed to swing in a full 180-degree arc and discharge into a 50.8-cm diameter pipeline. The discharge distance was about 792 meters. This dredge successfully removed 917,500 m^3 of sediment. Bjork (1974) indicated that the dredge head used on the Lake Trummen, Sweden, project had a specially designed "nozzle." The positive experience at Green Lake and at Lake Trummen indicates that dustpan types and other variations of conventional hydraulic suction heads should receive additional consideration for dredging highly flocculent freshwater lake sediments.

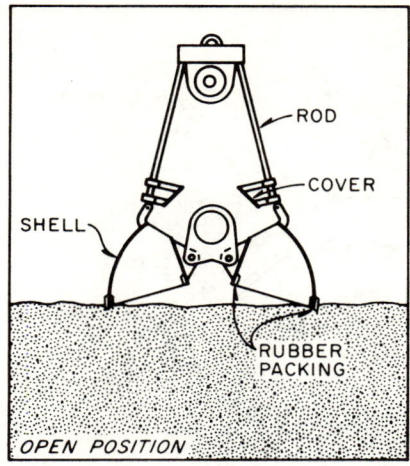 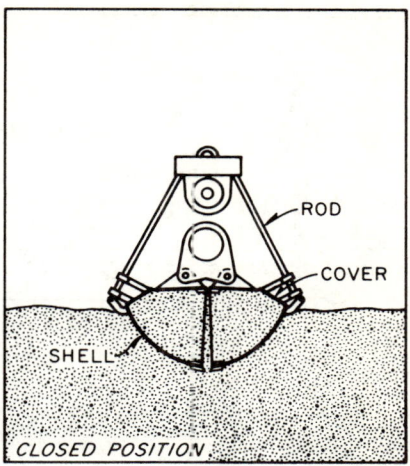

Figure 8-2 Open and closed positions of the watertight bucket (redrawn from Barnard, 1978).

Inland lake sediment removal is most commonly accomplished with a cutterhead hydraulic pipeline dredge. This type of equipment has become the workhorse of the dredging industry and has become relatively common to inland lake dredging projects with the advent of small, portable equipment. The primary components of any cutterhead dredge system include the hull, cutter head, ladder, pump, power unit, and a pipeline to distribute dredged material (Figure 8-3).

The dredge hull is made of steel and constructed to withstand the constant vibration created by the cutterhead when excavating consolidated material. The hull is the working platform that houses the main power plant, pump, lever room, and the assemblage of winches, wires, "A" frames, etc. that comprise the dredge.

At the bow is a steel boom or ladder with a cutter mounted at its distal end. Ladder length determines the practical dredge depth limitations. The ladder also supports the suction pipe and the cutter drive motor and shaft. In some cases there may be a submersible auxiliary suction pump mounted on the ladder. The ladder is raised and lowered by suspension cables attached at the outer end and to a hull-mounted winch.

The cutter or cutterhead typically consists of three to six smooth or toothed conical blades that rotate at 10-30 rpm to loosen compacted sediment (Bray, 1979). Cutter heads may be open-nosed, closed-nosed, straight-vaned, ribbon-screw-shaped, or auger-like. Most cutters have been designed specifically to loosen sand, silt, clay, or even rock material. Few, if any conventional hydraulic cutterheads have been designed to remove soft, flocculent lake sediment, so most of them probably are less efficient than they could be for lake dredging.

Spuds, vertically mounted pipes ranging from 25.4 cm to 127 cm in diameter, depending on the dredge size, are located at the stern of the dredge on both sides (Figure 8-4). They are used to "walk" the dredge forward by alternately raising and lowering them into the sediment.

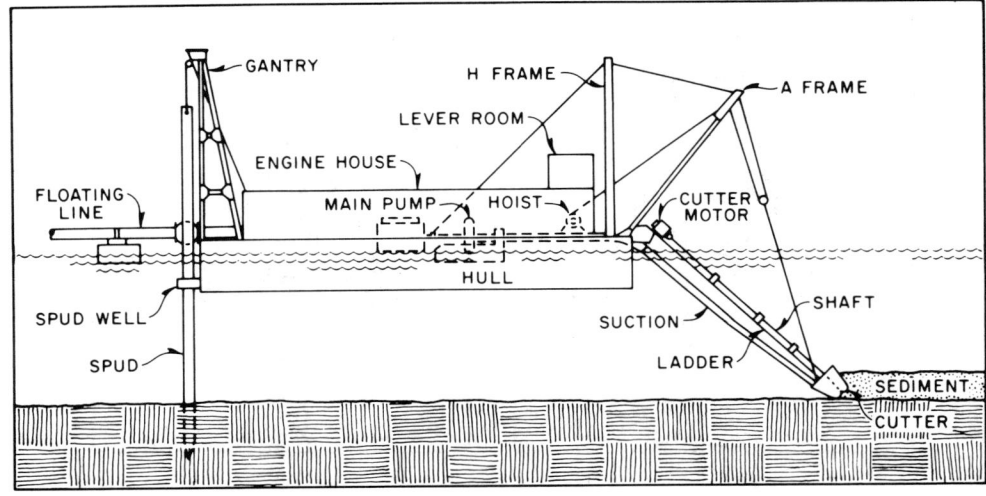

Figure 8-3 Configuration of a typical cutterhead dredge (from Barnard, 1978).

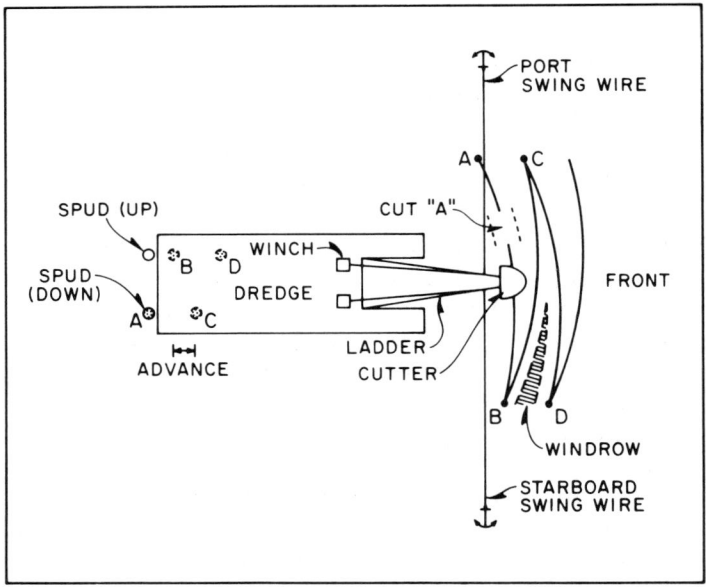

Figure 8-4 Spud-stabbing method for forward movement, and resultant pattern of the cut (from Barnard, 1978).

Operationally, sediment loosened by the cutter moves to the pickup head by suction from the dredge pump, which usually is of the centrifugal type. The sediment slurry is then discharged by pipeline to a remote disposal area. Cutterhead dredges are described by the diameters of their discharge pipes. Hydraulic dredges used for inland lake work usually range in size from 15 to 35 cm, although the one used recently at Vancouver Lake, Washington, was 66 cm (Raymond and Cooper, 1984). Figure 8-4 shows how the cutterhead is moved from side to side and how the cut path is created by pulling alternately on port and starboard swing wires. A major advantage of hydraulic cutter suction dredges over bucket types is that they are not confined in operation by the limitation of cable reaches. Another advantage is their continuous operating cycle. This cycle permits hydraulic dredges to produce large volumes of dredged material. This ability, however, is not without its negative aspects. Most slurries from hydraulic dredges contain only 10 to 20 percent solids; 80 to 90 percent of the slurry volume is water. This means that relatively large disposal areas, with adequate residence times, must be designed to precipitate solids from the dredge slurry. It also means that the large pumping capacity of hydraulic dredges might produce unplanned lake drawdowns, unless disposal-area overflow water is returned to the lake.

The amount of sediment supplied to the suction head is controlled by cutter rotation rate, thickness of the cut, and the swing rate (Barnard, 1978). Improper combination of any of these might cause the generation of excessive turbidity. Therefore, not only the configuration of the dredge equipment itself, but the skill of the dredge operator is important to minimizing turbidity.

Special-Purpose Dredges

Portable cutterhead dredges are essentially miniatures of large coastal waterway dredges. The cutterheads of coastal dredges were designed for cutting sand, clay, and silt; they were not intended for use in the fine, flocculent organic sediments (frequently 40% to 60% organics) encountered in lakes. Consequently, soft lake sediments have been a challenge to the dredging industry. Attempts to deal with the challenge have brought several dredging innovations. Among them is the cutterhead design used on Mud Cat® dredges. Mud Cat dredges utilize a horizontal auger to dislodge and move sediment to the center of a 2.4-m-wide dredge head, where it is sucked up by the pump and transported through a 20.3-cm discharge pipeline.

Figure 8-5 demonstrates the high mobility of the small Mud Cat dredges. Note the mud shield, which can be raised or lowered over the auger head to minimize resuspension of sediments. Nawrocki (1974) reported that turbidity plumes due to dredging with a Mud Cat machine were confined to an area no more than 6 m from the dredge, though operating conditions were not clearly defined. Suspended solids in the area of increased turbidity ranged from 39 to 1260 mg l^{-1}. Those near the bottom were approximately 100 mg l^{-1}. More turbidity is created by forward motion of the dredge than by backward motion. This appears to be associated with raising the mud shield while moving forward and lowering it when moving backward. Mallory and Nawrocki (1974) indicated the Mud Cat dredge should be capable of

Figure 8-5 The Mud Cat® dredge features a unique auger-type cutterhead. The size of the dredge makes it extremely portable (published with permission of National Car Rental).

producing slurry containing 30-percent to 40-percent solids. This represents nearly a doubling of the solids content commonly produced by conventional cutterhead dredges.

The Mud Cat guidance system is well suited to work on small water bodies. The dredge operates on a cable anchored at both shorelines. The guidance system permits uniform dredging of the bottom, with few missed strips. Mud Cat dredges have been used successfully at Collins Park and several other small lakes in New York State. The portability, guidance system, reduced turbidity, and increased solids content resulting from use of these dredges appears to make them ideally suited to small lake restoration projects.

Clark (1983) reported on a survey of portable hydraulic dredges available for use in the United States. The survey identified 46 models of portable equipment available from several different manufacturers. No attempt was made to critically analyze the features of one dredge relative to another, but tables are presented that describe the general dredge specifications, the pump characteristics, suction and discharge diameters, cutter type, and working capacity. The information should be useful to engineers for selecting dredges, since it includes dredging depth ranges from three to 18 m, production rate ranges from 15 to 1375 m^3 hr^{-1}, and a wide variety of cutterhead types.

Pneuma Dredging

There are a few other dredging systems that deserve mention because of their unique capabilities, although they are not commonly used for lake restoration and some are not currently available in the United States. Each of these systems was developed specifically for soft, viscous sediment similar to that found in lakes. The machines were originally created on a large scale for harbor maintenance, but it appears feasible to downsize them for lake dredging, as has been done in at least two instances (Matsubara, 1979; Murakami, 1984; Spencer Engineering, 1981).

The pneuma dredging system is unique. It was developed in Italy and employs air-driven pumps. The system commonly consists of three cylindrical pump bodies, a pickup head, a compressor, and air lines to the pump bodies. This system operates on the principle that sediment and water, under hydrostatic pressure, will be forced into the pump body, which is maintained at atmospheric pressure. Filling efficiency therefore increases with depth (increased hydrostatic pressure). At depths less than 10 m, however, hydrostatic pressure is insufficient to operate the system effectively. The Japanese Oozer® dredge system (a modified pneuma design) has overcome the problem of operating at shallow depths by creating partial vacuum in the pump bodies during the filling phase. Partial vacuum in the pump bodies also permits more rapid filling at greater depths. Figure 8-6 is a schematic diagram in the Oozer dredge system. The use of vacuum and air pressure to respectively fill and empty the pump body and drive sediment through the discharge pipe permits the removal of soft sediment at in situ density. This means the solids content of these dredge slurrys may run as high as 70 percent (Koba et al., 1975).

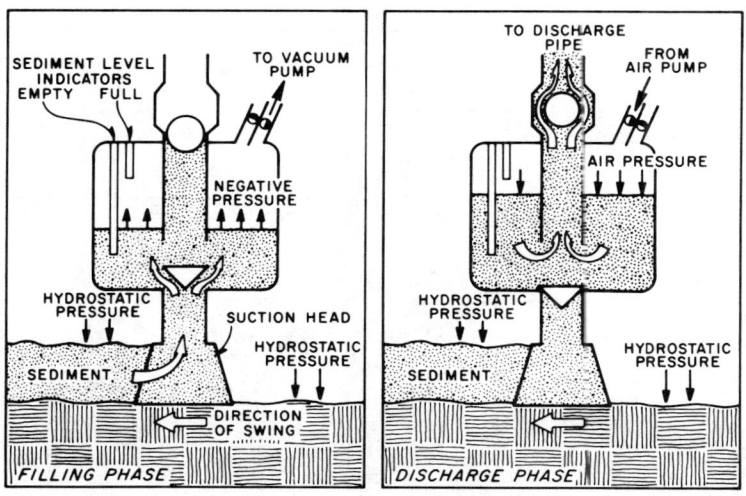

Figure 8-6 Schematic diagram of Oozer® dredge system.

The high solids content of the dredged material has several advantages, among them that solids production rates are increased, resuspension is minimized, and far less water volume must be contained at the disposal area.

Another dredge specialized for soft-sediment removal is the Cleanup System®, developed by TOA Harbor Works Company of Tokyo, Japan. This system employs a special suction head consisting of an auger, movable plates to direct sediment flow, a floating wing device to contain sediment as it is fed to the auger, and a large gas collection shroud (Figure 8-7). Methane and hydrogen sulfide gases, frequently liberated from organic sediment during dredging, are collected by the shroud, preventing release of foul odors.

The structural configuration of the Oozer and Cleanup dredges is similar to a conventional cutterhead dredge (Figure 8-3). Their cutpaths are likewise created by swinging the cutter from port to starboard and back. The major differences are in the pickup head configuration, the cutter (auger) apparatus, and the pumps. As with most hydraulic cutterhead dredges, the Cleanup System has a large centrifugal pump mounted within the hull, and in addition it has an auxiliary submerged centrifugal pump, mounted near the pickup head, which facilitates dredging at depths greater than 9 meters. The combination of pickup head configuration and pump equipment, as with the Oozer system, permits the removal of soft sediment at in situ density (20% to 70% solids in the slurry) (Sato, 1978). Both systems were compared with conventional cutterhead equipment to determine their relative ability to minimize secondary pollution from sediment resuspension (Suda, 1979). Suspended solids concentrations 1.5 m above the sediment-water interface and directly over the two special dredge heads averaged 6 to 8 mg l^{-1}. Under similar conditions the suspended solids concentrations around a conventional cutterhead averaged 40 to 80 mg l^{-1}

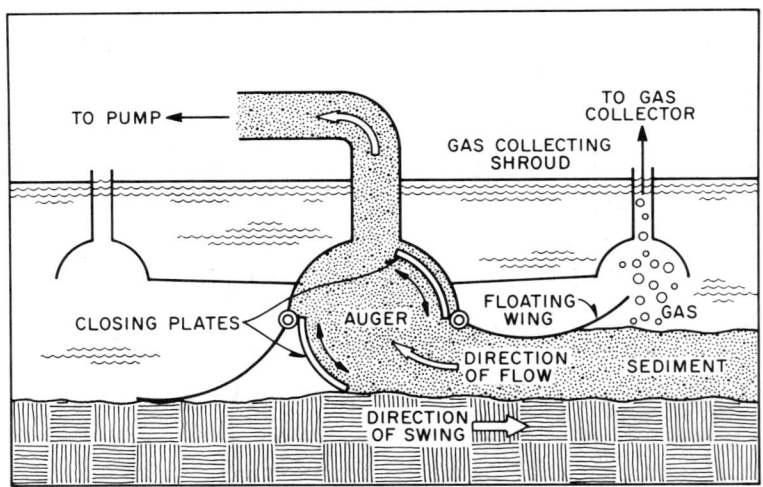

Figure 8-7 Suction head of the cleanup System® dredge.

at the same depth. Herbich and Brahme (1983), citing different references, reported average turbidity of 4.0 mg l^{-1} and 5.5 mg l^{-1} three and seven meters above bottom for the Cleanup and pneuma pump systems, respectively.

Although the Cleanup and Oozer dredges were developed by the Japanese to remove highly contaminated sediments from deep harbors, they demonstrate that extremely soft sediment can be removed efficiently with minimal resuspension if properly designed equipment is employed. A miniature, prototype Oozer system has been used successfully to remove polluted sediment from Lake Kasumigaura, Japan (Matsubara, 1979). An improved version of the original dredge has been in use on the lake since 1978. Through 1981 it had removed 265,000 m^3 of sediment Murakami, 1984).

A pneuma pump dredge was employed in 1981 at Gibraltar Lake, California, where mercury-contaminated sediment was removed. The pneuma dredge was suspended from an "A" frame on a stationary barge (Spencer Engineering, 1981). The plan called for dropping the dredge into the sediment and letting it dig a cone of depression (potholing method), but, as at Lilly Lake, Wisconsin, the plan failed when sediment did not flow.

A second method used was to tow the suspension barge so that the dredge was dragged over the sediment, forcing sediment into the pickup head. The major problem with this approach was maintaining proper speed and depth on the pickup head. In spite of this, it was reported that the dredge pumped an average of 500-600 m^3 hr^{-1} at an average velocity of 2-3 m sec^{-1} and transported it 700 m horizontally and 52 m vertically to the disposal area. The suspended solids concentration of the slurry averaged 40 to 50 percent by volume and occasionally reached 70 percent. The slurry was air-driven through the 35.5-cm discharge pipe by two compressors that produced air at the rate of 40 m^3 min^{-1} and a third one producing 45 m^3 min^{-1}.

The pneuma dredge performed well for a few months, after which it began to fail. Metal fatigue at the pump body seams was a major problem, and eventually all the seams required rewelding. Another major problem was encountered when rubber valve fittings were literally chewed up by debris. Both problems resulted in loss of air pressure and consequent loss of efficiency in the system. Replacement of rubber valves restored the dredge's efficiency, but the problem recurred in a few weeks. Others have reported similar problems concerning the rapid deterioration and plugging of pneuma pump valves.

The contractor finally devised a valving system that used a combination of urethane and rubber (actually machined bowling balls) that was more durable than the original equipment. Spencer Engineering (1981) reported this system to be much more satisfactory than the original equipment. Despite early problems, the final engineering report indicated that the pump performed well compared to more conventional dredging methods and that it was economical and efficient in transporting toxic sediments without disturbance to water supplies or the environment (Spencer Engineering, 1981). Goldman et al. (1981) confirmed the findings of Spencer Engineering. The chief concern in Gibraltar Lake was sedimented mercury. Goldman et al. (1981) found that elevated levels of mercury were not detected in the water at any station or at any depth during dredging. In fact, the dredging was so clean that no bathing beach areas in the 110.8-ha lake were forced to close during any phase of the dredging. Thus it appears that downsized and modified versions of the pneuma system, and possibly other special purpose dredges, may realize their greatest potential in lake restoration. Additional evaluation of these systems should be encouraged.

SUITABLE LAKE CONDITIONS

When a lake has filled in so that it is unusable for boating, swimming, and fishing activities, virtually nothing short of dredging will restore these uses. The high cost of dredging (Peterson, 1979), however, dictates that the feasibility of this treatment be examined closely in comparison to the intended use of the lake and alternative treatment methods (see Costs, later in this chapter).

Peterson (1981; 1982a) has described some of the problems in characterization of sediments that are unique to assessing the feasibility of dredging. Lake size might seem important to determining dredging feasibility, but, except for total cost, size appears not to be a constraint. Peterson's (1979) examination of 64 lake dredging projects showed that size ranged from less than 2 ha to over 1,050 ha and that volume of sediment removed ranged from a few hundred m^3 to over seven million m^3.

The one factor that might limit dredging of a large inland lake is the requirement for a commensurately large disposal area. Restoration most frequently is sought for lakes in high use areas, where sediment disposal space is scarce. It is also true, however, that the greatest user benefits will be derived from projects in densely populated areas (JACA, 1980). Therefore, it is important that disposal alternatives be explored for these situations.

Various productive uses of dredged material have been examined (Lunz et al., 1978; Spaine et al., 1978; Walsh and Malkasian, 1978). A practical solution was employed at Nutting Lake, Massachusetts, where approximately 153×10^3 m^3 of sediment was sold as a soil conditioner, at a price of \$1.40 m^{-3}. This reduced the total dredging cost by \$215,000 and the per-unit dredging cost to about \$1.00 m^{-3} (Worth, 1981). The selling and removal of dredged material does not negate the need for a "disposal area," but it may reduce the size requirement of what actually becomes a sediment rehandling area. A seven-hectare, two-basin handling area with a total holding capacity of approximately 98×10^3 m^3 was used at Nutting Lake. This area provided sufficient capacity for a season's dredging. Before the next dredging season (summer), the material was removed and sold, thus restoring the initial disposal capacity. This indicates that lack of an adequate disposal area to accommodate *all* of the dredged material need not prevent a project from going forward.

To be cost-effective, a sediment removal project should have reasonable assurance of longevity. An estimate of sedimentation rates should assist in determining the feasibility of dredging. Although dredging is expensive per unit of dredged material, when costs are amortized over the life expectancy of the project they may look much more reasonable. Sedimentation rates will help establish this "reasonableness." All other conditions being similar, lakes with relatively small watershed-to-surface ratios (nominally 10:1) will probably have lower sedimentation rates than those with large watersheds. Thus, a large lake with a small watershed should benefit more from dredging than will the reverse situation.

All of the above are considerations of physical features—depth, size, disposal, watershed area, and sedimentation rate. What of the chemical aspects of sediment and their influence on lake biota? Current information demonstrates that lakes with highly enriched surface sediments relative to underlying sediment might benefit from shallow dredging (Andersson et al., 1975; Bengtsson et al., 1975). Lake Trummen, Sweden, showed marked changes in water chemistry and biota when 40 cm of rich surface sediments were removed (Bjork, 1978). Similar changes were observed in Steinmetz Lake, New York, when 25 cm of organic sediments were removed and replaced by the same amount of clean sand (Snow et al., 1980). In both cases, extensive sediment surveys before dredging revealed that surface sediment was disproportionately rich in phosphorus and nitrogen relative to the deeper sediment. Open-water sediment is usually more important in these surveys than littoral sediment, since it becomes the sink. Littoral zones tend to be cleaned by wave action and, in the temperate zone, by spring ice scouring. The sediment survey should, at the minimum, determine the area of sediment to be removed and the depth. Horizontal sediment characteristics normally are more uniform than vertical sediment profiles. Sediment depth may vary considerably, depending on the basin configuration at the time of the lake formation or the transport of sediment to the lake via stream inlets. Vertical variation in the survey is important to note. Sediment profiles can be developed with the assistance of a Livingston piston corer. It will be important to note sediment color and texture differences with depth and to chemically characterize (phosphorus and nitrogen) the surface (0 to approximately 10 cm) sediment relative to deeper sediment if nutrient control is the intent (Peterson, 1981). Beyond this it will be quite useful to know the sediment particle size, settling rate, sediment volume, etc., in order to properly select a dredge for the job and design an adequate disposal area.

Several variables may determine the suitability of a lake for dredging, but generally the lakes most suitable have shallow depths, low sedimentation rates, organically rich sediments, long hydraulic residence times, and the potential for extensive use following dredging.

DREDGE SELECTION AND DISPOSAL AREA DESIGN

This section relies strongly on work by Pierce (1970). Implementation of lake dredging requires several decisions over the course of a project. Among the most important ones are what dredging equipment to select and what factors to consider in the disposal area design. Equipment selection will depend on several variables, including availability, project time constraints, slurry transport distances, discharge head, and the physical and chemical characteristics of the dredged material.

The primary factor controlling the design of disposal areas is the amount of dredged material that must be contained. In concert with this requirement, however, is the need to meet suspended solids requirements of the discharge permit. Therefore, factors such as sediment grain size, specific gravity, plasticity, and settling characteristics must be considered when designing the disposal area.

To illustrate the considerations most pertinent to dredge selection and disposal area design, the following example is offered. A feasibility study is conducted on the hypothetical Dead Lake, located in a rural area of the glaciated upper midwestern United States. The study reveals the following lake characteristics:

Lake area	= 120 ha
Maximum depth	= 5.5 m
Average depth	= 2.0 m
Normal water level	= 245 m
Sediment water content	= 30-60 %

Since total project cost is most likely calculated on the basis of sediment removal (payment is usually based on a measure of actual material removed), it is necessary to estimate the amount and type of sediment contained in the basin. The usual procedure for this is to collect hydrographic data suitable to developing a map showing the top of sediment depth and the configuration of the original basin, thereby describing the sediment volume. The accuracy of this map depends on the sampling interval and the relief of the original basin. Even relatively shallow glaciated lakes may have "deep holes," reinforcing the need for sediment depth mapping.

Sediment sampling frequency for the purposes of determining volume varies, depending on the basin configuration and the desired accuracy of the survey. Preliminary sampling stations should be broadly spaced, to provide a rough estimate of the solid bottom relief of the lake. This will help to define, and one hopes, limit the required number of stations for final mapping. Pierce (1970) suggested that small to medium-sized (< 40.5 ha) sediment removal projects should be mapped routinely by laying out sampling locations in a 15.25-m grid pattern. He suggested further that, for lakes with surface areas greater than 40.5 ha, the sample station grid size might be increased to 30.5 m without significant loss of accuracy. This reduces the station frequency by half without adverse effects on mapping accuracy if the bottom relief is relatively uniform. Individual lake characteristics ultimately dictate the required station frequency.

A common procedure for obtaining the necessary data is to make measurements at stations prescribed by the chosen grid size and relate the measurements to known elevation datum points on shore (topographic map, U.S. Geological Survey benchmark, etc.). The measurements can then be converted to elevations, thereby permitting the development of hydrographic maps and the calculation of sediment volume.

A simple means of obtaining the required data is to measure, at each station, the water depth to the sediment-water interface and the distance (depth) to which a probe can be pushed into the lake sediment before contacting hard bottom. Both measurements can be made at the same time by using a graduated probing ("sounding") rod. Lake sediment probes usually are steel rods measuring 0.95 cm to 1.6 cm in diameter. They can be bent. The investigator needs to develop "a feel" for the degree of resistance that determines hard lake bottom. Sediment depth is determined by calculating the difference between the rod interval reading at "hard bottom" and the reading at the sediment-water interface. Distinction of the sediment-water interface may be difficult in lakes with highly organic, flocculent sediments. In these cases it is advisable to use a lightweight disc or foot at the tip of the probing rod to establish water depth to the sediment surface. Alternatives to this are the use of a graduated line and Secchi disc, or an electronic depth sounder. It may be useful to employ divers to verify some of the depth determinations.

Depth determination should be made during calm periods on open waters. A pontoon boat makes the operation much easier. In cold climates the work can be accomplished even more easily by making the measurements through holes drilled in the ice. Winter lake mapping makes it much easier to locate your position accurately, and you need not contend with a pitching boat. Pierce (1970) indicated that a properly equipped crew working efficiently should be able to collect water and sediment data in this manner over 10 to 20 acres of lake surface per day. Efficiency is enhanced if data are collected during early winter, before ice has thickened to more than six or eight inches. Sediment depth measurements comprise an important aspect of the dredging project. Miscalculations in the sediment volume may lead to major errors in project cost estimates and to selection of improper dredging equipment, so accuracy should be stressed.

Sediment mapping on our hypothetical Dead Lake indicates that the deposits consist of highly organic silty material (muck). Water content of the sediment at the surface averages about 60 percent, while that at mid-depth and beyond ranges from 30 to 40 percent. Mapping data shows further that the thickness of sediment deposits is nearly 3.6 m at the south end, near the inlet, and decreases in depth to approximately 1.8 m on the north end. These sediment conditions are well suited to the use of a hydraulic cutterhead dredge. Three sediment disposal areas have been located around the lake. The desire to minimize pumping distances thus makes it convenient to divide the lake surface area into three pieces and identify them with the nearest disposal areas. Figure 8-8 shows how the lake might be divided to best utilize the available upland disposal areas.

The feasibility study shows that sedimentation rates in the lake have been reduced significantly over the past 15 years as a result of shifts from row crop to small grain and hay crop farming in the watershed. The accumulated sediment is not contaminated, and recent accumulations result mostly from autochthonous organic material decomposition. Therefore, it appears that deepending at least 15 percent of the lake to about 6.0 m, while leaving a fish spawning and wildlife area intact, will have a positive effect toward restoring the fishery and other beneficial uses. The study indicated further that water depth 60 m from shore should be a minimum of 2.5 m, and that the bottom should then slope at a five-percent grade to a maximum depth of 8.5 m. This depth would be highly desirable for maintaining adequate dissolved oxygen concentrations. The major portion of the lake should average about 3.5 m deep.

The maximum depth calculations based on these recommendations indicate that approximately 1,530,000 m^3 of sediment needs to be removed. It is desirable to complete the project as rapidly as possible, to minimize lake use disruption, so project duration is targeted for two years (mid-April through mid-November: ice-free months, over two consecutive seasons).

160 *Physical and Chemical Methods*

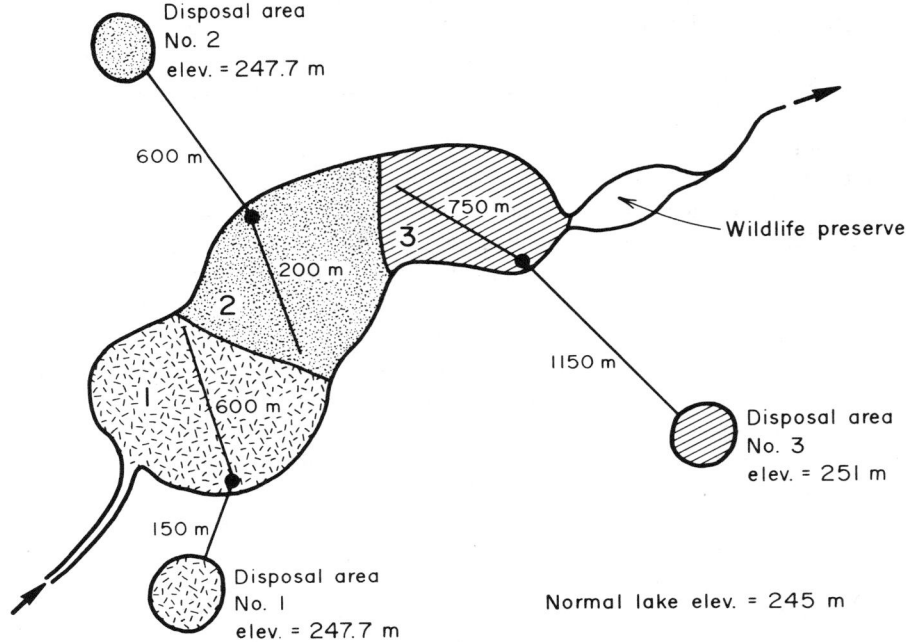

Figure 8-8 Dead Lake (hypothetical), showing the planned dredging areas, pipeline distances to disposal areas, and the wildlife area that will remain undredged.

Dredge Selection

Proper selection and use of hydraulic dredging equipment will permit implementation of the feasibility recommendations. Pierce (1970) suggested a series of considerations when selecting a hydraulic cutterhead dredge for any particular project. They are presented in the remainder of this section.

1. Plan to optimize the available disposal area.

Long pumping distances to disposal areas should be minimized to the extent possible, since energy requirements will increase with pumping distances. In Figure 8-8, disposal area No. 1 is the closest, at 750 m when pumping, from lake area No. 1. Disposal area No. 2 is 800 m and disposal area No. 3 is 1900 m, when pumping, from the respective lake areas. It has been calculated that areas 1, 2, and 3 will hold 574,000 m^3, 413,000 m^3, and 918,000 m^3 of dredged material, respectively. Therefore, area 1 would receive 574,000 m^3, area 2 would receive 413,000 m^3, and area 3 would receive the remainder of the total dredge volume (543,000 m^3), to optimize disposal efficiency by minimizing pipeline length. The lake configuration relative to the disposal locations indicates the shortest pipeline length will be about 300 m when working lake area No. 1. It will be nearly the full 800 m in lake area No. 2. In lake area No. 3 the maximum pipeline length will be 1900 m.

2. Analyze the production capacity of available dredging equipment.

It will be necessary to analyze the production of various sizes of dredges to determine which equipment might complete the job within the planned two-year period. A quick survey of equipment in the immediate area reveals that 20-cm, 25-cm, and 30-cm dredges are available, so production analysis will be limited to these sizes.

Dredge pump production rate ranges, rather than specific numbers, are usually listed by the manufacturers, since dredging conditions and thus production rates vary considerably. Production ranges for the available dredges (20, 25, and 30 cm) are taken from Figure 8-9 to illustrate the method. Similar dredge capacity charts are available from various dredge pump manufacturers. Charts for the specific equipment in question should be used whenever they are available. Figure 8-8 and the feasibility study show that the greatest sediment volume is located near the center of the lake, and that transport from this area to the disposal cells will require pipeline transport distances in excess of 600 m. Based on that information, the following dredge pump production range analysis was developed, using the minimum, a medium, and the maximum pipeline lengths:

300-meter length of discharge pipeline:
20 cm pump = 50 to 110 m^3 per hour, average 80 m^3/hr
25 cm pump = 80 to 190 m^3 per hour, average 135 m^3/hr
30 cm pump = 310 to 420 m^3 per hour, average 365 m^3/hr

600-meter length of discharge pipeline:
20 cm pump = beyond effective discharge length; booster pump required
25 cm pump = 60 to 120 m^3 per hour, average 90 m^3/hr
30 cm pump = 220 to 290 m^3 per hour, average 255 m^3/hr

800-meter length of discharge pipeline:
20 cm pump = beyond effective discharge length; booster pump required
25 cm pump = 50 to 80 m^3 per hour, average 65 m^3/hr
30 cm pump = 190 to 250 m^3 per hour, average 220 m^3/hr

The analysis reveals that using the 25-cm system for distances of 600 to 800 m is marginally efficient. It also indicates that any disposal in cell No. 3 from lake area No. 3, even with the largest system available (30 cm), will require the use of a booster pump.

3. Compute the number of dredging days required to complete the job.

Approximately 1,530,000 m^3 of sediment must be removed from Dead Lake. A hydraulic dredge normally is operated 24 hours a day, in three shifts, for efficiency.

162 Physical and Chemical Methods

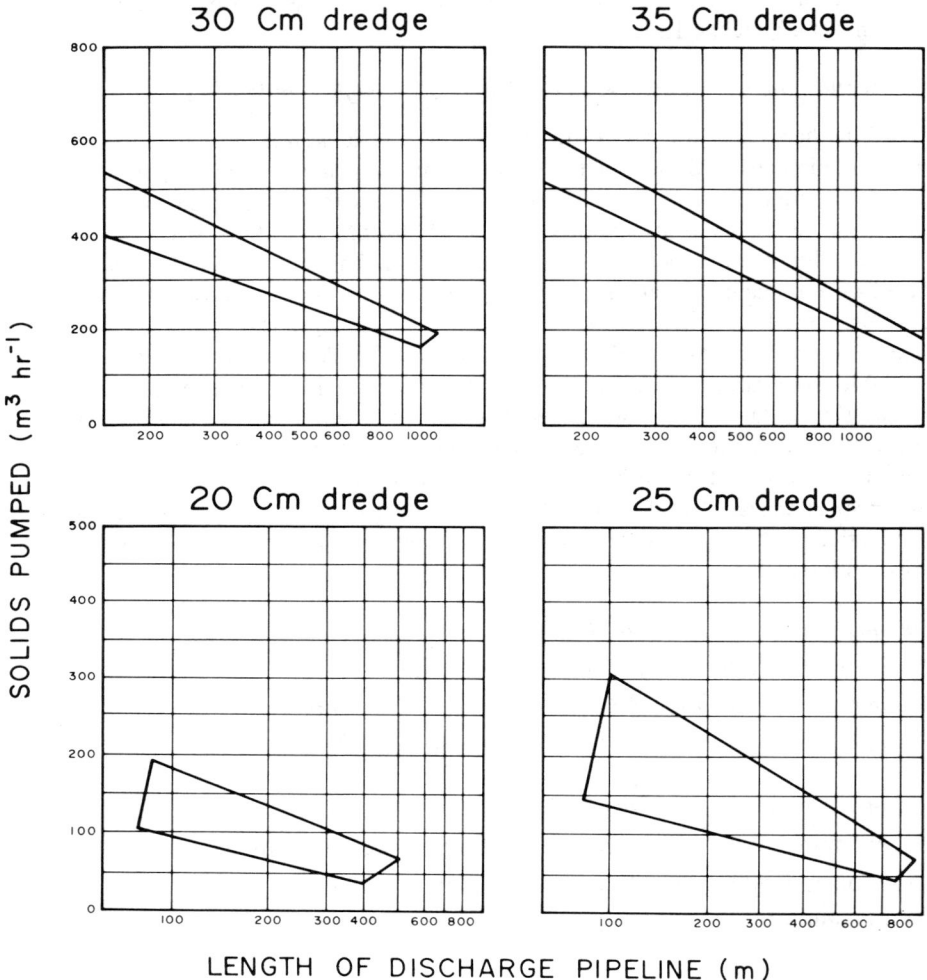

Figure 8-9 Representative production characteristics for various sizes of dredge systems (modified from Pierce, 1970).

There is always some down time for maintenance and pipeline relocation, so we can assume normal productive dredge time at approximately 20 hours per day, six days per week.

It was already observed from the above production analysis that the 20 cm dredge would not operate efficiently without a booster pump at discharge distances exceeding 600 meters. Since most of the pumping will require discharge at 600 meters and beyond, the 20-cm pump should not be given any further serious consideration. Using the average (rounded for illustrative convenience) discharge rates of the 25-cm and 30-cm systems for 600 m and 800 m pipeline lengths, we can calculate the number of days required to complete the Dead Lake dredging project.

25-cm pump system:

$$\frac{90 + 65}{2} = 77.5 \cong 75$$

$$\frac{1{,}530{,}000 \text{ m}^3}{(75 \text{ m}^3 \text{ hr}^{-1})(20 \text{ hr/day})} = 1020 \text{ days}$$

30-cm pump system:

$$\frac{255 + 220}{2} = 237.5 \cong 230$$

$$\frac{1{,}530{,}000 \text{ m}^3}{(230 \text{ m}^3 \text{ hr}^{-1})(20 \text{ hr/day})} = 237 \text{ days}$$

Dead Lake is located in a north temperate zone of the United States that is subject to freezing conditions from approximately mid-November until mid-April. This reduces annual open-water work days to about 185. If the 25-cm pump system was used on this project, completion time would be in excess of five years (1020 ÷ 185 = 5.5 yr). If the project is to be completed within the targeted time of two years (two open-water seasons), the 30-cm dredge will be required (237 ÷ 185 = 1.3 yr). Disposal in area No. 3 will require the use of an appropriately placed booster pump, since the pipeline length to that area exceeds the efficient pumping distance (approximately 1000 m) of the 30-cm pump system.

4. **Determine the required head discharge characteristics of the main pump when pumping material with the specific gravity of lake sediment (approximately 1.20).**

The required head discharge characteristics of a pump will be dependent on the discharge pipe length; i.e., the longer the pipeline, the higher the total head required. Pump head discharge characteristics must be analyzed for both minimum and maximum discharge distances. In the case of Dead Lake these distances are about 300 m when pumping from lake area 1 to disposal area 1 and about 1920 m when pumping from lake area 3 to disposal area 3.

The sum of the total suction lift and total discharge head is defined by Pierce (1970) as the total dynamic head against which a pump must work. The heads commonly are computed from basic hydraulic formulas, corrected for specific gravity of the pumped material. Suction lift incorporates suction elevation head, suction velocity head, and friction head in the suction pipe. The total discharge head is calculated by summing the pump velocity head, the discharge elevation head, and the friction head in the pipeline. Minor head losses usually are not considered.

164 *Physical and Chemical Methods*

Suction Head. Since the weight of dredged material (specific gravity of lake sediment is approximately 1.20) is greater than that of water, the surface of a column of water equal to the depth of Dead Lake would always be at greater elevation than the surface of an equal-sized (diameter) column of dredged material of the same weight. The resultant difference in column heights is the suction elevation head. The suction elevation head is always referred to the horizontal center line of the main pump and is computed as

$$h_{ss} = S_1 A - S_2 B, \qquad (8.2)$$

where h_{ss} = the suction elevation head (meters of fresh water)
S_1 = specific gravity of lake water (1.0)
S_2 = specific gravity of material being pumped (1.2)
A = distance from the bottom of the cut to the water surface (m)
B = distance from the pump center to the bottom of the cut (m).

Assuming that the dredge pump is mounted on the hull at lake level and that maximum dredged depth is 8.5 meters, the static suction head is

$$h_{ss} = 1.00 \, (8.5) - 1.20 \, (8.5)$$

$$h_{ss} = -1.7 \text{ m}.$$

The minus sign indicates that a suction head exists. This number must be added positively to other heads computed for the suction system.

The suction velocity head is the energy required to start the movement of dredge material into the suction pipe. It can be computed as

$$h_{sv} = S_2 \frac{V_s^2}{2g}, \qquad (8.3)$$

where h_{sv} = velocity head (meters of fresh water)
S_2 = specific gravity of the material being pumped
V_s = velocity of the mixture in the suction pipe (m sec^{-1})
g = acceleration rate of gravity (m sec^{-2}).

Normal suction pipe velocity should be maintained at 3.0 to 4.0 m sec^{-1} to assure that solids are carried into the pump. If we assume an upper-midrange suction pipe velocity of 3.6 m sec^{-1}, the velocity head in the suction pipe will be

$$h_{sv} = (1.20) \frac{(3.6)^2}{2(9.82)}$$

$$h_{sv} = 0.8 \text{ m}.$$

Friction losses caused by aqueous flow characteristics in pipes create the major portion of the head that a dredge pump must overcome. Pipeline friction loss is influenced by a variety of variables. Among them are the type of pipe, diameter of pipe, flow velocity in the pipe, the pipeline length, the pipeline configuration, and the percentage and type of solids in the pumped mixture. Since friction losses are magnified as the diameter of the suction pipe decreases, many small dredges utilize suction pipes one size (usually 5-cm increments) larger than the discharge pipe. For example, a 30-cm dredge (size of discharge) might have a 35-cm suction line. Since the volume entering the larger suction pipe (35 cm) must be accommodated by the smaller discharge pipe (30 cm), the rate of velocity in the discharge pipe will be increased over that in the suction pipe. The velocity in the discharge pipe varies as the ratio of the square of the diameter of the larger pipe divided by the square of the diameter of the smaller one $[(35)^2 \div (30)^2 = 1.36]$. Therefore, the 3.6 m sec^{-1} velocity in the suction will be increased to approximately 4.9 m sec^{-1} in the discharge. All influences affecting pipeline friction losses must be considered and applied to an acceptable equation for formulating friction losses. One of the most commonly used formulas is the Darcy-Weisbach pipe friction formula.

Pierce (1970) defines the suction friction head as the energy required to overcome friction losses in the pump suction line. The suction friction head can be computed using the Darcy-Weisbach formula:

$$h_{sf} = f \left[1 + \frac{(P - 10)}{100} \right] \frac{LV_s^2}{2gD} , \qquad (8.4)$$

where h_{sf} = friction head (meters of fresh water)
f = the friction factor
P = solids in dredge slurry (% by volume)
L = equivalent length of suction pipe (m)
V_s = velocity of the mixture in the suction pipe (m sec^{-1})
g = acceleration rate of gravity (m sec^{-2})
D = inside diameter of the suction pipe (m).

The friction factor f is a dimensionless number that is a function of the Reynolds number and the relative roughness (absolute roughness ÷ diameter of pipe in m) of different types of pipe. The functions have been obtained experimentally for clear water and expressed graphically (Figure 8-10) by Moody (1944). Use of f as described by the Moody diagram for computing dredged material pipeline transport necessarily becomes an approximation at best, since solids in the slurry will affect the number. Despite this apparent problem, Moody f values are commonly used to estimate various hydraulic pipeline dredging figures. The Reynolds number can be calculated from the formula

$$R = \frac{VD}{\nu}, \qquad (8.5)$$

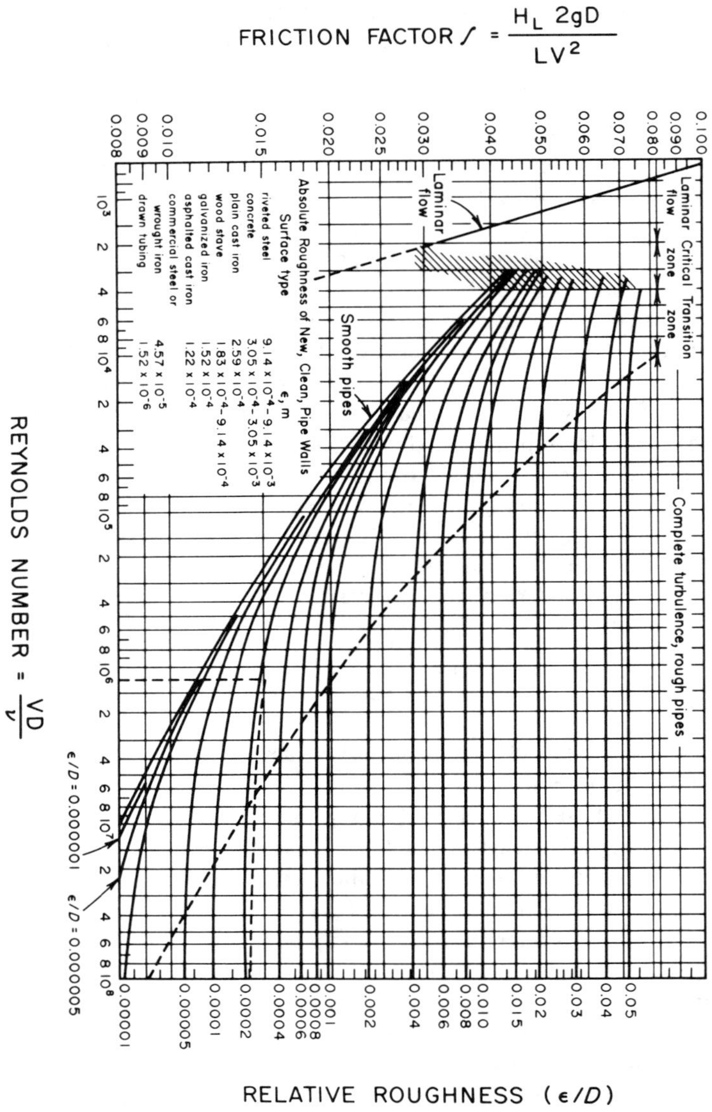

Figure 8-10 Moody diagram showing friction factors for pipe flows (redrawn from Randall, 1977).

where V = velocity in the pipeline (m sec^{-1})
       D = inside diameter of the pipeline (m)
       ν = temperature-corrected kinematic viscosity of water (m^2 sec^{-1} × 10^{-6})
       (See Table 8-1).

Table 8-1 Selected Physical Properties of Water at Various Temperatures

Temperature T (deg C)	Density p (gm cm^{-3})	Viscosity μ (gm cm^{-1} sec^{-1} × 10^{-2})	Kinematic Viscosity ν (cm^2 sec^{-1} × 10^{-2})*
0	0.9999	1.787	1.787
5	1.0000	1.514	1.514
10	0.9997	1.304	1.304
15	0.9991	1.137	1.138
20	0.9982	1.002	1.004
25	0.9971	0.891	0.894
30	0.9957	0.798	0.802
35	0.9941	0.720	0.725
40	0.9923	0.654	0.659
50	0.9881	0.548	0.554
60	0.9832	0.467	0.475
70	0.9778	0.405	0.414
80	0.9718	0.355	0.366
90	0.9653	0.316	0.327
100	0.9584	0.283	0.295

*cm^2 sec^{-1} × 10^{-4} = m^2 sec^{-1}

As stated above, the velocity of suction pipeline slurries commonly ranges from 3.0 to 4.0 m sec^{-1} or greater to maintain the suspension of solids (turbulent flow). If we use the previous assumed slurry velocity of 3.6 m sec^{-1} and a kinematic viscosity of water at 20°C (1.0 × 10^{-6} m^2 sec^{-1}) and apply these figures to a 35-cm (0.35 m) suction pipe, the following Reynolds number can be calculated from equation 8.5:

$$R = \frac{3.6\,(0.35)}{1.0 \times 10^{-6}}$$

$$R = 1.3 \times 10^6.$$

If we assume a pipe roughness of 8.7 × 10^{-5} m the relative roughness will be

$$rr = \frac{\epsilon}{D}, \qquad (8.6)$$

where rr = relative roughness
 ϵ = absolute roughness (m)
 D = inside diameter of pipe (m).

168 Physical and Chemical Methods

Then

$$rr = \frac{8.7 \times 10^{-5}}{0.35}$$

$$rr = 2.5 \times 10^{-4}$$

Applying relative roughness and the Reynolds number to the Moody diagram (Figure 8-10), the resultant friction factor is 0.015 for these conditions. Note that the intersection of these two variables falls within the transition zone between laminar flow and complete turbulence in rough pipes.

Friction head losses in suction pipes vary with configuration, valving, suspended solids concentration, and cutterhead types. Cutterhead losses are highly variable, and losses associated with fine-grain dredge material are not well defined. (Note the comment above concerning the use of f values.

Correction factors for these variables are not readily available in tabular form. Therefore, engineering best judgment based on a combination of practical experience and laboratory tests frequently is applied to actual suction pipe lengths to calculate "the equivalent length of suction pipe." In effect, the equivalent length is a correction for suction pipe head loss. The suction pipe "correction factor" commonly is within the range of 1.3 to 1.7 (D. Hayes, personal communication). To dredge to a depth of 8.5 m (maximum lake depth after dredging), the dredge ladder (suction pipe length) will need to be approximately 15 m long. Applying a suction-pipe-equivalency correction factor of 1.7, the equivalent suction pipe length is 25.5 m (15 × 1.7 = 25.5). Substituting the required figures (assume 20% solids) into equation 8.4 determines the suction friction head:

$$h_{sf} = 0.015 \left[1 + \frac{(20-10)}{100} \right] \frac{25.5 \, (3.6)^2}{2(9.82)(0.35)}$$

$$h_{sf} = 0.8 \text{ m}.$$

The total suction head (H_s) on the dredge pump is the sum of the suction elevation head, the velocity head, and the friction head:

$$H_s = h_{ss} + h_{sv} + h_{sf}$$

$$= 1.7 + 0.8 + 0.8$$

$$H_s = 3.3 \text{ m}.$$

Discharge Head. The discharge elevation head is represented by the difference in elevation (vertical distance) between the pump center line and the end of the discharge pipe, corrected for the specific gravity of the dredge slurry. As mentioned previously, the specific gravity of lake sediment dredge slurry for this example is given as 1.20.

The pump center line of dredges being considered for this job is at the water line of the dredge hull (from Figure 8-8, normal water level is 245 m). The top of the dike elevation at disposal sites 1 and 2 is determined to be 247.7 m. This information yields the discharge elevation head, using the equation

$$h_{de} = S_2 (E_D - E_p), \qquad (8.8)$$

where h_{de} = discharge elevation head (meters of fresh water)
S_2 = specific gravity of the mixture being pumped
E_D = elevation of the center line of the discharge pipe at the point of discharge (m)
E_p = elevation of the center line of the dredge pump (m).

Therefore, when pumping to disposal areas 1 and 2, the discharge elevation head will be

$h_{de} = 1.20 (247.7 - 245.0)$

$h_{de} = 3.2$ m.

The discharge friction head is the energy needed to overcome friction losses in the discharge pipe; it can be computed using the form of equation 8.4. The dredge pump will have to overcome maximum friction head when pumping from lake area 2 to disposal area 2 (greatest discharge distance without a booster pump). The pipeline length in this case will be about 200 m of floating pipe and 600 m of shore pipe. The two pipes differ considerably in joint configuration, since the floating pipe must be flexible enough to accommodate wave action and relocation of the dredge. Therefore, the factor applied to the two types of pipe to calculate the equivalent length is different. Pierce (1970) indicates that the floating pipe factor typically ranges from 1.35 to 1.5 (more bends than shore pipe), while that for shore pipe usually lies between 1.1 and 1.25. If we use the maximum factor of 1.5 for floating pipe (200 m) and the minimum of 1.1 for shore pipe (600 m) the factors will tend to normalize the pipeline equivalent lengths. Therefore,

Floating pipe length = 200 (1.5) = 300
Shore pipe length = 600 (1.1) = 660
Total equivalent length = 960 m.

This total equivalent length, substituted into the form of equation 8.4 with the calculated discharge pipeline velocity (4.9 m sec^{-1}), results in a discharge pipeline friction head loss of

$$h_{df} = 0.015 \left[1 + \frac{(20-10)}{100} \right] \frac{960(4.9)^2}{2(9.82)(0.30)}$$

$h_{df} = 65.0$ m.

170   Physical and Chemical Methods

The same value can be obtained from Figure 8-11 by entering the velocity scale at 4.9 m sec^{-1} and reading vertically to the 0.30-m pipeline intersection and then reading left to 2.05 on the friction head loss scale. The friction head loss scale is in meters per 30.5 m of pipe, so the scale reading must be multiplied by the number of times that 30.5 can be divided into the equivalent pipe length (960 ÷ 30.5 = 31.47; 2.05 × 31.47 = 64.5 ≅ 65 m).

The pump velocity head is the energy required to increase the pump suction line velocity to the discharge pipeline velocity. It can be computed from equation 8.9:

$$h_{dv} = S_2 \frac{V_d^2 - V_s^2}{2(g)}, \tag{8.9}$$

where h_{dv} = pump velocity head (meters of fresh water)
       S_2 = specific gravity of the dredged material
       V_d = velocity of dredged material in the discharge pipeline (m sec^{-1})
       V_s = velocity of dredged material in the suction pipeline (m sec^{-1}).

The suction velocity for this example is 3.6 m sec^{-1}, and the discharge velocity is 4.9 m sec^{-1}. Therefore, the pump velocity head is

$$h_{dv} = (1.20) \frac{(4.9)^2 - (3.6)^2}{2(9.82)}$$

$$h_{dv} = 0.7 \text{ m}.$$

The total discharge head on the main pump is the sum of the discharge heads, per equation 8.10:

$$\begin{aligned} H_d &= h_{de} + h_{df} + h_{dv} \\ &= 3.2 + 65.0 + 0.7 \\ H_d &= 68.9 \text{ m}. \end{aligned} \tag{8.10}$$

The total dynamic head on the main pump is the sum of the total suction head and the total discharge head, per equation 8.11:

$$\begin{aligned} H_{TDH} &= H_s + H_d \\ &= 3.3 + 68.9 \\ H_{TDH} &= 72.2 \text{ m}. \end{aligned} \tag{8.11}$$

Once the total dynamic head is known, the power necessary to operate the pump against the resistance in the system can be calculated. First, however, it is necessary to know the theoretical pump output, which can be calculated as follows:

$$Q = \frac{\pi}{4} \times 3600 \, D^2 \, V_d, \tag{8.12}$$

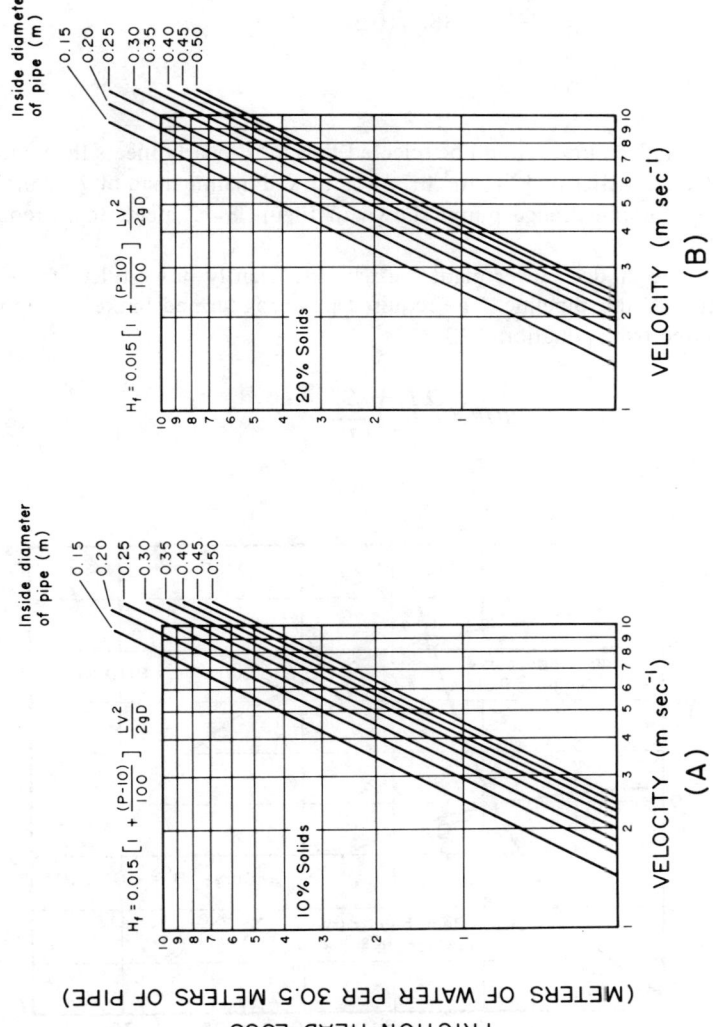

Figure 8-11 Friction head loss for 10-percent and 20-percent solids in various diameter pipelines as a function of slurry velocity (modified from Pierce, 1970).

where Q = output of the dredge pump (m^3 hr^{-1})
D = inside diameter of the discharge pipe (m)
V_d = velocity of slurry in the discharge pipe (m sec^{-1}).

The 30-cm dredge pump output, when pipeline velocity is 4.9 m sec^{-1}, is

$$Q = \frac{3.14}{4} \times 3600 \, (0.30)^2 \, (4.9)$$

$$Q = 1246 \, m^3 \, hr^{-1}.$$

Therefore, a dredge pump should be selected that most nearly meets the required head discharge characteristics of 1246 m^3 hr^{-1} at a total dynamic head of 72.2 m. The performance curve for the dredge pump shown in Figure 8-12 meets these requirements at Point C.

In addition, the dredge power plant must be sufficiently powerful to force the pump output through the pipeline. The required power is termed brake horsepower and can be computed from equation 8.13:

$$BHP = \frac{QH_{TDH} S_2}{2.737 \, E}, \qquad (8.13)$$

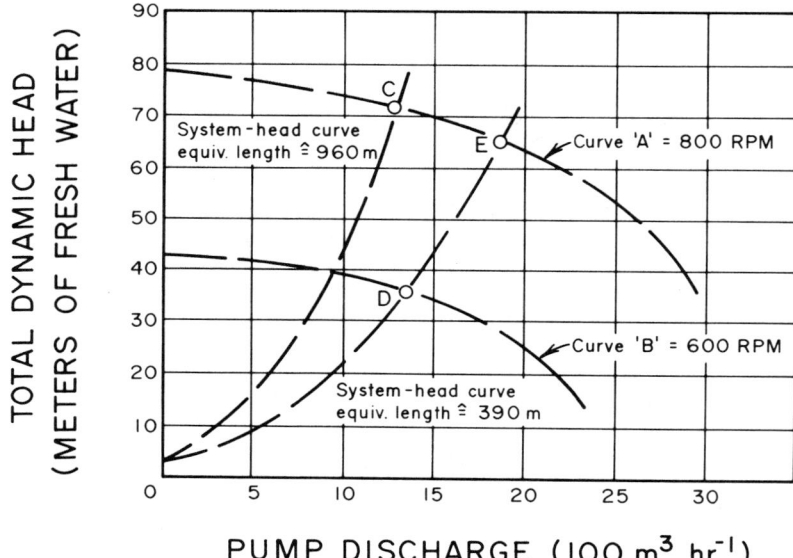

Figure 8-12 System head curve for a 30-cm dredge pump (modified from Pierce, 1970).

where BHP = continuous brake horsepower at the pump
Q = dredge output (m^3 hr^{-1})
H_{TDH} = total dynamic head the pump works against (meters of fresh water)
S_2 = specific gravity of the dredged material
E = dredge pump efficiency (%).

Pierce (1970) notes that the pump efficiency of smaller dredges ranges from 50 percent to 65 percent and decreases with wear. It is therefore recommended that a conservative figure be adopted for efficiency. Pierce (1970) recommends an efficiency figure of 55 percent. Therefore,

$$BHP = \frac{(1248)(72.2)(1.20)}{(2.737)(55.0)}$$

$$BHP = 718 \text{ hp}.$$

An additional consideration for dredge power plants, according to Pierce (1970), is that the manufacturer's rated continuous-duty capacity at any RPM should be discounted by at least ten percent. This will assure that the power plant is adequately sized to rotate the pump at the required RPM (800 in this case), and at the same time it will provide a longer, more trouble-free engine lifetime. In this case 718 × 1.10 = 790 hp, at 800 RPM, that will be required. Engine selection should be made based on a 1.5 to 1.0 reduction gear between the engine and the pump. Therefore, an engine of at least 790 horsepower at 1200 RPM should be used.

From the above dredge pump system analysis it can be concluded that:

1. The minimum dredge size that should be considered for this job is 30 cm, with sufficient ladder length to dredge to a depth of 8.5 m.
2. The average production rate of a 30-cm dredge pumping distances of 600 m will be approximately 230 m^3 hr^{-1}.
3. The job can be completed in two summer seasons using a 30-cm dredge.
4. The closest disposal sites should be filled first.
5. There will be a maximum head of 72.2 m on the dredge pump when pumping from lake area 2 to disposal area 2.
6. The dredge pump power plant should have a minimum continuous horsepower rating of 790 at 1200 RPM.

This represents an analysis of maximum head conditions on the pump.

5. Determine minimum head conditions when pumping to the nearest disposal area.

The total head on the pump decreases as the pumping distance decreases. This means the pump output increases and the velocity of dredged material in the pipeline increases. As noted previously, a minimum pumping distance of about 300 m (150 m of shore pipe and 150 m of floating pipe) will be encountered when dredging and disposing in area 1. By doing the same series of computations as were done for the 800-m pipeline, the following can be concluded for the 300-m pipeline:

1. Average production rate for a 30-cm dredge with a 300-m discharge pipeline length is about 360 m^3 hr^{-1}.
2. The system head curve for the 300-m pipeline is shown in Figure 8-12 as having an equivalent length of approximately 390 m. Point E in Figure 8-12 shows that at 800 RPM the pump discharge would exceed 1800 m^3 hr^{-1}. This would increase the discharge pipeline velocity beyond 6.5 m sec^{-1}, creating excessive pump and pipeline wear, possible pump cavitation, and extreme taxing of the engine, all of which would create an inefficient operation.
3. There are two possible solutions to the problem: Install a smaller pump impeller or reduce the engine speed. Figure 8-12 shows that if the 30-cm pump is operated at 600 RPM in conjunction with a 390-m equivalent pipeline length (point D), the pump will deliver about 1340 m^3 hr^{-1} at a head of 36 m. According to Pierce (1970), at this capacity the discharge velocity would be reduced to an acceptable 5.1 m sec^{-1}.
4. The continuous horsepower required for the 36-m head and 600-RPM operating conditions is reduced to 385 hp. Using the 1.5-to-1 reduction would require the engine to operate at 900 RPM, to turn the pump at 600 RPM.

It is apparent that the head discharge curves for a pump over its recommended speed range will be very helpful in selecting a dredge system and in determining the optimum pump speed at various discharge distances, to maximize production. It should be noted that pump system performance will change as the system components wear through use, and it is a good policy to periodically check the performance against the manufacturer's rating curves. This is most easily accomplished by running tests on clear water when the system is first mobilized and rerunning the tests periodically after the pump has been in service. Results of these tests will permit the dredge operator to modify operating procedures as necessary to maintain optimum production.

6. Analyze booster pump requirements for pumping to distances beyond the capacity of the main pump.

The pipeline transport distance (1900 m) from lake area 3 to disposal site 3 (elevation = 251 m) exceeds the efficient capacity of the 30-cm dredge (see Figure 8-9), due to the increased friction head and reduced output from the pump. The booster pump selected must be capable of increasing the total discharge rate to maintain a minimum discharge pipeline velocity of 4.9 m sec^{-1}.

Figure 8-13 shows the head discharge for the main pump operating alone and the curve for the main pump and an identical booster pump operating in series. The "operating point" on the two-pump curve can be determined by calculating the system head curve for the 1900-m discharge pipeline. For this system the equivalent length of the discharge line is

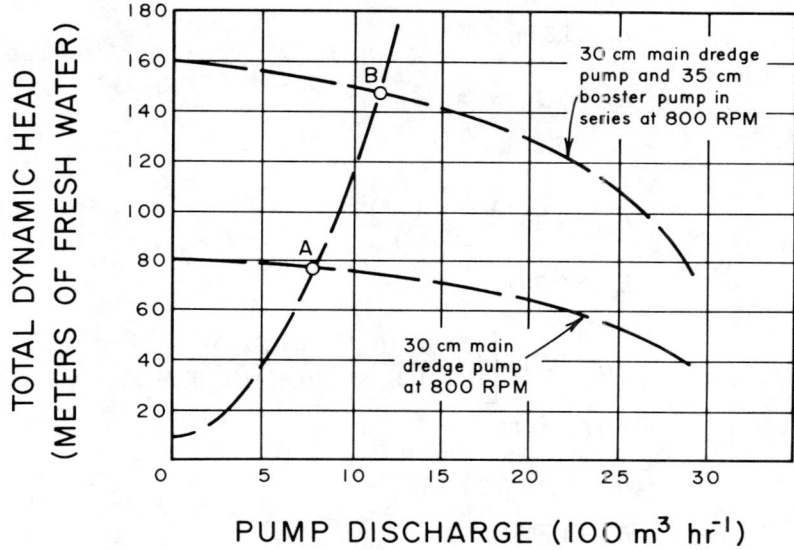

Figure 8-13 Head discharge relationships for a 30-cm dredge pump and a 30-cm booster pump (modified from Pierce, 1970).

$$\text{Floating pipe length} = 760\,(1.5) = 1140$$

$$\text{Shore pipe length} = 1160\,(1.1) = 1276$$

$$\text{Total equivalent length} = 2416 \text{ m}$$

The system head curve, as shown in Figure 8-13 for the 1900-m discharge pipeline, can be computed as follows:

$$h_{ss} = 1.0(8.5) - 1.20(8.5)$$

$$h_{ss} = -1.7 \text{ m};$$

$$h_{sv} = 1.2 \frac{(3.6)^2}{2(9.82)}$$

$$h_{sv} = 0.8 \text{ m};$$

$$h_{sf} = 0.015 \left[1 + \frac{(20-10)}{100} \right] \frac{25.5(3.6)^2}{2(9.82)(0.35)}$$

$$h_{sf} = 0.8 \text{ m};$$

$H_s = 1.7 + 0.8 + 0.8$

$H_s = 3.3$ m;

$h_{de} = 1.2 (251 - 245)$

$h_{de} = 7.2$ m;

$h_{dv} = 1.20 \dfrac{(4.9)^2 - (3.6)^2}{2(9.82)}$

$h_{dv} = 0.7$ m;

$h_{df} = 0.015 \left[1 + \dfrac{20-10}{100} \right] \dfrac{2416(4.9)^2}{2(9.82)(0.30)}$

$h_{df} = 162$ m;

$H_d = 7.2 + 0.7 + 162$

$H_d = 169.9$ m;

$H_{TDH} = H_s + H_d$

$\phantom{H_{TDH}} = 3.3 + 169.9$

$H_{TDH} = 173.2$ m;

$Q = 2826(0.30)^2 (4.9)$

$Q = 1246$ m^3 hr^{-1}.

The calculated head discharge relationship determines one point on the pump system head curve (Figure 8-13). Other points on the curve can be calculated to develop the system curve for plotting. Point A in Figure 8-13 shows that the dredge pump alone would produce about 815 m^3 hr^{-1} at a head of 75 m. By rearrangement of equation 8.12 to

$$V_d = \dfrac{Q}{\dfrac{\pi}{4} \times 3600 D^2}$$

it can be determined that the discharge pipeline velocity under these conditions would be reduced to slightly more than 3 m sec^{-1}, which is at the lower edge of the efficient operating range (3.0 to 4.0 m sec^{-1}). Point B in Figure 8-13 shows that the dredge pump, operated in series with a second, identical booster pump would increase the discharge to about 1180 m^3 hr^{-1} at a head of 145 m. Under these conditions the velocity in the discharge pipeline would be increased to a more acceptable 4.6 m sec^{-1}.

Use of a 30-cm booster pump would be quite acceptable. However, if the 30-cm pump is unavailable, what does one do? Let us assume that the only pump available is a 35-cm, high head model. The head discharge curve for this booster pump, together with the 30-cm dredge pump curve, is shown in Figure 8-14. Figure 8-14 also shows the 30 cm-35 cm series operation pump curve and the system head curve for the 1900-m discharge pipeline. Point A on Figure 8-14 represents the capacity of the two pumps in series when delivering through the 30-cm-diameter, 1900-m-long discharge pipeline. The discharge at point A is about 1281 m³ hr⁻¹, at a head of 172 m. The discharge pipeline velocity at this discharge rate is about 4.9 m sec⁻¹.

Figure 8-14 can be used to determine what portion of the total head is contributed by each pump: Simply construct a vertical line from Point A downward. The total head for the main pump is 72 m (Point C), and that for the booster pump is 100 m (Point B). Added together, the two heads equal the total head of 172 m at operating point A. The continuous brake horsepower needed to operate the two pumps will be

Dredge pump:

$$BHP = \frac{(1281)(72.0)(1.20)}{(2.737)(55.0)}$$

$$BHP = 735 \times 1.10 = 809 \text{ hp}$$

Booster pump:

$$BHP = \frac{(1281)(100.0)(1.20)}{(2.737)(55.0)}$$

$$BHP = 1021 \times 1.10 = 1123 \text{ hp}.$$

The horsepower requirement for the dredge pump is slightly more than the 790 hp that was computed previously, but it is not enough to pose any problem for operating, if one considered the ten-percent factor that was used in the initial selection of the power plant. The booster pump power plant should be selected accordingly, in conjunction with a speed reduction gear, so that the booster pump runs at 800 rpm.

An energy diagram (Figure 8-15), depicting the heads developed throughout the length of the pipeline is useful in determining the maximum and minimum allowable distances between the dredge pump and the booster pump. Since there are several variables in the dredging process, a positive suction head (H_s) of 10.6 m at the booster pump will be assumed. Also, since there is a positive suction head (H_s) of about 3.3 m on the main dredge pump, that amount must be subtracted from the total dynamic head (H_{TDH}) on the dredge pump to obtain the dredge pump head (H_d): 72.0 - 3.3 = 68.7. From this and the discharge friction head loss per 30.5 m, for 30-cm discharge pipe at a velocity of 4.9 m sec⁻¹ (from Figure 8-11), we can compute the maximum pump spacing in equivalent length of pipe:

$$\frac{(68.7 - 10.6)\, 30.5}{2.05} = 864 \text{ m (equivalent length).}$$

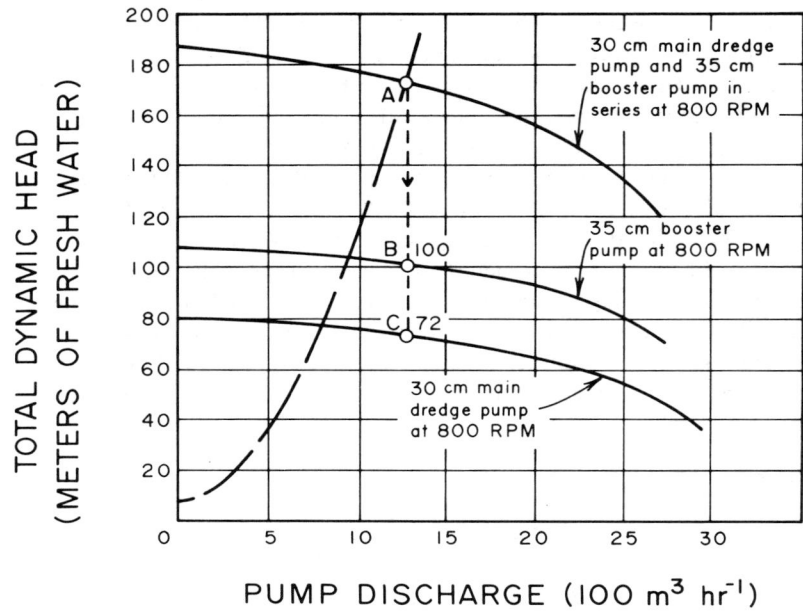

Figure 8–14 Head discharge relationships for a 30-cm dredge pump and a 35-cm booster pump (modified from Pierce, 1970).

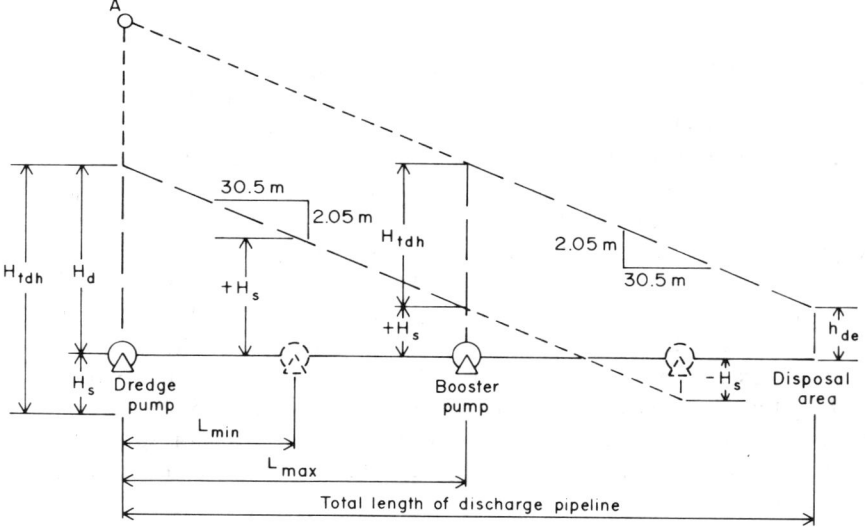

Figure 8–15 Energy diagram for a 30-cm dredge pump operating in series with a 35-cm booster pump (modified from Pierce, 1970).

The equivalent pipeline length can be converted to actual pipeline length by dividing by the floating pipeline equivalent length correction factor (1.5), which gives an actual pipeline length of 576 m. Since the floating pipeline length is about 760 m, this means the booster pump will have to be barge-mounted on the lake a maximum distance from the dredge of no more than 576 m. Plus and minus signs used with H_s in Figure 8-15 indicate the presence of a suction head or suction lift, respectively. Therefore, it can be seen that a booster pump located at a distance greater than L_{max} from the dredge pump will operate under a suction lift. This condition should be avoided.

The minimum spacing between the dredge pump and the booster pump can be computed if the discharge pipeline working pressure is known. We will assume a working pressure of 1.40×10^5 kg m^{-2}, which is equivalent to 140 m of waters. The two pumps placed immediately adjacent to one another would produce a discharge head of 168.7 m [(dredge pump head H_d - dredge pump suction head H_s) + booster pump total dynamic head H_{TDH}], indicated by point A in Figure 8-15. By proportion, the 168.7-m head would result in a discharge pipeline pressure of 1.68×10^5 kg m^{-2}, which exceeds the required working pressure of 1.40×10^5 kg m^{-2}. Therefore, it will be necessary to locate the booster pump some minimum distance from the dredge, so that the pipeline friction will reduce the pipeline pressure to a value below the working pressure. This distance should be calculated so that the discharge pressure of the dredge pump plus the booster pump, minus the pipeline friction head loss between the two pumps, is less than the pipeline working pressure. The slope of the energy gradient in Figure 8-15 is nearly constant throughout the length of the discharge pipe. The friction head loss created by the minimum distance between pumps can be computed from formula 8.14:

$$H_1 = H_{TDH} + H_d - W_p, \qquad (8.14)$$

where H_1 = head loss in the discharge pipe between the dredge pump and the booster pump (meters of water)
H_{TDH} = booster pump total dynamic head (m)
H_d = dredge pump total head minus the positive suction head (m)
W_p = discharge pipe working pressure (meters of water).

Therefore

$$H_1 = 100 + 68.7 - 140$$

$$H_1 = 28.7 \text{ meters of water.}$$

Using the above information, the length of pipeline necessary to create this friction head loss can be computed. Recall from Figure 8-11 that 30-cm discharge pipeline head loss is 2.05 m per 30.5 m of pipeline length at a discharge velocity of 4.9 m sec^{-1}. Therefore

$$\frac{(30.5)\ 28.7}{2.05} = 427 \text{ m equivalent pipeline length.}$$

The equivalent pipeline length can be converted to actual pipeline length by dividing by the equivalent pipeline length correction factor (1.5):

$$\frac{427}{1.5} \cong 284 \text{ m}.$$

Pierce (1970) used this type of example to develop two formulas for calculating the maximum and minimum distances required between the dredge pump and the booster pump to operate within the prescribed discharge pipeline pressure limits. The formulas should be quite useful in determining these distances when working with the smaller dredges commonly employed in lake restoration. Pierce pointed out that the approach described above could be used to determine spacing between pumps if more than one booster pump was required. The two formulas for calculating maximum and minimum spacing requirements, as modified from Pierce (1970), are

$$L_{max} = \frac{30.5 \, (H_d - H_s)}{1.5 \, h_{df}}, \tag{8.15}$$

$$L_{min} = \frac{30.5 \, (H_d + H_{TDH} - W_p)}{1.5 \, h_{df}}, \tag{8.16}$$

where L_{max} = maximum distance between the main dredge pump and the booster pump (meters of discharge pipeline)
L_{min} = minimum distance between the main dredge pump and the booster pump (meters of discharge pipeline)
H_d = dredge pump discharge head (meters of fresh water, which equals the discharge pressure gauge reading at the dredge pump)
H_s = booster pump suction head (m)
h_{df} = friction loss in the discharge pipeline (meters per 30.5 meters of discharge pipeline; see Figure 8-11)
H_{TDH} = booster pump total dynamic head (meters of water; equals the discharge pressure gauge reading at the dredge pump)
W_p = discharge pipe working pressure (meters of water).

From this analysis, for pumping from lake area 3 to disposal site 3, the following can be concluded:

1. The disposal area is beyond the efficient pumping capacity of the dredge pump, so a booster pump will be required.
2. Discharge pipeline velocities could be increased to an acceptable level by employing either an identical 30-cm booster pump or the available 35-cm model.
3. Head discharge characteristic graphs are highly desirable, to assist in booster pump selection, since the required head discharge characteristics of the booster are highly dependent on those of the dredge pump.

4. The booster pump should not be located too close to or too far away from the dredge pump. Excessive pipeline pressures will result if they are too close. If the booster is too far, it may operate under a suction lift, perhaps resulting in pump cavitation, reduced output, and excessive equipment wear.

This information on hydraulic dredge selection should be helpful to the dredge-plan designer, for selecting the proper equipment, and to the lake manager, to assure that the designer has selected the proper equipment. How dredging is actually conducted on a project will depend on the type of equipment selected and on site-specific constraints, all of which must be considered when developing the dredge operating plan. The example above can be used as a general guide, though, as stated by Pierce (1970), there is no substitute for actual dredge discharge relationship curves for dredge pumps actually being considered for selection. It is from these relationships or from pump manufacturers' specifications that the pump suction heads can be obtained.

Disposal Area Design

Once dredge equipment is selected, the critical concern of the dredge plan designer is the disposal area. Upland disposal has become a common practice. The challenge is to design and construct a containment and disposal area of adequate size and retention time to hold the dredged material volume and to reduce suspended solids concentrations to meet effluent requirements. Procedures were developed by Montgomery (1978; 1979) for designing disposal areas for retention of suspended solids based on settling characteristics of the sediments. These procedures are presented in the following paragraphs.

Field investigation of the dredge site must be conducted, to obtain disposal area design information. The most basic piece of field information is an estimate of the in-place sediment volume, as described previously. Two types of laboratory tests also need to be conducted to develop information for the disposal area design. The first of these is sediment characterization tests, including natural water content, Atterberg limits, organic content, and specific gravity for fine-grain sediments. Grain size analyses are adequate for coarse-grained sediment. The second type of test is to determine sedimentation rate. Montgomery (1978) demonstrated that most freshwater dredge slurries could be characterized by flocculent settling tests, where particles agglomerate during settling, with different physical properties and settling rates. Hydraulically dredged lake sediments could therefore be characterized by this test procedure.

Montgomery (1978) prefaces the flocculent settling test procedures with the statement that, "if an interface forms near the top of the settling column during the first day of the test, sedimentation is governed by zone settling, and that test procedure should be initiated." He indicates that zone settling might prevail at high solids concentrations or if sediments are contaminated with high levels of organics.

182   *Physical and Chemical Methods*

Therefore, the zone testing procedure, in which the flocculent suspension forms a lattice structure and settles as a mass, is also described.

Flocculent Settling Procedure

   1. A settling column such as that shown in Figure 8-16 is used. The test column depth should approximate the effective settling depth of the proposed containment area. A practical test depth is 2 m. The column should be at least 20 cm in diameter, with sample ports at 0.3-m intervals. The column should have provisions to bubble air from the bottom, to keep the slurry mixed during the column-filling period.
   2. Mix the sediment slurry to the desired suspended solids concentration in a container with sufficient volume to fill the test column.
   3. Pump or pour the slurry into the test column, using air to maintain a uniform concentration during the filling period.
   4. While the column is completely mixed, draw off samples at each sample port and determine the suspended solids concentration. Average these values and use the results as the initial concentration. After the initial samples are taken, stop the air bubbling and begin the test.
   5. Allow the slurry to settle, then withdraw samples from each sampling port at regular time intervals, and determine the suspended solids concentrations. Sampling intervals depend on the settling rate of the solids—usually at 30-minute intervals for the first three hours and then at four-hour intervals until the end of the test. Continue the test until the interface of solids can be seen near the bottom of the column and the suspended solids level in the fluid above the interface is < 1 g l^{-1}.
   6. If an interface has not formed within the first day on any previous tests, run one additional test with a suspended solids concentration sufficiently high to induce zone settling behavior. This test should be carried out according to the procedures outlined below. The exact concentration at which zone settling behavior occurs depends upon the sediments being used to estimate the volume required for dredged material storage.

Zone Settling Test Procedure

This test procedure consists of placing a slurry in a column similar to that in Figure 8-16 and recording the fall of the liquid–solid interface over time. The depth to the interface is then plotted as a function of time. From this plot, the slope of the constant settling zone of the curve represents the zone settling velocity, which is a function of the initial test slurry concentration. Information needed to design a containment area when zone settling characteristics prevail can be obtained by using the following procedure, from Montgomery (1978):

   1. Use a settling column such as that shown in Figure 8-16. It is important that the column diameter be sufficient to reduce wall effects, and that the test be performed at a slurry depth near that expected in the field. A one-liter graduated cylinder should never be used to perform a zone settling test for sediment slurries representing dredging and disposal activities.

Sediment Removal 183

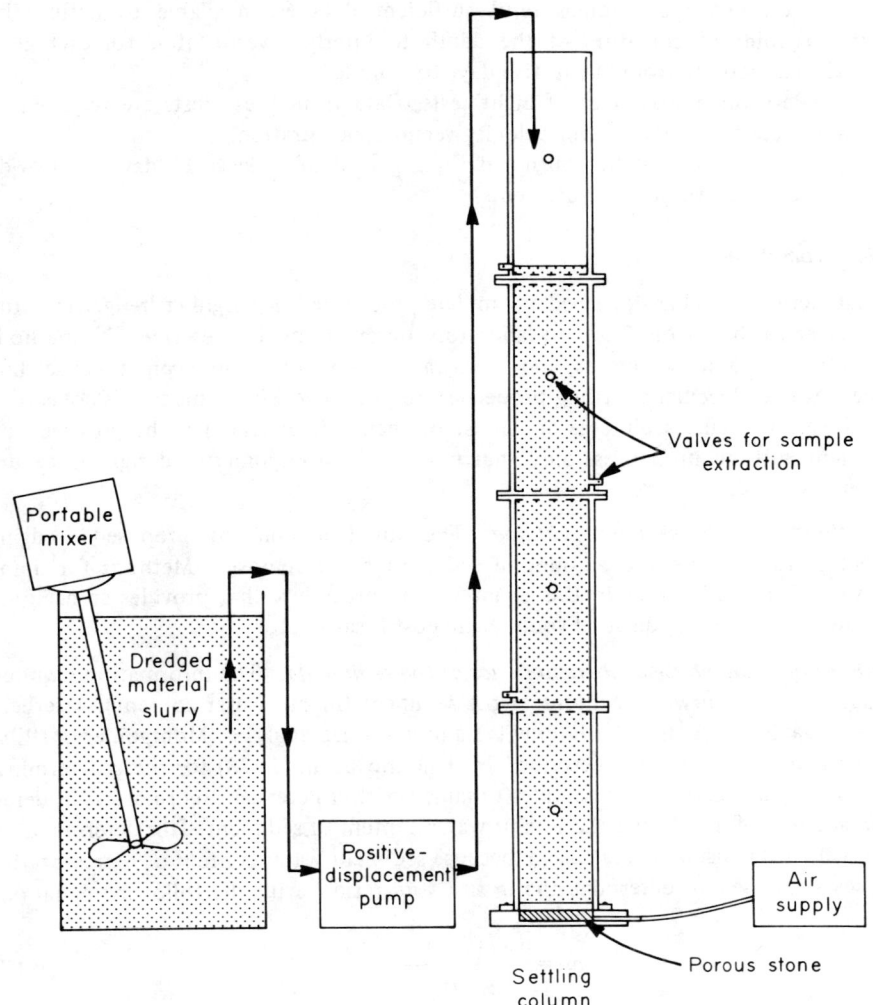

Figure 8-16 Schematic of fine-grained sediment settling test equipment (modified from Montgomery, 1978).

2. Mix the slurry to the desired concentration and pump or pour it into the test column. Test concentrations should range from about 60 g l to 200 g l^{-1}. Air may not be necessary to keep the slurry mixed if the filling time is less than one minute.

3. Record the depth to the solid-liquid interface with respect to time. Observations must be made at regular intervals to gain data for plotting the curve of depth to interface versus time. It is important to make enough observations to clearly define this curve for each test.

4. Continue the readings until sufficient data are available to define the maximum point of curvature of the depth to interface versus time for each test. The tests may require from one to five days to complete.

5. Perform a minimum of eight tests. Data from these tests are required to develop the curve of zone settling velocity versus concentration.

6. The test should be continued for a period of at least 15 days to provide data for estimating volume requirements.

Design Procedures

Montgomery (1978) has described a complete procedure for designing dredged material disposal areas, based on field and laboratory observations. He describes the methods for both saltwater sediments and freshwater sediments, based on zone settling properties and flocculent settling properties, respectively. Since this text is concerned with freshwater lake sediments and most of them are expected to be governed by flocculent settling properties, only that portion of Montgomery's design procedure will be described.

1. Estimate in situ sediment volume. The initial step in any proposed dredging project is to estimate the amount of sediment to be removed. Methods for doing this were described earlier in this chapter. Any procedure that provides an accurate estimate for computing disposal volume and cost is satisfactory.

2. Determine the physical characteristics of the sediments. The information required includes the natural water content of the sediment (in situ water content), Atterberg limits, organic content, specific gravity, and grain-size analyses. Montgomery (1978) recommended the methods described by Palermo et al. (1978) for these determinations. Fine-grained sediment samples obtained with a Petersen dredge are considered to be adequate for defining the in situ water content of sediment. This determination is critical to the disposal area design, because sediment water content of representative samples w is used to determine the in situ void ratio e with the following equation:

$$e_i = \frac{(w/100) G_s}{S_d/100}, \qquad (8.17)$$

where w = water content of sediment (%)
G_s = specific gravity of sediment solids
S_d = the degree of saturation (equal to 100% for sediment).

3. Analyze the proposed dredging and disposal data and laboratory sedimentation tests. Once tentative dredge selection has been made, this information can be used in the disposal area design. For example, the designer must estimate the containment area influent rate, influent suspended solids concentration, effluent rate (for weir sizing), effluent suspended solids concentration allowed, and time required to complete the disposal activity. If information, as previously described in this chapter, is unavailable, one must assume the largest size dredge that could be used on the project to assure the disposal area is not undersized. Montgomery (1980) has recommended the use of a suspended solids concentration of 145 g l^{-1} (13% by weight) to be used for design purposes if the actual concentrations are unknown. That figure, however, is derived from experience in maintenance dredging of waterways (freshwater), not from dredging lake-type sediments, so it should be used with caution when dealing with flocculent lake sediments. Sedimentation tests for fresh water should be performed at this point, according to the method described previously.

Design Method for Freshwater Sediments

According to R.L. Montgomery (personal communication), the following design method should be sufficient to remove suspended solids from freshwater disposal area discharges, down to a level of < 1 g l^{-1}. If that level is insufficient to meet discharge requirements it will be necessary to treat the discharge.

Because larger-particle flocs settle at faster rates, overtaking finer flocs in their descent, the increased contact increases the floc size and enhances settling rates. Therefore, the greater the ponding depth in the disposal area, the greater the probability for contact among sediments and flocs and the more efficient the settling.

Montgomery (1978) tested laboratory settling rates on freshwater sediments and arranged the data as shown in Table 8-2. From this point he prescribed the following procedure:

1. Arrange data from the laboratory tests illustrated in Table 8-2 into the form in Table 8-3.
2. Plot these data as shown in Figure 8-17. The percent of initial concentration, by weight, for each depth and time is given in Table 8-3. The solid curved lines represent the concentration depth profile at various times during settling (refer to Figure 8-17). Numbers appearing along the horizontal depth lines indicate area boundaries.
3. Compute a design concentration using data from the 15-day zone settling test.

The design concentration, C_D, is defined as the average concentration of the dredged material in the containment area at the end of the disposal activity. The following steps can be used to compute C_D:

1. Compute concentration versus time for the 15-day settling test. Assume zero solids in the water above the solids interface, to simplify calculation.
2. Plot concentrations versus time on log-log paper (Figure 8-17).
3. Draw a straight line through the data points. This line should be drawn through the points representing the consolidation zone.
4. Estimate the time of dredging by dividing the dredge production rate into volume of sediment to be dredged.
5. Estimate the concentration at time $t_{1/2}$ (half the time required for the disposal activity, determined in Step 4), using the figure developed in Steps 2 and 3. This time is an approximation of the average time of residence for the dredged material in the containment area. Since concentration is a function of time, one-half the dredging time would represent a period during which one-half of

Table 8-2 Observed Flocculent Settling Concentrations with Depth, g/l (modified from Montgomery, 1978); data from actual test on freshwater sediments.

	Depth From Top of Settling Column (m)					
Time (min)	0.3	0.6	0.9	1.2	1.5	1.8
0	132.00	132.00	132.00	132.00	132.00	132.00
30	46.00	99.00	115.00	125.00	128.00	135.00
60	25.00	49.00	72.00	96.00	115.00	128.00
120	14.00	20.00	22.00	55.00	78.00	122.00
180	11.00	14.00	16.00	29.00	75.00	119.00
240	6.80	10.20	12.00	18.00	64.00	117.00
360	3.60	5.80	7.50	10.00	37.00	115.00
600	2.80	2.90	3.90	4.40	14.00	114.00
720	1.01	1.60	1.90	3.10	4.50	110.00
1020	0.90	1.40	1.70	2.40	3.20	106.00
1260	0.83	1.14	1.20	1.40	1.70	105.00
1500	0.74	0.96	0.99	1.10	1.20	92.00
1740	0.63	0.73	0.81	0.85	0.94	90.00

Table 8-3 Percent Initial Concentration* with Time (modified from Montgomery, 1978).

	Depth From Top of Settling Column (m)		
Time (min)	0.3	0.6	0.9
0	100.0	100.0	100.0
30	35.0	75.0	87.0
60	19.0	37.0	55.0
120	11.0	15.0	17.0
180	8.0	11.0	12.0
240	5.0	8.0	9.0
360	3.0	4.0	6.0
600	2.0	2.2	3.0
720	1.0	1.2	1.4

*Initial suspended solids concentration = 132 g/l.

the dredged material would have been in the area longer than and the other half less than a time equal to one-half the dredging time.

This value is the design solids concentration, C_D.

4. **Compute detection time required for sedimentation.**

1. Calculate removal percentage at depths of 0.3, 0.6, and 0.9 m for various times, using the plot illustrated in Figure 8-17. The removal percentage for depth

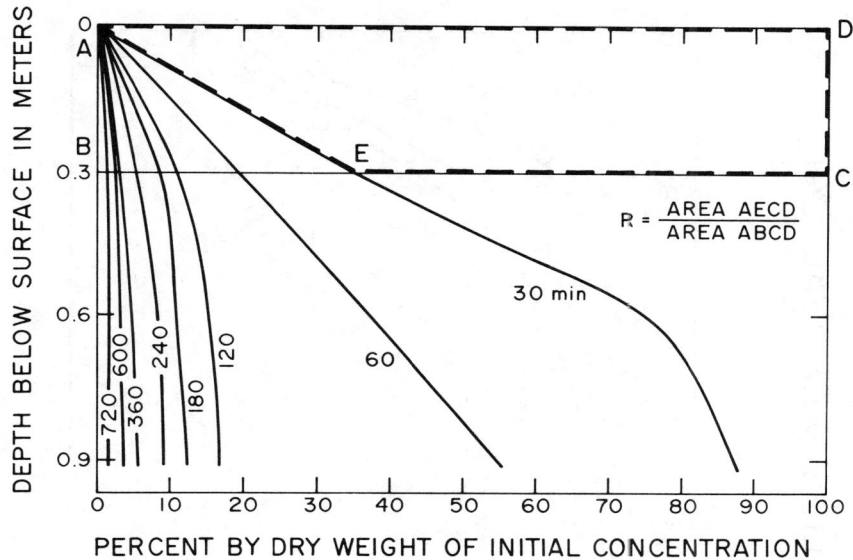

Figure 8-17 Profile of percent initial concentration versus depth (modified from Montgomery et al., 1983).

d_1 (0.3 m) and t = 30 min is computed from formula 8.18:

$$R = \frac{\text{Area A, E, C, D}}{\text{Area A, B, C, D}} \times 100, \qquad (8.18)$$

where R = the removal percentage, and the area numbers correspond to the numbers used in Figure 8-17 to indicate the area boundaries for the total area down to depth, 0.3 m (A, B, C, D), and the area to the right of the t = 30-min line (A, E, C, D).

2. Plot the solids removal percentages versus time, as shown in Figure 8-18.
3. Theoretical detention times can be selected from Figure 8-18 for various solids removal percentages. Select the detection time T that gives the desired removal percentage for the design ponding depth.
4. Montgomery recommends that a correction factor of 2.25 be applied to the theoretical detention time T, because of short-circuiting and dispersion of flows through the disposal basin:

$$T_d = 2.25\ T, \qquad (8.19)$$

where T_d = design detention time
T = theoretical detention time.

188 Physical and Chemical Methods

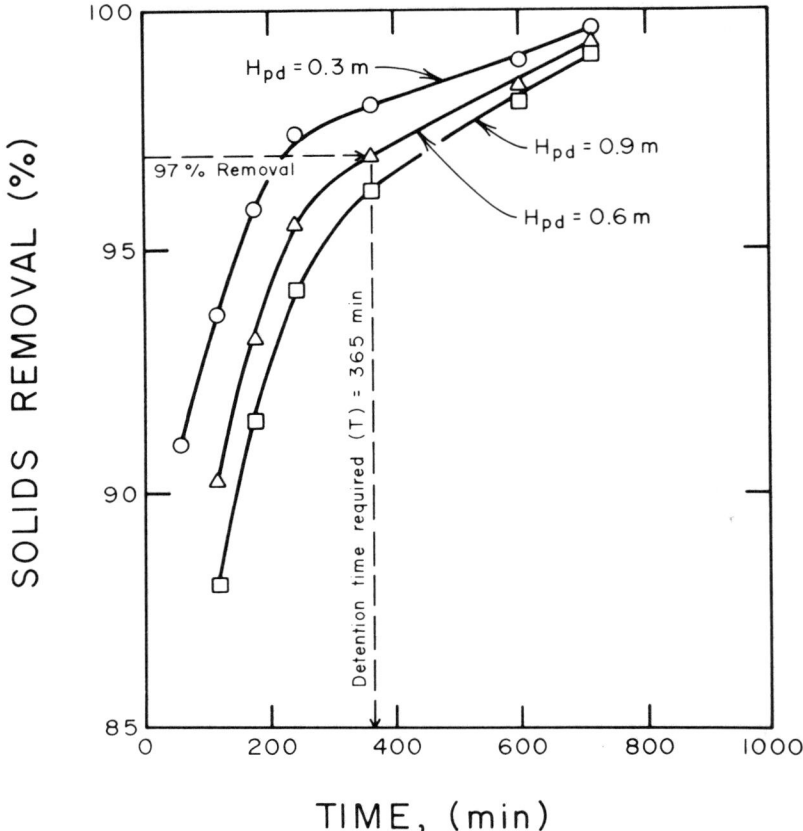

Figure 8-18 Solids removal versus time, as a function of depth (modified from Montgomery, 1978).

These procedures are aimed at providing sedimentation basins with sufficient areas and detention times to accommodate continuous hydraulic dredge disposal activities while also meeting requirements for effluent suspended solids. The basins must also be designed to meet the volume requirements of the job. The sedimentation basin total volume requirement includes volume for storage of dredged material, volume for sedimentation (ponding depths), and free-board volume (volume above water surface). Coarse-grained material storage volume (> 200 sieve) must be determined separately, since this material behaves differently from the fine-grained (< 200 sieve) material.

5. *Estimate the volume occupied by dredged material in the sedimentation basin.*

1. Compute the average void ratio of fine-grained dredged material in the sedimentation basin at the completion of the dredging operation, using the design

concentration determined in earlier steps as dry density of solids. Use the following equations to determine void ratio:

$$e_0 = \frac{G_s \gamma_w}{\gamma_d} - 1, \qquad (8.20)$$

where e_0 = average void ratio of dredged material in the sedimentation basin at the completion of the dredging operation
G_s = specific gravity of sediment solids
γ_w = density of water (g l^{-1})
γ_d = dry density of solids at design concentration ($C_D = \gamma_d$), (g l^{-1}) (from Figure 8-19).

2. Compute the changes in volume of fine-grained sediments after disposal in the sedimentation basin:

$$\Delta V = V_i \frac{e_0 - e_i}{1 + e_i}, \qquad (8.21)$$

where ΔV = change in volume of fine-grained sediments after disposal in the sedimentation basin (m^3)
e_i = average
V_i = volume of fine-grained sediments (m^3).

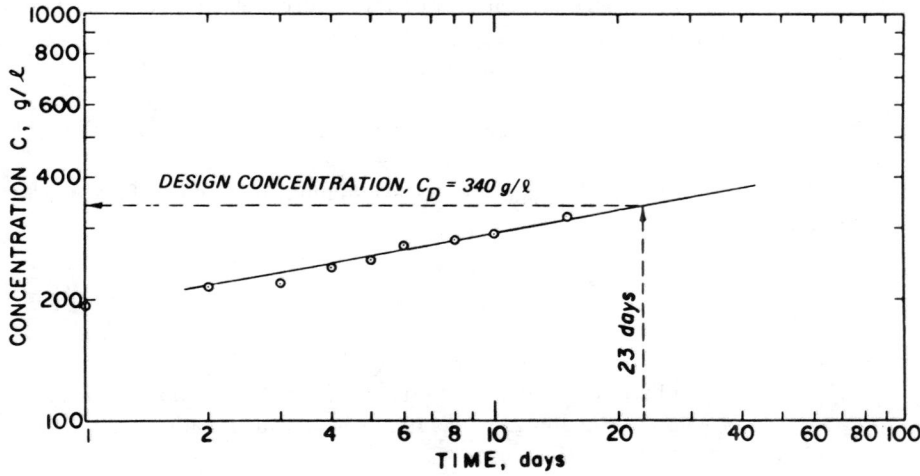

Figure 8-19 Concentration determined from column test versus time (from Montgomery et al., 1983).

3. Compute the volume required in the sedimentation basin for dredged material:

$$V = V_i + \Delta V + V_{sd}, \tag{8.22}$$

where V = volume of dredged material in the sedimentation basin at the end of the dredging operation (m³)
V_{sd} = volume of sand (compute using 1:1 ratio), (m³).

6. Estimate basin depth. The above procedures have provided a design detention time, T_d, required for sedimentation of fine-grained dredged material. Equations 8.20-8.22 are used to estimate the volume and corresponding depth requirements for the storage of solids in the containment area. The existing topography and surficial geology of the proposed containment must be kept in mind throughout the design process, since they can have a significant effect on the average depth of the containment area. The following procedures were recommended by Montgomery (1978) to estimate the thickness of dredged material at the end of disposal, when freshwater sediments are being disposed.

1. Compute the volume required for sedimentation:

$$V_B = Q_i T_d, \tag{8.23}$$

where V_B = sedimentation basin volume required to meet requirements for suspended solids in effluent (m³)
Q_i = influent rate (m³ hr⁻¹), $Q_i = A_p V_d$; assume V_d = 4.6 m sec⁻¹ in the absence of data, and convert Q_i calculated in m³ sec⁻¹ to m⁻³ hr¹; A_p is the cross-sectional area of the dredge discharge pipe (m²), and V_d is the velocity of dredge discharge (m sec⁻¹)
T_d = required design detention time, from equation 8.19.

2. Consult with soils design engineers to determine maximum height allowable for confining dikes, D.

3. Compute the required design area as a minimum required surface area for solids storage:

$$A_d = \frac{V}{H_{dm(max)}}, \tag{8.24}$$

where A_d = the design basin surface area (m²)
V = volume of dredged material in the sedimentation basin at the end of the dredging operation (m³), from equation 8.22
$H_{dm(max)}$ = $D - H_{pd} - H_{fb}$ (m)

where D = dike height (m),
H_{pd} = average ponding depth over the area (m)
H_{fb} = freeboard above basin water surface to prevent wave overtopping and subsequent damage to the confining earth dikes (m).

$H_{dm\,(\max)}$ is the thickness of dredged material in the basin corresponding to a known volume of material. Thickness decreases as surface area increases. Minimums of 0.6 m are recommended for H_{pd} and H_{fb}, to account for fetch and wind.

4. Evaluate volume available for sedimentation near the end of the disposal operation:

$$V^* = H_{pd} A_d, \qquad (8.25)$$

where V^* = volume available for sedimentation near the end of the disposal operation (m^3)
H_{pd} = average ponding depth over the area (m)
A_d = design basin surface area (m^2).

5. Compute V^* and V_B. If the volume required for sedimentation is larger than V^*, the sedimentation basin will not meet the suspended solids effluent requirements throughout the disposal operation. If that is the case, one of the following measures can be taken to ensure that effluent suspended solids concentrations are met:

a. Increase the design area A_d.
b. Operate the dredge intermittently when V^* becomes less than V_B, or use a smaller dredge.
c. Provide for post-treatment of effluent to remove solids (see Barnard and Hand, 1978).

Sediment that exhibits zone settling characteristics requires a different disposal area design method. If this should be encountered, the reader is referred to Montgomery (1978; 1979) and Montgomery et al. (1983). The above disposal area design guidance is the best available at this time, though some have expressed the opinion that it may be inadequate to address the extremely flocculent organic sediments found in lakes. Dunst et al. (1984) reported that the specific gravity of sediment from Lilly Lake, Wisconsin, was only 1.02. One should be extremely cautious about relying completely on the above disposal area design procedures.

CASE STUDIES

Dunst et al. (1974) identified 33 lake-dredging projects completed or in progress in 1974. Seven years later Peterson (1981) listed 64 such projects. Both the 1974 and 1981 listings probably were incomplete, since many small lake-dredging projects are completed but never reported in the open literature. If records are available from these projects there is little data collected beyond the completion date, so few well-documented long-term evaluations are available. The following case studies represent relatively well-documented projects. They are presented for that reason and because each has achieved its objective, through somewhat different approaches. Peterson (1981) has described additional case studies.

It should not be construed from the following presentations that all dredging projects to control internal nutrient recycling are successful. On the contrary, there have been failures, but almost all failures or marginal successes can be traced to inaccurate predredging assessment of the problem and to inadequate amounts of sediment removed (Brashier et al., 1973; Churchill et al., 1975; George et al., 1981; Ryding, 1982). The need for accurate assessment of the problem cannot be overemphasized.

Lake Trummen, Sweden

The most thoroughly documented long-term evaluation of lake dredging was of the project at Lake Trummen, Sweden. The lake, located near the city of Växjö, began receiving domestic waste discharge from the kitchen of St. Sigfrid's Hospital in 1895 (Sjön Trummen i Växjö, 1977). Modern toilets, served by septic tanks near the lakeshore were installed in 1936. In 1943, a flax mill began discharging to the 400-ha lake. From that time winter fish kills became common, and the quality of the lake declined rapidly. Sewage discharge to the lake stopped in 1959, after the flax-dressing plant closed and wastewater from the surrounding area was connected to a municipal treatment system. Despite the wastewater diversion, no apparent recovery of the lake followed. In fact, its condition became so bad in the early 1960s that citizens of Växjö considered filling the basin (Bjork, 1974). Limnological investigations conducted by the University of Lund from 1966 to 1969 revealed that the upper one meter of sediment was extremely rich in nutrients relative to that underlying it.

One-half meter of sediment was dredged from the main lake basin in 1970, and another one-half meter was removed in 1971 (Bjork, 1974). This removal increased the mean depth of the lake from 1.1 m to 1.75 m and the maximum depth from 2.1 m to 2.5 m. The total volume of the sediment removed was about 30×10^5 m^3. By any standard, the lake was still extremely shallow and could have remained eutrophic, except that the sediment skimming treatment reduced the phosphate phosphorus content in the surface layer of sediment from approximately 2.4 mg l to 0.1 mg l^{-1} (Sjön Trummen i Växjö, 1977).

Part of the dredged material was disposed of in shallow, diked-off bays. The remainder went to upland diked ponds, where return flow was treated with aluminum sulfate to reduce the total phosphorus concentration from about 1 mg l^{-1} to 30 µg l^{-1}. Dried dredge material was sold as topsoil dressing for about $2 (U.S.) m^{-3}.

Bengtsson et al. (1975) indicated that the role of the sediment in recycling nutrients was reduced substantially following dredging. The data in Figure 8-20 demonstrate an approximate 90-percent reduction of total phosphorus in the surface water of Lake Trummen after dredging (from 600 µg l^{-1} to a range of 70-100 µg l^{-1}). Phosphorus reductions are especially noticeable in summer. Total nitrogen concentrations were also reduced approximately 80 percent, from 6.3 mg l^{-1} to 1.3 mg l^{-1} (G. Andersson, personal communication). The temporary increase of phosphorus in 1975 (Figure 8-20) was associated with a large influx of planktivorous cyprinid fish (Andersson et al., 1978). This type of fish population is generally considered to be undesirable and associated with reduced water quality. During 1976 about two metric

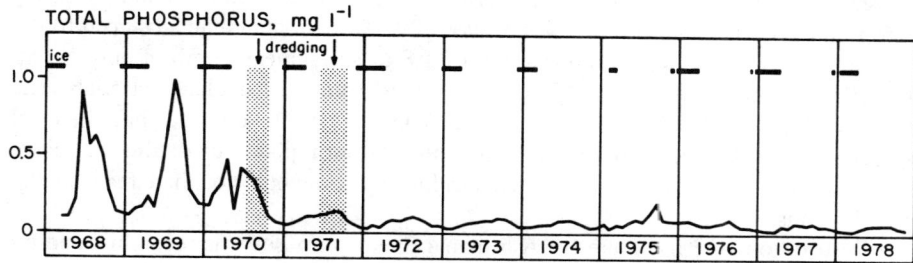

Figure 8-20 Total phosphorus concentration in Lake Trummen, Sweden, before and after dredging (courtesy of Gunnar Andersson, Department of Limnology, University of Lund, Lund, Sweden).

tons (30 kg ha^{-1}) of fish were taken from the lake. The practice continued through 1979 and phosphorus concentrations remained low. When fishing stopped in 1979 the P levels began to rise again (Andersson, personal communication). This suggests that Shapiro et al. (1975) may have been correct when they recommended that symptoms of eutrophication in lakes might be controlled by fish populations.

The significant nutrient reduction in the lake resulted in equally significant biological changes. The Shannon diversity index for phytoplankton rose from 1.6 in 1968 to 3.0 in 1973 (Cronberg et al., 1975). Secchi disc transparency went from 23 cm to 75 cm during the same period. The blue-green algal biomass was reduced dramatically, and the nuisance species *Oscillatoria agardhii* disappeared completely. Phytoplankton productivity went from a level of 370 g C m^{-2} in 1968-69 to about 225 g C m^{-2} in 1972-73.

The effect of dredging on the benthic community of Lake Trummen was negligible. A year after dredging tubificid oligochaetes and chironomids were more numerous than before dredging, but the total number of benthic organisms changed little (Andersson et al., 1975). Rapid recolonization was attributed to the mobility and constant swarming of chironomids.

Surface sediment skimming at Lake Trummen was successful in rolling back the effects of eutrophication. The results remained relatively constant for nine years, which attests to thorough pretreatment evaluation and a rational plan for restoration. These results are highly encouraging, because if surface sediment dredging is successful in a shallow lake like Trummen, it also should be successful in deeper ones with similarly enriched surface sediment.

Lilly Lake, Wisconsin

Lilly Lake is a 37-ha closed-basin lake in southeastern Wisconsin. Its agricultural watershed is only 155 ha. The lake suffered for several years from infilling by floating leafed pond weeds (*Potamogeton amplifolius*). By 1977 shoaling had advanced to the point that the lake had a maximum depth of 1.8 m and a mean depth of only 1.4 m. The basin contained more than 10 m of partially decomposed plant materials. The Wisconsin Department of Natural Resources (WDNR) (1969) reported that chemical eradication and restocking of centrarchid and northern pike failed, due to severe winter kill problems. The rate of basin infilling by sedimentation had reached 0.5 cm yr^{-1} (Dunst, 1981).

Restoration of Lilly Lake for fish management purposes was recommended by the Wisconsin DNR (1969). The plan called for deepening at least 10 percent of the basin to a depth of 6 m. This required removing 665×10^3 m^3 of sediment. Use of a hydraulic suction dredge was proposed, with the idea that sediment would flow to the dredge head as a cone of depression was created. The lightweight sediment (specific gravity = 1.02), however, had much greater cohesiveness than the predredging assessment indicated, and a cutterhead had to be employed.

Dredging began in July 1978 and continued until near freeze-up at the end of October. It began again in May 1979 and was completed at the end of August. Most of the dredged material was pumped approximately 3 km to an abandoned gravel pit. Although the groundwater table rose temporarily due to this disposal, it resumed normal levels shortly after disposal ceased in both 1978 and 1979, and monitoring wells revealed no adverse impact due to chemicals (Dunst et al., 1984). Figures 8-21, 8-22, 8-23, and 8-24 show in-lake changes for selected variables before, during, and after dredging. Prior to dredging in 1977 the total inorganic nitrogen (TIN) concentrations were near detection limits. Although total phosphorus (TP) increased to more than 40 μg l^{-1} during July, August, and September before dredging, the increase was not due to excessive phytoplankton concentrations, as evidenced in Figure 8-23. The increases may have been due to decomposition of sloughed macrophyte parts and filamentous algae on the bottom. Much of the increase during dredging was due to resuspended sediment. Despite previously reported winter kills of fish, the dissolved oxygen (DO) concentrations remained above 6 mg l^{-1} throughout the study, except for February and March of 1979 (Figure 8-24). These DO declines were preceded by large increases in TIN and TP late in the fall of 1978. The DO depletion may have resulted from oxygen demand from organics suspended as a result of dredging into the late fall. There was no similar oxygen decline the following year, when dredging stopped in August. The water quality per se of Lilly Lake was not exceptionally poor prior to dredging, but there was so little water in the lake (1.4 m), due to infilling by macrophytes and partially decomposed macrophytes, that recreational use was severely impaired.

Adverse, although not severe short-term changes occurred during the active dredging period. TIN increased rapidly (Figure 8-21) when dredging began in July 1978, due mostly to liberation of ammonium-nitrogen from the sediment (Dunst, 1981). It remained in excess of 3.5 mg l^{-1} during most of the 1978-1979 ice-cover

Sediment Removal 195

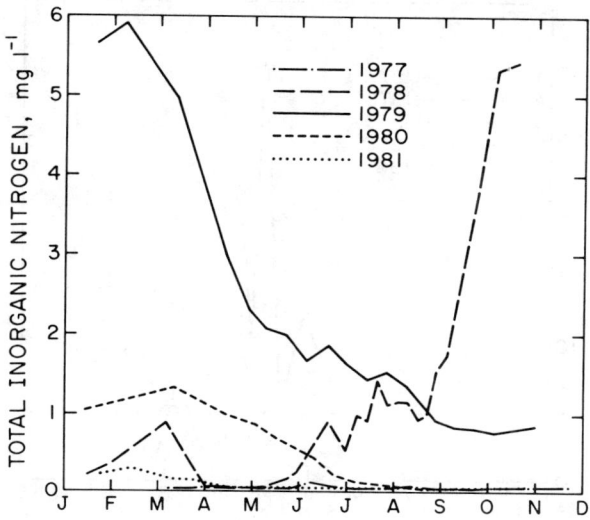

Figure 8-21 Total inorganic nitrogen in Lilly Lake, Wisconsin (data courtesy of Russell Dunst, Wisconsin Department of Natural Resources, Madison, Wisconsin).

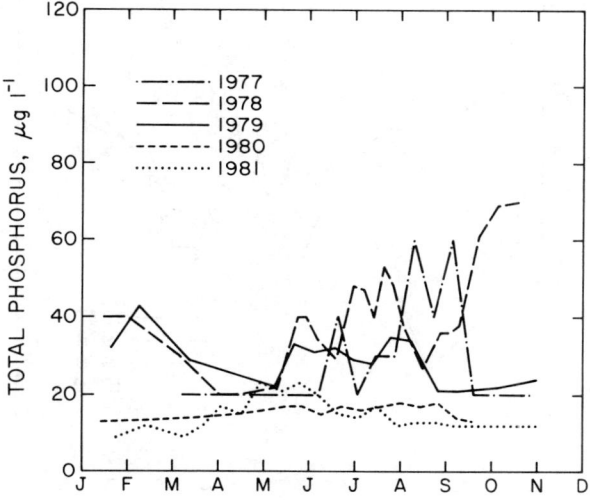

Figure 8-22 Total phosphorus in Lilly Lake, Wisconsin (data courtesy of Russell Dunst, Wisconsin Department of Natural Resources, Madison, Wisconsin).

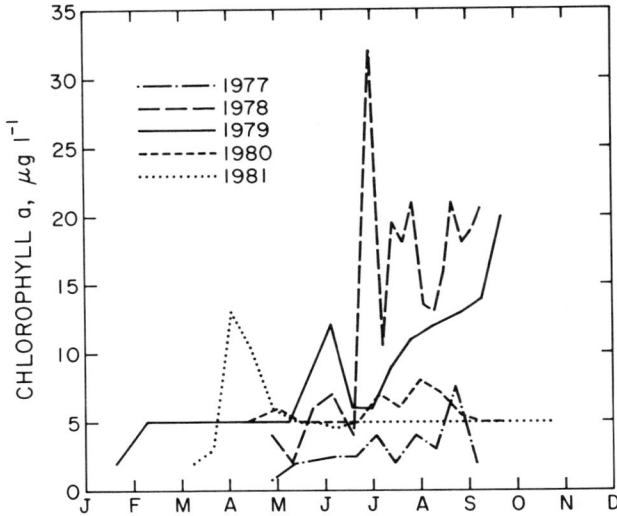

Figure 8-23 Chlorophyll *a* in Lilly Lake, Wisconsin (data courtesy of Russell Dunst, Wisconsin Department of Natural Resources, Madison, Wisconsin).

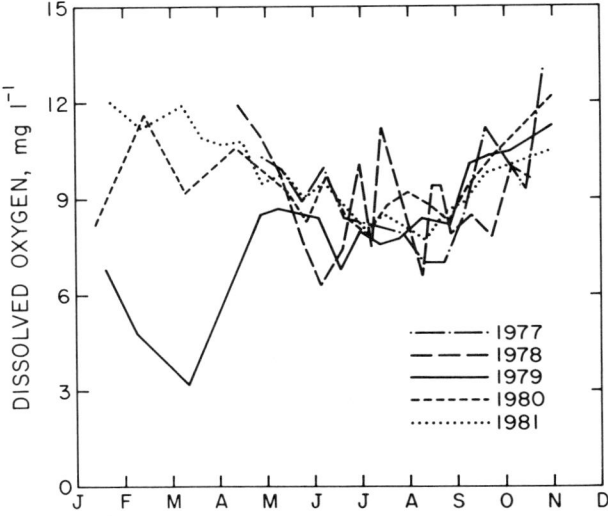

Figure 8-24 Dissolved oxygen in Lilly Lake, Wisconsin (data courtesy of Russell Dunst, Wisconsin Department of Natural Resources, Madison, Wisconsin).

period, then declined steadily throughout the spring and summer, even though dredging ended in November 1978 and began again in May 1979.

The TP concentrations behaved in a fashion similar to TIN, except there were minor peaks (lower than predredging, 1977) in TP during the summer of 1978. The TP remained relatively high during the fall and winter of 1978-79, probably due to increased levels of suspended solids in the water column from the dredging. This is confirmed in part by elevated turbidity and BOD levels during the same period (Dunst et al., 1984).

Phytoplankton responded in a predictable manner to the increased nutrient concentrations (Figure 8-23). The July-September average gross primary productivity increased from 185 mg C m^{-3} day^{-1} in 1976 and 140 mg C m^{-3} day^{-1} in 1977 to over 1000 mg C m^{-3} day^{-1} in 1978 (Dunst, 1981).

With the completion of dredging in September 1979, all of the above variables returned to near or below predredging levels. The 1980 and 1981 TP concentrations were significantly lower than concentrations prior to or during dredging. Chlorophyll a concentrations were slightly higher in 1980 and 1981 than in the predredging year of 1977, but the increase was relatively insignificant (approximately 3-4 μg l^{-1}). More importantly, the water storage capacity (basin volume) of the lake was increased 128 percent (Dunst, 1981), and the overall objective of increasing the recreational potential of Lilly Lake appeared to have been realized. The two-summer inconvenience of dredging resulted in what lake users hoped would be a long-term benefit.

Dredging has not resulted in complete success in creating a weed-free lake, however. High water transparency (2.2 m) in 1981 permitted plant growth to a depth of approximately 3.7 m. This is in total agreement with the maximum depth of growth (3.5 m) predicted by equation 8.1 (Dunst et al., 1984). Plant growth over 75 percent of the lake basin was dominated by *Chara* (a macrophytic alga) and *Myriophyllum* spp. The plant growth was particularly offensive because it grew to the surface in many areas. In 1982 plant growth covered the same area but was less offensive, since it remained 1.2 and 1.8 m below the water surface except in the nearshore areas. While the 1980 and 1981 macrophyte biomass was considerably less (about 100 g m^{-2}) than during the predredging period (685 and 335 g m^{-2} in 1976 and 1977, respectively), this rapid reinvasion poses a major concern regarding the effectiveness of dredging to control macrophytes. The project at Lilly Lake should receive intermittent monitoring to determine if macrophytes will continue to be a problem.

Additional case studies have been described by Peterson (1982a). Most are not as clearly successful as the Lake Trummen project, but dredging to control algal productivity seems more likely to succeed than macrophyte control projects unless depth requirements are better defined and adhered to. The high cost of dredging projects dictates that their objectives be well defined and that plans be well thought out and executed in the best fashion possible.

COSTS

The relevance of cost comparisons for sediment removal is questionable because a large number of variables affect the cost figures. These include the type of equipment used, the project size (volume of material to be removed), availability of disposal site, the density of the material being removed, the distance to the disposal area, and the ultimate use of the removed material. Data for 64 lake sediment removal projects in the United States indicated a cost range from $0.24 m^{-3} to $14.00 m^{-3} (Peterson, 1981). Given such wide variability, an average cost figure is not helpful in estimating costs for a single project. Hydraulic dredging costs ranging from $1.25 m^{-3} to $1.75 m^{-3} are relatively common, however, and probably can be considered "reasonable." If sediment contaminated with toxic substances is encountered and special dredges or treatment methods are required, dredging costs may exceed $25.00 m^{-3} (Barnard and Hand, 1978; Koba et al., 1975; Matsubara, 1979). In general, the per-unit-volume cost of sediment removal will be inversely related to the total volume of material removed. Any time dredged material is of sufficient quality for use as potting soil or topsoil dressing, as was the case at Lake Trummen, Sweden (Sjön Trummen i Växjö, 1977), Nutting Lake, Massachusetts (Worth, 1981), Lilly Lake, Wisconsin (Dunst et al., 1984), and Paradise Lake, Illinois (Lembke et al., 1983), the overall project costs might be reduced significantly.

With cost comparisons being extremely difficult among sediment removal projects themselves (Peterson, 1981), it may seem impossible to make realistic cost comparisons between sediment removal and any other lake treatment technique. Peterson (1982b) attempted to do so by making the basic assumptions that both treatment approaches are aimed at the control of in-lake nutrient cycling and that nutrient inactivation and dredging treatments are based on criteria thought necessary to achieve the necessary reduction in nutrient cycling. Thus, treatment costs should reflect the amount of phosphorus removed (dredging) or sealed into the lake system (nutrient inactivation) to prevent excessive internal cycling.

Direct cost comparisons of the two treatments cannot be made, since nutrient inactivation costs usually are reported on the basis of materials and labor required to treat a hectare of lake surface (area), while dredging costs are computed on the basis of cost to remove a cubic meter of sediment (volume). Peterson (1982b) used per-cubic-meter dredging costs, lake area, percent of basin area dredged, and dredge volume to calculate the dredging cost on a per hectare basis. If the assumption is correct that only sufficient sediment material was removed to control the internal nutrient cycling (lake deepening projects per se were not included in the calculation), then dredging costs and nutrient inactivation costs should be comparable. Calculations were made using nutrient inactivation cost data from Cooke and Kennedy (1981) and dredging cost data from Peterson (1981). Results of the calculations are shown in Table 8-4.

The nutrient inactivation costs in Table 8-4 do not include equipment costs, but that expenditure for this type of treatment is generally considered to be small relative to other costs (about 9% of total costs at Medical Lake, Washington). Adding equipment costs for the Medical Lake project would increase the total by $256 ha^{-1}, bringing the most expensive (per hectare) nutrient inactivation project cost close to the least expensive dredging project cost. Labor costs for nutrient inactivation in Table 8-4 assumed an eight-hour work day and labor charge of $5.00 per hour, which may be conservative. Even if doubled, however, the labor cost would not make dredging generally competitive with nutrient inactivation, if cost alone is considered.

Table 8-4 Per-Hectare Cost Comparison of Dredging and Alum Treatment to Control In-Lake Phosphorus Dynamics in 20 Treated Lakes. Costs were calculated from selected data of Peterson (1981) and Cooke and Kennedy (1981).

Lake	Treatment Type	Physical and Chemical Data	Chemical and Dose	Sediment Removed (m^3)	Treatment Cost ($/ha)
Horseshoe Lake, Wisconsin	Liquid alum	A_0 = 8.9 ha V = 3.6 × 10^5 m^3 Z_{max} = 16.7 m $\bar{Z}$ = 4.0 m Alk = 218–278 mg/l pH = 6.8–8.9 dimictic	2.6 g Al/m^3	—	101
Lake San Marcos	Liquid alum	A_0 = 18.2 ha V = 4.3 × 10^5 m^3 Z_{max} = 2.3 m $\bar{Z}$ = 2.3 m Alk = 190–268 mg/l pH = 7.3–9.1	6.0 g Al/m^3	—	126
Welland Canal, Ontario, Canada	Liquid alum, surface application	A_0 = 74 ha V = 6.2 × 10^6 m^3 Z_{max} = 9.0 m $\bar{Z}$ = 9.0 m Alk = 109 mg/l dimictic	2.5 g Al/m^3	—	204

Table 8-4 (continued)

Lake	Treatment Type	Physical and Chemical Data	Chemical and Dose	Sediment Removed (m^3)	Treatment Cost ($/ha)
Mirror Lake, Wisconsin	Liquid alum and aeration	A_0 = 5.1 ha V = 4×10^5 m^3 Z_{max} = 13.1 m $\bar{Z}$ = 7.8 m Alk = 222 mg/l pH = 7.6 monomictic	6.6 g Al/m^3	—	400[a]
Shadow Lake, Wisconsin	Liquid alum	A_0 = 17.1 ha V = 9.1×10^5 m^3 Z_{max} = 12.4 m $\bar{Z}$ = 5.3 m Alk = 188 mg/l pH = 7.4 dimictic	5.7 g Al/m^3	—	400[a]
Cline's Pond, Corvallis, Oregon	Liquid sodium aluminate and HCl	A_0 = 0.4 ha V = 9600 m^3 Z_{max} = 4.9 m $\bar{Z}$ = 2.4 m Alk = 30–50 mg/l pH = 7.0–7.7 monomictic	10.0 g Al/m^3	—	420

Table 8-4 (continued)

Lake	Treatment Type	Physical and Chemical Data	Chemical and Dose	Sediment Removed (m^3)	Treatment Cost ($/ha)
West Twin Lake, Ohio	Liquid alum	A_0 = 34 ha V = 14.2 × 10^4 m^3 Z_{max} = 11.5 $\bar{Z}$ = 4.4 m Alk = 102–149 mg/l dimictic	26.0 g Al/m^3	—	425
Dollar Lake, Ohio	Liquid alum	A_0 = 2.2 ha V = 0.86 × 10^5 m^3 Z_{max} = 7.5 m $\bar{Z}$ = 3.9 m Alk = 101–127 mg/l pH = 6.7–8.6 dimictic	20.9 g Al/m^3	—	504
Medical Lake, Washington	Liquid alum	A_0 = 64 ha V = 6.4 × 10^6 m^3 Z_{max} = 18 m $\bar{Z}$ = 10 m Alk = 750 mg/l pH = 8.5–9.5	12.2 g Al/m^3	—	1,740[b]

Table 8-4 (continued)

Lake	Treatment Type	Physical and Chemical Data	Chemical and Dose	Sediment Removed (m^3)	Treatment Cost ($/ha)
Half Moon Lake, Wisconsin	Dredging (~30% of basin)	A_0 = 53.4 ha V = 8.9×10^5 m^3 Z_{max} = 2.7 m $\bar{Z}$ = 1.7 m	—	25×10^3	2,137
Lilly Lake, Wisconsin	Dredging (100% of basin)	A_0 = 35.6 ha V = 5.3×10^8 m^3 Z_{max} = 1.8 m $\bar{Z}$ = 1.4 m	—	680×10^3	4,584
Commonwealth Lake, Oregon	Dredging (100% of basin)	A_0 = 2.6 ha $\bar{Z}$ = 0.9 m	—	19×10^3	5,769
Steinmetz, New York	Draining and bulldozing (75% of basin)	A_0 = 1.2 ha V = 8.1×10^5 m^3 Z_{max} = 2.1 m $\bar{Z}$ = 1.5 m	—	2×10^3	7,233
Carnegie Lake, New Jersey	Dredging (75% of basin)	A_0 = 110 ha Z_{max} = 3 m	—	765×10^3	12,054
Lenox Lake, Iowa	Dredging (100% of basin)	A_0 = 13.4 ha Z_{max} = 3.4 m $\bar{Z}$ = 0.9 m	—	76×10^3	13,328

Table 8-4 *(continued)*

Lake	Treatment Type	Physical and Chemical Data	Chemical and Dose	Sediment Removed (m^3)	Treatment Cost ($/ha)
Nutting Lake, Massachusetts	Dredging (56% of basin)	A_0 = 31.6 ha Z_{max} = 2.1 m $\bar{Z}$ = 1.3 m	—	275×10^3	18,858
Sunshine Springs, Wisconsin	Dredging (100% of basin)	A_0 = 0.4 ha V = 1.7×10^3 m^3 Z_{max} = 1.2 m $\bar{Z}$ = 0.5 m	—	5.1×10^3	26,832
Krause Springs, Wisconsin	Dredging (100% of basin)	A_0 = 0.3 ha V = 0.98×10^3 m^3 Z_{max} = 1.0 m $\bar{Z}$ = 0.34 m	—	4.9×10^3	31.754
Collins Park Lake, New York	Dredging (15% of basin)	A_0 = 24.3 ha V = 6.5×10^5 m^3 Z_{max} = 9.75 m $\bar{Z}$ = 2.7 m	—	52×10^3	39,178
59th Street Pond, New York	Draining and bulldozing (100% of basin)	A_0 = 1.8 ha $\bar{Z}$ = 0.2 m	—	13×10^3	100,605[c]

[a]Cost per hectare, spread over Mirror Lake and Shadow Lake as a unit (5.1 ha + 17.1 ha = 22.2 ha).
[b]From Gasperino et al. (1980).
[c]All sediment from 59th Street Pond had to be trucked out of New York City, thus escalating the costs.

To put the costs in Table 8-4 in proper perspective, it should be noted that the Medical Lake project costs are abnormally high. They represent real costs, but are high because the lake has steep banks, it is relatively deep (Z = 10 m; Z_{max} = 18 m), has a large volume (6.2 × 10^6 m^3), and it has an uncommonly high total alkalinity (approximately 750 mg l^{-1} as $CaCO_3$). The large volume and high alkalinity required greater amounts of aluminum sulfate to reduce the dissolved reactive phosphorus concentration by the targeted 87 percent. These factors make Medical Lake somewhat unusual among the alum-treated lakes. The 59th Street Pond dredging costs also were unusually high, for the reasons noted in Table 8-4. If these two outliers are eliminated from Table 8-4 and the median cost per hectare of the remaining projects is computed, it can be seen that nutrient inactivation ($400 ha^{-1}) is quite attractive relative to dredging ($12,691 ha^{-1}) when cost alone is examined.

It might be argued that dredging and nutrient inactivation costs should not be compared, since dredging addresses one set of lake problems while nutrient inactivation addresses another. Peterson (1982b) concluded that was not necessarily the case. He found considerable overlap between the lake conditions suitable to dredging and those suited to nutrient inactivation. A possible exception to the similarities was depth and rooted plant growth. There is still some question about the effectiveness of nutrient inactivation in shallow lakes, which are generally ideal candidates for dredging, but the experience of Welch et al. (1982) at Long Lake, Washington, adds credence to the idea that nutrient inactivation can be effective in shallow as well as deeper lakes.

The costs in Table 8-4 were calculated as "end-of-treatment" costs, i.e., what it actually cost to complete the treatment. Costs of this nature are misleading. To be meaningful the treatment costs should be amortized over the effective life expectancy of a project. Unfortunately, lake restoration is new enough that this kind of cost figure cannot be generated with any confidence. Continued monitoring of lake restoration projects will improve this situation. Preliminary evidence indicates that initially successful dredging projects may have greater durations of effectiveness than nutrient inactivation, but the evidence is not definitive. If this proves over time to be the case, then it should become possible to determine if the initial high cost of dredging, compared to nutrient inactivation treatments, is warranted.

If the amortization of cost rationale is applied to the Lake Trummen dredging costs (not shown in Table 8-4), they look much more reasonable than the end-of-project costs. The 1971 cost of the entire project was $572,222 (using a conversion factor of 4.5 Sw. Kr. per dollar). This translates to a cost of approximately $1,430 ha^{-1}. If the cost is amortized over the post-treatment years the lake has shown benefits (9 years), the costs translate to approximately $159 ha^{-1} yr^{-1}. This figure will continue to decline as long as the lake maintains its current level of quality. A figure such as the cost ha^{-1} yr^{-1} makes it much easier to compare the "real cost" of treatment across different treatment types, since the cost can be calculated on the basis of effectiveness. For example, compare the costs for dredging Lake Trummen to the cost of treating West Twin Lake, Ohio: West Twin Lake's area is 34 ha. The cost to treat it with alum in 1975 was $425.00 ha^{-1}. Cooke et al. (1981) have shown the treatment benefits held for five years beyond treatment (not measured since then), so the costs would be about $85 ha^{-1}.

From this brief, rough analysis, nutrient inactivation still holds the edge over dredging in terms of cost, but a much larger sample of cost comparisons is needed to draw any valid conclusions. Only time (determination of restorative effects) will tell the real story. This type of assessment applies only to cases where treatment is directed toward limiting internal nutrient cycling. Dredging is still the only practical method of restoration if deepening a shoaled lake is the objective, or if littoral-zone removal of macrophytes is targeted, as in Lake Trummen. Alum treatment apparently has limited or no effect for reducing macrophyte biomass, and may even enhance plant growth by increasing light penetration.

SUMMARY

There are four major reasons for removing sediment from a lake. They include deepening, limitation of nutrient recycling, reduction of macrophyte nuisances, and the removal of sedimented toxic substances. Environmental concerns associated with lake sediment removal appear to be less negative than might be expected. Case studies reveal that while adverse impacts do occur, they generally are not severe and are of short duration. Another possible long-term impact of dredging is the effect dredged material has on the area where it is disposed.

Dredging depth will depend on the purpose of the project and the sediment characteristics of the lake. Two different approaches are cited for determining dredging depth for the control of in-lake nutrient cycling from the sediments. A third method has been proposed when the intent is to control macrophytes. An important aspect of any sediment-removal project is a thorough pretreatment evaluation of nutrient and sediment mass balance. It will help determine the feasibility of dredging and provide insight to the expected longevity of effectiveness.

Almost all projects designed to deepen a lake are successful. Those designed to control internal nutrient cycling have had mixed results, but, most failures can be traced to incomplete preevaluation or to simply removing too little sediment. Macrophyte control projects have not been evaluated extensively enough to permit generalization. Toxic sediment removal projects have been completed successfully, but generally require use of special dredges and special disposal practices. Numerous specialized techniques are available.

Cost comparisons among dredging projects are difficult, due to the large number of variables involved. Cost comparisons of dredging with other treatment techniques are not commonly made because of the many variables and the nonuniform units for measuring costs—$ ha^{-1} of treated area for nutrient inactivation; $ m^{-3} of sediment dredged. A few basic assumptions permit conversion of the different costing units into common terms. All of the other variables still exist, but it appears that dredging is several times more expensive than nutrient inactivation to accomplish the same short-term results (minimization of internal nutrient recycling) if cost alone is considered. "Real cost" determination must be deferred until the longevity of treatment effectiveness for each technique is evaluated and the cost can reasonably be put in terms of $ $ha^{-1} yr^{-1}$.

There is little doubt that dredging can be a successful lake restoration technique if there is a thorough preimplementation evaluation of the lake setting, if proper equipment is selected, if the disposal areas are designed for end-of-treatment effectiveness, and if the dredging is conducted by conscientious and competent operators. Successful examples of both surface-sediment skimming and deep (6 to 8 m) dredging can be cited. Thorough preimplementation evaluation is the most important aspect of the project.

REFERENCES

Andersson, G., Berggren, H., Cronberg, C., and Gelin, C. 1978. Effects of planktivorous and benthivorous fish on organisms and water chemistry in eutrophic lakes. *Hydrobiologia* 59:8-15.

Andersson, G., Berggren, H., and Hambrin, S. 1975. Lake Trummen restoration project III. Zooplankton macrobenthos and fish. *Verh Int Ver Limnol* 19:1097.

Barko, J.W., and Smart, R.M. 1979. The role of *Myriophyllum spicatum* in the mobilization of sediment phosphorus. In J.E. Breck, R.J. Prentki, and O.L. Loucks, eds., *Aquatic Plants, Lake Management, and Ecosystem Consequences of Lake Harvesting*, Madison WI: Center for Biotic Systems, Univ Wisconsin, 435.

Barko, J.W., and Smart, R.M. 1980. Mobilization of sediment phosphorus by submerged freshwater macrophytes. *Freshwater Biol* 10:229-238.

Barnard, W.D. 1978. *Prediction and Control of Dredged Material Dispersion Around Dredging and Open-Water Pipeline Disposal Operations*, Tech. Report DS-78-13, Vicksburg, MS: U.S. Army Corps of Engineers, Waterways Experiment Station.

Barnard, W.D., and Hand, T.D. 1978. *Treatment of Contaminated Dredged Material*, Tech. Report DS-78-14, Vicksburg, MS: U.S. Army Corps of Engineers, Waterways Experiment Station.

Belonger, B. 1969. *Aquatic Plant Survey of Major Lakes in the Fox-Illinois Watershed*, Madison, WI: Wisconsin Dept. Nat. Resour., Res. Rept. No. 39.

Bengtsson, L., Fleischer, S., Lindmark, G., and Ripl, W. 1975. Lake Trummen restoration project I. Water and sediment chemistry. *Verh Int Ver Limnol* 19:1080.

Bjork, S. 1972. Ecosystem studies in connection with the restoration of lakes. *Verh Int Ver Limnol* 18:379-387.

Bjork, S. 1974. European lake rehabilitation activities. Plenary lecture of the Conference on Lake Protection and Management, Madison, Wisconsin.

Bjork, S. 1978. Restoration of degraded lake ecosystems. Lecture at MAB Project 5 Regional Workshop, *Land Use Impacts on Lake and Reservoir Ecosystems*, Warsaw, Poland, May 26-June 2, 1978. CODEN LUNBDS/(NBLI-3008)/1-24 (1978)/ISSEN 0348-0798. Univ. Lund, Sweden.

Born, S.M., Wirth, T.L., Brick, E.M., and Peterson, J.O. 1973. *Restoring the Recreational Potential of Small Impoundments*, Tech. Bull. No. 71, Madison, WI: Wisconsin Dept. of Nat. Resour.

Brannon, J.M. 1978. *Evaluation of Dredged Material Pollution Potential*, Tech. Rept. DS-78-6, Vicksburg, MS: U.S. Army Corps of Engineers, WES.

Brashier, C.K., Churchill, L., and Leidahl, G. 1973. *Effect of Silt Removal in a prairie Lake*, EPA Ecol. Res. Series R3-73-037, Corvallis, OR: USEPA, CERL.

Bray, R.N. 1979. *Dredging: A Handbook for Engineers*. London: Arnold Press.

Bremer, K.E. 1979. PCB contamination of the Sheboygan River, Indiana Harbor, and Saginaw River and Bay. In S.A. Peterson and K.K. Randolph, eds., *Management of Bottom Sediments Containing Toxic Substances: Proceedings of the 4th U.S./ Japan Experts Meeting*, EPA-600/3-79-102, Corvallis, OR: USEPA, CERL.

Canfield, D.E. Jr., Langeland, K.A., Linda, S.B. and Haller, W.T. 1985. Relations between water transparency and maximum depth of macrophyte colonization in lakes. *J Aquatic Plant Manage* 23:25–28.

Carline, R.F., and Brynildson, O.M. 1977. *Effects of Hydraulic Dredging on the Ecology of Native Trout Populations in Wisconsin Spring Ponds*, Tech. Bull. No. 98, Madison, WI: Wisconsin Dept. Nat. Resour.

Carignan, R., and Kalff, J. 1980. Phosphorus sources for aquatic weeds: Water or sediment? *Science* 207:987.

Chen, K.Y., Eichenberger, B., Mang, J.L., and Hoeppel, R.E. 1978. *Confined Disposal Area Effluent and Leachate Control (Laboratory and Field Investigations)*, Tech. Rept. DS-78-7, Vicksburg, MS: U.S. Army Corps of Engineers, WES.

Churchill, C.L., Brashier, C.K., and Limmer, D. 1975. *Evaluation of a Recreational Lake Rehabilitation Project*, OWRR Comp. Report No. B-028-SDAK, Brookings, SO: Water Resources Inst., South Dakota State University.

Clark, G.R. 1983. *Survey of Portable Hydraulic Dredges*, Tech. Rept. HL-83-4, Vicksburg, MS: U.S. Army Engineers Waterways Experiment Station.

Collett, L.C., Collins, A.J., Gibbs, P.J., and West, R.J. 1981. Shallow dredging as a strategy for the control of sublittoral macrophytes: A case study in Tuggerah Lakes, New South Wales. *Aust J Marine Freshwater Res* 32:563–571.

Cooke, G.D., and Kennedy, R.H. 1981. *Precipitation and Inactivation of Phosphorus as a Lake Restoration Technique*, EPA-600/3-81-012, Corvallis, OR: USEPA, CERL.

Cooke, G.D., Heath, R.T. Kennedy, R.H. and McComas, M.R. 1982. Change in lake trophic state and internal phosphorus release after aluminum sulfate application. *Wat Res Bull* 18:699–705.

Cronberg, G., Gelin, C., and Larsson, K. 1975. Lake Trummen restoration project II. Bacteria, phytoplankton, and phytoplankton productivity. *Verh Int Ver Limnol* 19:1088.

Crumpton, J.E., and Wilbur, R.L. 1974. *Habitat Manipulation*, Dingell–Johnson Job Completion Report, Proj. No. F-26-5, Florida Game and Freshwater Fish Comm.

Digerfeldt, G. 1972. The post-glacial development of Lake Trummen, regional vegetation history, water level changes, and paleolimnology. *Folia Limnol Scand* 16:1.

Dunst, R. 1980. Sediment problems and lake restoration in Wisconsin. In S.A. Peterson and K.K. Randolph, eds., *Management of Bottom Sediments Containing Toxic Substances: Proceedings of the 5th U.S./Japan Experts Meeting*, EPA Ecol. Res. Ser. Rept. EPA-600/9-8-044, Corvallis, OR: USEPA, CERL.

Dunst, R.C. 1981. Dredging activities in Wisconsin's lake renewal program. In *Restoration of Lake and Inland Waters: International Symposium on Inland Waters and Lake Restoration*, EPA-440/5-81-010, Washington, DC: USEPA, Office of Water Regulations and Standards.

Dunst, R. 1982. Sediment problems and lake restoration in Wisconsin. *Environ Int* 7:87–92.

Dunst, R.C., Born, S.M., Uttormark, P.O., Smith, S.A., Nichols, S.A., Peterson, J.O., Knauer, D.R., Serns, S.R., Winters, D.R., and Wirth, T.L. 1974. *Survey of Lake Rehabilitation Techniques and Experiences*, Tech. Bull. 75, Madison, WI: Wisc. Dept. Nat. Resour.

Dunst, R.C., Vennie, J.G., Corey, R.B., and Peterson, A.E. 1984. *Effect of Dredging Lilly Lake, Wisconson*, EPA-600/3-84-097, EPA cooperative agreement No. R804875, Corvallis, OR: CERL.
Fujiki, M., and Tajima, S. 1973. The pollution of Minamata Bay and the neighbouring sea by factory wastewater containing mercury. In F. Coulston, F. Korte, and M. Goto, eds., *New Methods in Environmental Chemistry and Toxicology*, papers presented at the International Symposium on Ecological Chemistry, Susono, Totsuka, Tokyo: International Academic Printing Co.
Gambrell, R.P., Kincaid, R.A., and Patrick, W.H. Jr. 1978. *Disposal alternatives for Contaminated Dredged Material as a Management Tool to Minimize Adverse Environmental Effects*, Tech. Rep. DS-78-8, Vicksburg, MS: U.S. Army Corps of Engineers, WES.
Gasperino, A.F., Beckwith, M.A., Keizur, G.R., Saltero, R.A., Nichols, D.G., and Mires, J.M. 1981. Medical Lake improvement project: Success story. In *Restoration of Lakes and Inland Waters: International Symposium on Inland Waters and Lake Restoration*, EPA-440/5-81-010, Washington, DC: USEPA, Office of Water Regulation and Standards.
George, C., Tobiessen, P., Snow, P., and Jewell, T. 1981. *The Monitoring of the Restorational Dredging of Collins Lake, Scotia, New York. Final Project Report*, Grant No. R804572, Corvallis, OR: USEPA, CERL.
Goldman, C.R., Gersberg, R.M., Axler, R.P. 1981. *Gibraltar Lake Restoration Project, City of Santa Barbara. Final Report on Limnological Monitoring During Dredging*, Davis, CA: Ecological Research Associates.
Herbich, J.B., and Brahme, S.B. 1983. *Literature Review and Technical Evaluation of Sediment Resuspension During Dredging*, Report No. COE-266, College Station, TX: Ocean and Hydraulic Engineering Group, Texas A&M University.
Higginson, F.R. 1970. Ecological effects of pollution in Tuggereh Lakes. *Proc Ecol Soc Aust* 5:143–152.
Horn, E., and Hetling, L. 1978. Hudson River PCB study description and detailed work plan. In S.A. Peterson and K.K. Randolph, eds., *Management of Bottom Sediments Containing Toxic Substances: Proceedings of the 3rd U.S./Japan Experts Meeting*, EPA-600/3-78-084. Corvallis, OR: USEPA, CERL.
Huston, J. 1970. *Hydraulic Dredging: Theoretical and Applied*, Cambridge, MD: Cornell Maritime Press.
Hutchinson, G.E. 1975. *A Treatise on Limnology – Vol. III – Limnological Botany*, New York: Wiley.
JACA Corp. 1980. *Economic Benefits Assessment of the Section 314 Clean Lakes Program*.
Koba, H., Shinohara, K., and Sato, E. 1975. Management techniques of bottom sediments containing toxic substances. Paper presented at *1st U.S./Japan Experts Meeting on the Management of Bottom Sediments Containing Toxic Substances*, Nov. 17–21, 1975, Corvallis, OR: USEPA, CERL.
Larsen, D.P., Schults, D.W., and Malueg, K.W. 1981. Summer internal phosphorus supplies in Shagawa Lake, Minnesota. *Limnol Oceanogr* 26:740–753.
Lembke, W.D., Mitchell, J.K., Fehrenbacher, J.B., Barcelona, M.J., Garske, E.E., and Heffelfinger, S.R. 1983. *Dredged Sediment for Agriculture: Lake Paradise*, Research Rept. No. 175, Water Resources Center, University of Illinois: Urbana–Champaign.

Lie, G.B. 1979. The influence of aquatic macrophytes on the chemical cycles of the littoral. In J.E. Breck, R.T. Prentki, and O.L. Loucks, eds., *Aquatic Plants, Lake Management, and Ecosystem Consequences of Lake Harvesting*, Center for Biotic Systems, Univ. Wisconsin-Madison.

Livingston, D.A., and Boykin, J.C. 1962. Distribution of phosphorus in Linsley Pond mud. *Limnol Oceanogr* 7:57-62.

Lunz, J.D., Diaz, R.J., and Cole, R.A. 1978. *Upland and Wetland Habitat Development with Dredged Material; Ecological Considerations*, Tech. Rept. DS-78-15, Vicksburg, MS: U.S. Army Corps of Engineers, WES.

Mackenthun, K.M., Brossman, M.W., Kohler, J.A., and Terrell, C.R. 1979. Approaches for mitigating the kepone contamination in the Hopewell/James River area of Virginia. In S.A. Peterson and K.K. Randolph, eds., *Management of Bottom Sediments Containing Toxic Substances: Proceedings of the 4th U.S./Japan Experts Meeting*, EPA-600/3-79-102, Corvallis, OR: USEPA, CERL.

Mallory, C.W., and Nawrocki, M.A. 1974. *Containment Area Facility Concepts for Dredged Material Separation, Drying, and Rehandling*, DMRP Contract Rep. D-74-6, Vicksburg, MS: U.S. Army Corps of Engineers, WES.

Maristo, L. 1941. Die Seetypen Finnlands auf floristicher und vegetations-physiognomischer Grundlage. *Ann Bot Soc Zool Bot Vanamo* 15:314-310.

Matsubara, M. 1979. The improvement of water quality at Lake Kasumigaura by the dredging of polluted sediments. In S.A. Peterson and K.K. Randolph, eds., *Management of Bottom Sediments Containing Toxic Substances: Proceedings of the 4th U.S./Japan Experts Meeting*, EPA-600/3-79-102, Corvallis, OR: USEPA, CERL.

Modlin, R. 1970. *Aquatic Plant Survey of Major Lakes in the Milwaukee River Watershed*, Res. Rept. No. 52, Madison, WI: Wisconsin Dept. Nat. Resour.

Moody, L.F. 1944. Friction factors for pipe flow. *Trans ASME* 56:51-61.

Montgomery, R.L. 1978. *Methodology for Design of Fine-Grained Dredged Material Containment Areas for Solids Retention*, Tech. Rept. D-78-56, Vicksburg, MS: U.S. Army Engineers Waterways Experiment Station.

Montgomery, R.L. 1979. Development of a methodology for designing fine-grained dredged material sedimentation basins. Ph.D. thesis, Vanderbilt University, Nashville, Tennessee.

Montgomery, R.L. 1980. Containment area sizing for sedimentation of fine-grained dredged material. In S.A. Peterson and K.K. Randolph, eds., *Management of Bottom Sediments Containing Toxic Substances: Proceedings of 5th U.S./Japan Experts Meeting*, EPA-600/9-80-044, Corvallis, OR: USEPA, CERL.

Montgomery, R.L. 1982. Containment area sizing for disposal of dredged material. *Environ Int* 7:151-161.

Montgomery, R.L., Thackston, E. and Parker, F.L. 1983. Dredged material sedimentation basin design. *J Environ Eng Div ASCE* 109:466-484.

Murakami, K. 1984. Dredging for controlling eutrophication of Lake Kasumigaura, Japan. In *NALMS Proceedings of International Symposium on Lake and Reservoir Management*, Knoxville, Tennessee, October 1983.

Murakami, K., and Takeishi, K. 1977. Behavior of heavy metals and PCBs in dredging and treating of bottom deposits. In S.A. Peterson and K.K. Randolph, eds., *Management of Bottom Sediments Containing Toxic Substances: Proceedings of the 2nd U.S./Japan Experts Meeting*, EPA-600/3-77-083, Corvallis, OR: USEPA, CERL.

Nawrocki, M.A. 1974. *Demonstration of the Separation and Disposal of Concentrated Sediments*, EPA-660/2-74-072, USEPA.
Palermo, M.R., Montgomery, R.L., and Poindexter, E. 1978. *Guidelines for Designing, Operating, and Managing Dredged Material Containment Areas*, Tech. Rept. DS-78-10, Vicksburg, MS: U.S. Army Engineers, WES.
Peterson, S.A. 1979. Dredging and lake restoration. In *Lake Restoration: Proceedings of a National Conference*, EPA-400/5-79-001, Washington, DC: USEPA, Office of Water Planning and Standards.
Peterson, S.A. 1981. *Sediment Removal as a Lake Restoration Technique*, EPA-600/3-81-013, Corvallis, OR: USEPA, CERL.
Peterson, S.A. 1982a. Lake restoration by sediment removal. *Water Res Bull* 18:423–435.
Peterson, S.A. 1982b. Dredging and nutrient inactivation as lake restoration techniques: A comparison. In *Management of Bottom Sediments Containing Toxic Substances: Proceedings of the 6th U.S./Japan Experts Meeting*, Feb. 1981, Tokyo, Japan, Vicksburg, MS: U.S. Army Engineers, Dredging Operations Technical Support Program.
Pierce, N.D. 1970. *Inland Lake Dredging Evaluation*, Tech. Bull. 46, Madison, WI: Wisconsin Dept. Nat. Resour.
Randall, R.E. 1977. *Notes from the Fifth Dredging Engineering Short Course*, College Station, TX: Texas A&M University.
Raymond, R.B., and Cooper, F.C. 1984. Vancouver lake: Dredged material disposal and return flow management in a large lake dredging project. In *NALMS Proceedings of International Symposium on Lake and Reservoir Management*, Knoxville, Tennessee, October 1983.
Reimold, R.J. 1972. The movement of phosphorus through the salt marsh cord grass, *Spartina alterniflora* Loisel. *Limnol Oceanogr* 17:606–611.
Ryding, S.O. 1982. Lake Trehörningen restoration project. Changes in water quality after sediment dredging. *Hydrobiologia* 92:549–558.
Sakakibara, A., and Hayashi, O. 1979. Lake Suwa water pollution control projects. In S.A. Peterson and K.K. Randolph, eds., *Management of Bottom Sediments Containing Toxic Substances: Proceedings of the 4th U.S./Japan Experts Meeting*, EPA-600/3-79-102, Corvallis, OR: USEPA, CERL.
Schults, D.W., and Malueg, K.W. 1971. Uptake of radiophosphorus by rooted aquatic plants. In *Proceedings of the 3rd National Symposium on Radioecology*, Oak Ridge, Tennessee, May 10–12, 1971, 417–424.
Shapiro, J., Lamarra, V., and Lynch, M. 1975. Biomanipulation—An ecosystem approach to lake restoration. In P.L. Brezonik and J.L. Fox, eds., *Proceedings of a Symposium on Water Quality Management through Biological Control*, Univ. of Florida, Gainesville.
Sjön Trummen i Växjö. Förstörd, Restaurerad, Pånyttfödd. 1977. *Länsstyrelsen I Kronobergs Län, Växjö Kommun.*
Snow, P.D., Cook, W., and McCauley, T. 1980. *The Restoration of Steinmetz Pond, Schenectady, New York*, EPA Final Project Report, Grant No. NY-57700108, Washington, DC: USEPA.
Spaine, P., Llopis, L., and Perrier, E.R. 1978. *Guidance for Land Improvement Using Dredged Material*, Tech. Rept. DS-78-21, Vicksburg, MS: U.S. Army Corps of Engineers, WES.
Spencer Engineering. 1981. *Gibraltar Lake Restoration Project Final Report*. Santa Barbara, CA: Spencer Engineering.

Spitler, F.J. 1973. *Dredging Long Lake, Michigan to Improve Boating and Fishing,* Tech. Bull. 73-17, Lansing, MI: Mich. Dept. Nat. Resour.

Stauffer, R.E., and Lee, G.F. 1973. The role of thermocline migration in regulating algal blooms. In E.J. Middlebrooks, D.H. Falkenburg, and T.E. Maloney, eds., *Modeling the Eutrophication Process,* Proceedings of a workshop at Utah State University, Logan.

Stefan, H., and Ford, D.E. 1975. Temperature dynamics in dimictic lakes. *J Hydraul Div ASCE* 101(HY1), Proc. Paper 11058:97–114.

Stefan, H., and Hanson, M.J. 1979. *Fairmont Lakes Study: Relationships between Stratification, Phosphorus Recycling, and Dredging,* Project Report No. 183, Univ. Minnesota, St. Anthony Fall's Hydraulic Laboratory.

Suda, H. 1979. Results of the investigation of turbidity generated by dredges at Yokkaichi Port. In S.A. Peterson and K.K. Randolph, eds., *Management of Bottom Sediments Containing Toxic Substances: Proceedings of the 4th U.S./Japan Experts Meeting,* EPA-600/3-79-102, Corvallis, OR: USEPA, CERL.

Toubier, J., and Westmacott, R. 1976. *Lakes and Ponds,* Tech. Bull. No. 72, Washington, DC: Urban Land Institute.

Twilley, R.R., Brinson, M.M., and Davis, G.J. 1977. Phosphorus absorption, translocation, and secretion in *Nuphar luteum. Limnol Oceanogr* 22:1022–1032.

U.S. Department of Agriculture. 1971. *Ponds for Water Supply and Recreation,* Handbook No. 387, Washington, DC: U.S. Government Printing Office.

Walsh, M.R., and Malkasian, M.D. 1978. *Productive Land Use of Dredged Material Containment Areas: Planning and Implementation Consideration,* Tech. Rept. DS-78-020, Vicksburg, MS: U.S. Army Corps of Engineers, WES.

Wechler, B.A., and Cogley, D.R. 1977. *Laboratory Study Related to Predicting the Turbidity-Generation Potential of Sediments to be Dredged,* Tech. Rept. D-77-14, Vicksburg, MS: U.S. Army Corps of Engineers, WES.

Welch, E.B., Michaud, J.P., and Perkins, M.A. 1982. Alum control of internal phosphorus loading in a shallow lake. *Water Res Bull* 18:929–936.

Welch, E.B., Lynch, P.D., and Hufschmidt, D. 1979. Internal phosphorus related to rooted macrophytes in a shallow lake. In J.E. Breck, R.T. Prentki, and O.L. Loucks, eds., *Aquatic Plants, Lake Management, and Ecosystem Consequences of Lake Harvesting,* Center for Biotic Systems, Univ. Wisconsin–Madison.

Wetzel, R.G. 1975. *Limnology.* Philadelphia: Saunders.

Wisconsin Department of Natural Resources. 1969. *Lilly Lake, Kenosha County, Wisconsin,* Lake Use Rept. No. FX-34, Madison, WI: Wisconsin Dept. Nat. Resour.

Worth, D.M. Jr. 1981. Nutting Lake restoration project: A case study. In *Restoration of Lakes and Inland Waters: International Symposium on Inland Waters and Lake Restoration,* EPA-440/5-81-010, Washington, DC: USEPA, Office of Water Regulation and Standards.

9

Hypolimnetic Aeration

PRINCIPLE

The objectives of hypolimnetic aeration are usually threefold. The first and usually most attainable is to raise the oxygen content of the hypolimnion without destratifying the water column or warming the hypolimnion. The second is to provide an increased habitat and food supply for cold-water fish species. Third, the internal loading of P should decline by establishing aerobic conditions at the sediment–water interface. Other constituents that reach high and possibly undesirable concentrations under anaerobic conditions, such as NH_4^+, Mn, and Fe, should also be diminished by hypolimnetic aeration. Pastorak et al. (1980; 1982) have provided a thorough review of this process and its effects.

There are many designs for hypolimnetic aerators. Fast and Lorenzen (1976) have reviewed 21 and were able to group them into three categories: mechanical agitation, which involves removal, treatment, and return of the hypolimnetic water; injection of pure oxygen; and injection of air, either through a full or partial air-lift design or through a down-flow injection design.

Mechanical agitation involves drawing off water from the hypolimnion, aerating it on shore or on the lake surface by means of a splash basin, and returning the water to depth with minimal increase in temperature. This has not been a popular system because of the poor gas exchange efficiency (Pastorak et al., 1982). To improve the efficiency of oxygenation, pure oxygen has been used to aerate hypolimnetic water drawn from depth, treated at the shore and returned. Pure oxygen can be introduced into the hypolimnion and forced downward with a pump or released at depth and allowed to rise through the hypolimnion. Although gas exchange would be favorable with this method, there are potential problems, especially with N_2-filled bubbles escaping to the lake surface and associated mixing of hypolimnetic water with epilimnetic water (Fast and Lorenzen, 1976).

Injection of air via air-lift systems has been the most popular for hypolimnetic aeration. Full air-lift brings bottom water to the surface and returns it to the hypolimnion after eliminating air bubbles. Partial air-lift aerates hypolimnetic water in place, with water and air bubbles being separated at depth. Fast et al. (1976) and Lorenzen and Fast (1977) found that the full air-lift design is least costly and more efficient at delivering oxygen than the other systems (Pastorak et al., 1982). The partial

214 *Physical and Chemical Methods*

air-lift design, however, is probably the most frequently used system and is marketed by Atlas Copco, Aquatec, as the "Limnox" (Figure 9-1). A new, flexible "Limno" constructed mainly of non-corrosive PVC-coated polyester fabric is the unit commonly used.

SIZING

The partial air-lift device successfully aerates hypolimnia by lifting bottom water vertically through a pipe as a result of injecting compressed air at the bottom. The aerated water is recirculated within the hypolimnion, but excess air and reduced gases are vented to the surface. Pastorak et al. (1982) have stressed the importance of adequately sizing the aeration system, and even oversizing, to compensate for unpredicted variation in oxygen consumption, hypolimnetic volume, or mechanical breakdown. Sizing is determined by estimating the loading of oxygen needed and the time duration. This can probably be obtained from existing data on hypolimnetic volume, O_2 deficit rate, and the period of stratification. The largest portion of the cost is the compressed air delivered from the shore.

The oxygen deficit rate (ODR) can be determined from a plot of mean hypolimnetic oxygen concentrations over time:

$$\text{ODR mg m}^{-2} \text{ day}^{-1} = \frac{\overline{X}DO_{t_1} - \overline{X}DO_{t_2}}{t_2 - t_1} \cdot \overline{Z}_h, \qquad (9.1)$$

where DO_{t_1} = dissolved oxygen at the beginning of stratification
DO_{t_2} = DO prior to any concentrations less than 1 mg l^{-1}
$t_2 - t_1$ = elapsed time in days
$\overline{Z}_h$ = mean depth of hypolimnion.

Estimating the rate from the slope of a least squares regression line is preferable to an estimate based on only two points. The amount of free air required could be estimated by

$$\text{Air flow, m}^3 \text{ day}^{-1} = \frac{ODR \cdot A_h \cdot t_s \cdot 2 \cdot 10^{-6}}{1.205 \cdot 0.2}, \qquad (9.2)$$

where A_h = area of hypolimnion, m
t_s = duration of stratification, days
2 = factor to allow for unmeasured O_2 demand. Factor of 4/10°C may be necessary if temperature increases significantly (Steinberg and Arzet, 1984)
0.2 = fraction of air as O_2
10^{-6} = kg mg^{-1}
1.205 = kg m^{-3} air at 1 atm. and 20°

Hypolimnetic Aeration 215

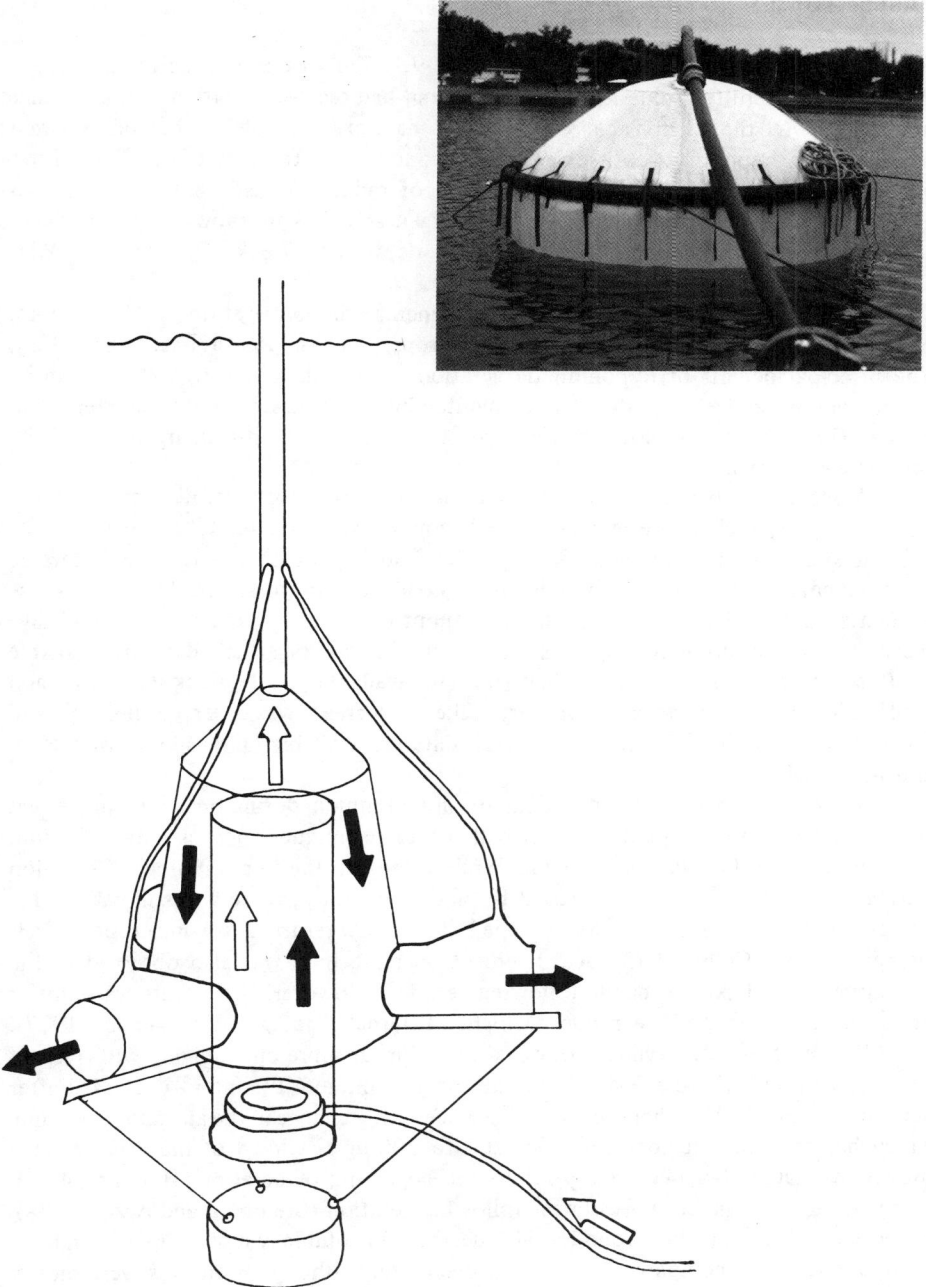

Figure 9-1 Atlas Copco's Limnox 20 partial air-lift hypolimnetic aerator. The open arrows indicate the direction of air and the solid arrows the direction of water. Insert photo shows installation of the new flexible Limno.

CASES STUDIES

Fifteen cases of aeration are shown in Table 9.1. This list is not inclusive, because Björk (1974) identified nine lakes in Europe that had received aeration in this manner up to 1973. Of the 13 in Table 9.1 for which data are available, all but one showed an increase in hypolimnetic dissolved oxygen, and usually to at least 7 mg l^{-1}, following aeration. At the same time, the content of reduced constituents, such as H_2S and NH_4^+, decreased. Results for dissolved oxygen and H_2S are shown for Brunsviken, where four units have operated since 1973, delivering 77.0 kg O_2 per day (Atlas Copco, 1976; Figures 9-2 to 9-4).

In only two cases did the hypolimnion increase in temperature by 4°C or more, but in one of those cases the design was inappropriate (Fast, 1971a; Fast et al., 1973). The principal benefit of hypolimnetic aeration is that anaerobic hypolimnia can be readily and effectively aerated while maintaining their normal cold-water environments. The beneficial biological effects are associated with this enlargement of the aerobic environment.

A secondary benefit of hypolimnetic aeration is a potential decrease in P and trace elements, such as Fe and Mn, which could result in a reduction of internal P loading and an improved water supply. By changing the sediment-water interface environment from reducing to oxidizing, the release of dissolved forms of those elements should decrease. This potential improvement, however, has not always reached expectations following aeration. Of nine lakes for which data are available on P, some decrease was seen in eight (no data available for Hemlock, Ottoville, and Ghirla). TP data were available for Tory Lake—a decrease was observed there. Fe and Mn decreased in all of the three for which data are available (Jarlasjön, Spruce Run, and Wahnbach).

Although P has been reduced in the hypolimnion during aeration, the effect is not as great nor as permanent as with other techniques such as alum addition, hypolimnetic withdrawal, or dredging. While total P in the hypolimnion of Jarlasjön during the stratified period decreased by about one-third, there was still 500 µg l^{-1} TP present. As soon as aeration stopped, P rapidly returned to preaeration levels (Bengtsson and Gelin, 1975). A 30-percent decrease in P was also observed in the hypolimnion of Lake Waccabuc following aeration. However, the effects were not as pronounced the second year due to higher external loading (Garrell et al., 1977).

Results from Brunsviken, in Stockholm, appear more encouraging with respect to a decrease in P (Figure 9-4). While the spring minimum P content decreased after aeration began in 1972, there seems to have been little change in the autumn maximum, which has remained at approximately 200 to 300 µg l^{-1}. Most of the decrease in P shown in Figure 9-4 occurred prior to the beginning of aeration. Hypolimnetic P decreased by 55% in Lake Wesslinger following aeration (Steinberg and Arzet, 1984). McQueen and Lean (1984) found a 64% decrease in soluble reactive P in a 15 m deep aerated enclosure compared to a control column, although iron was periodically added to their column.

The benefits of hypolimnetic aeration, as well as its limitations, are clearly seen in Kolbotnvatn, Norway. Prior to aeration, dissolved oxygen was depleted throughout

Table 9-1 Lakes Receiving Hypolimnetic Aeration and Associated Characteristics (modified from Pastorak et al., 1980).

Lake	Depth Max.	Depth Mean	Device	Volume ($m^3 \times 10^6$)	Area (ha)	Q_{air} ($m^3\ min^{-1}$)	References
Brunsviken,* Sweden	50		13		100	15.5	Atlas Copco, 1976
Caldonazzo, Italy			11		700	44	Atlas Copco, 1980
Hemlock,* Michigan	18.6		18.6		2.4	2.8	Fast, 1971a
Jarlasjön,* Sweden	24	9.3	24	7.8	84	22.8	Bengtsson and Gelin, 1975
Kolbotnvatn,* Norway	18.5	10.3	18.5	3.1	30.3	5.5	Atlas Copco, 1980; Holton and Holton, 1978
Larson,* Wisconsin	11.9	4.0	11.9	0.188	4.8	0.45	Smith et al., 1975
Mirror,* Wisconsin	13.1	7.6	12.8	0.4	5.3	0.45	Smith et al., 1975
Ottoville Quarry,* Ohio	18		18	0.063	0.73	0.11	Fast, 1973; Overholtz et al., 1977
Spruce Run, New Jersey	13.1		12.2			0.15	Whipple et al., 1975
Tegeler See, W. Germany	16		12/16		420	12	Atlas Copco, 1980
Waccabuc,* New York	13		13	4.053	53.6	7.93	Fast et al., 1975; Garrell et al., 1977
Wahnbach,* W. Germany	43	19.2		41.63	214.5	9	Bernhardt, 1967; 1974
Spruce Knob,* W. Virginia	5.7	2.1	5.2	0.224	10.5	1.3	Hess, 1977; LaBaugh, 1980
Ghirla,* Italy	14.0	8.0	14.0	2.0	24.5		Bianucci and Bianucci, 1979
Tory,* Ontario	10.0	4.5	9.0	0.055	1.23	3.54	Taggart and McQueen, 1981

*Data on lake response available.

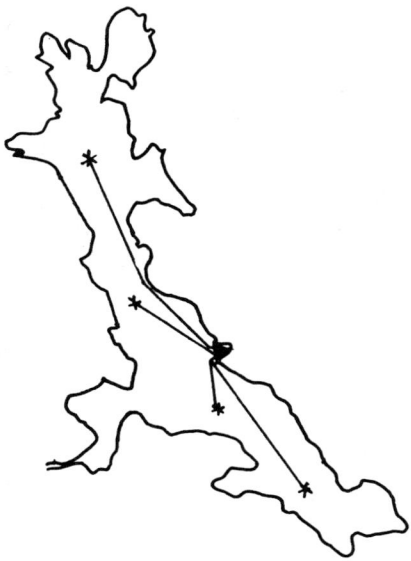

Figure 9-2 Locations of hypolimnetic aerators (Limnox units) and air feed lines from the main compressor in Brunsviken Lake, Sweden (Atlas Copco, 1980).

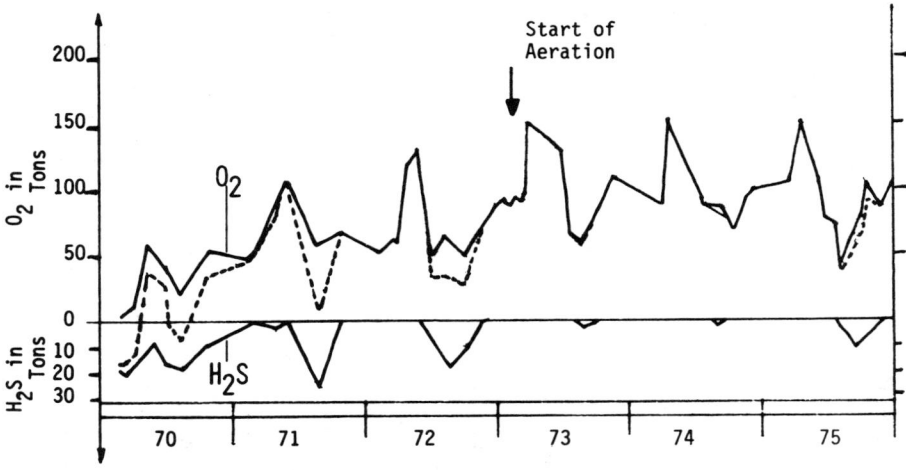

Figure 9-3 Oxygen and hydrogen sulfide (and resulting net O_2, assuming 1.9 moles/liter H_2S = 1.0 moles/liter O_2) in Brunsviken Lake, Sweden, before and after hypolimnetic aeration (from Atlas Copco, 1980).

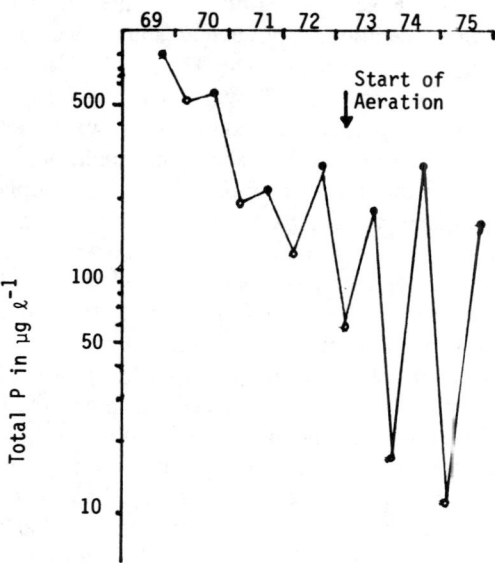

Figure 9-4 Total phosphorus concentration (lake weighted means) in Brunsviken Lake, Sweden, before and after hypolimnetic aeration (from Atlas Copco, 1980).

the 18-m water column from February until iceout in April. The hypolimnion was devoid of dissolved oxygen from June through October. With aeration the hypolimnetic dissolved oxygen has remained between 30 percent and 70 percent saturation, and stratification has persisted. Although P content near the sediment-water interface has decreased from 1600 to 500 $\mu g\ l^{-1}$ during winter stagnation and from about 900 to 600 $\mu g\ l^{-1}$ in summer, the lake P content still reaches 300 $\mu g\ l^{-1}$ and chlorophyll at times exceeds 100 $\mu g\ l^{-1}$. In this case, the high P and Chl a are maintained by high external loading to the lake (Holton et al., 1981; Holton and Holton, 1978). Nevertheless, the oxygen environment has been greatly improved for fish, especially cold-water species.

There has apparently been little change in epilimnetic content of phytoplankton, or in transparency, in other lakes (Smith et al., 1975) treated by hypolimnetic aeration. This is not surprising, in view of the rather minor impact on P release from sediments, and presumably on internal loading as well. There could be a significant indirect effect on phytoplankton control, however, by furnishing a greatly increased aerobic habitat as a refuge for herbivorous zooplankton. The hypolimnion is dark to dimly lit, which gives further protection against sight predators during daylight, in addition to having a greater dilution of the populations (Shapiro, 1979 and Fast, 1979). If the lake is highly eutrophic, metalimnetic minima may occur in spite of an aerated hypolimnion, and could serve as a barrier to anmeal migration (Taggart and McQueen, 1981).

In support of this, the large *Daphnia pulex* became very abundant in Hemlock Lake following hypolimnetic aeration. It reached an abundance nearly 90 times greater than during the pre-dilution time (Fast, 1971b). The smaller cladocerans (*Bosmina* and *Diaphanasoma*) also increased, to a lesser extent than *Daphnia,* and there was little change in *Diaptomus*. Before aeration all zooplankton were excluded from depths between 11 m and the bottom (18.5 m) by anaerobic conditions. Although documentation is mostly lacking on the impact of more distributed zooplankton populations for controlling phytoplankton in lakes with aerated hypolimnia, the significance of zooplankton grazing is well known. One can assume that encouragement of the large zooplankters should lead to greater phytoplankton loss rates and lower biomass. If that effect can be consistently realized, hypolimnetic aeration can offer much broader benefits in restoring lakes than are presently expected.

Growth and survival of cold-water fishes, as well as reproduction of some lake-spawning species (e.g., whitefish), can be severely limited in eutrophic lakes with anaerobic hypolimnia. Preferred temperatures for cold-water species are usually less than 18°C (Welch, 1980). When epilimnetic temperatures exceed that level, as is common even in cool-temperate areas, growth is reduced and fish seek cooler waters. If the cool hypolimnetic waters are devoid of oxygen, that habitat is not available and cold-water fish production may be impaired. In addition, the production of benthic fish-food organisms, especially chironomids, *Chaoborus*, and tubificids, could be expected to increase following a change from anaerobic to aerobic conditions (Fast, 1971a).

In three of the lakes listed in Table 9-1 (Hemlock, Ottoville Quarry, and Waccabuc), planted rainbow trout were observed to be distributed throughout the hypolimnion (Fast, 1973; Garrell et al., 1977; Overholtz et al., 1977). Before aeration, trout were excluded from the hypolimnion. Analyses of trout stomachs showed that they utilized food organisms that had increased in abundance in the hypolimnion as a result of aerobic conditions in the hypolimnion (e.g., *Chaoborus* and chironomids) or, like *Daphnia*, had migrated into the hypolimnion during daylight hours (Fast, 1973; Garrell et al., 1977).

UNDESIRABLE SIDE EFFECTS

Supersaturation of hypolimnetic water with N_2 was suggested as a possible problem that might lead to gas-bubble disease in fish. Although this apparently has not occurred as a result of hypolimnetic aeration, N_2 content can reach potentially damaging levels; 150-percent saturation relative to surface temperature and pressure occurred after 80 days of aeration in Lake Waccabuc (Fast et al., 1975).

Hypolimnetic aeration may increase eddy diffusion of nutrients into the epilimnion even though stratification is maintained. Metalimnetic P content increased by a factor of two in Lake Wesslinger during hypolimnetic aeration, which doubled the phytoplankton biomass, composed mostly of blue-greens (Steinberg and Arzet, 1984). One of the principal reasons for maintaining stratification during aeration is to prevent the recirculation of sedimented P. If this is not accomplished, hypolimnetic aeration loses some of its appeal over complete circulation as an enhancement to fisheries-usable habitat.

COSTS

The largest lake thus far treated by hypolimnetic aeration is Tegeler See, which is 420 ha (Table 9-1). Fifteen Limnox units were installed in 1980 to aerate 32×10^6 m^3 of hypolimnion, delivering 4.5 tons of O_2 per day. Cost of the Limnox aeration system was $2,118,000 (1983 dollars). This is about $2.40 per kg of O_2, or $5,040 ha^{-1} for six months of operation per year. Atlas Copco has reported the costs for six projects (R.S. Geney, personal communication). At a power rate of $0.07 kw/hour, the average operating cost/kg O_2 · day was $0.056 ± 0.02, and the average installed cost/kg O_2 · day was $354 ± 216.

REFERENCES

Atlas Copco. 1976. *Aeration of Lake Brunsviken,* Communications Dept., Wilrijk, Belgium.
Atlas Copco. 1980. Communications Dept., Wilrijk, Belgium.
Bengtsson, L., and Gelin, C. 1975. *Artificial Aeration and Suction Dredging Methods for Controlling Water Quality. Proc. Symposium on Effects of Storage on Water Quality,* Water Res. Center, Medmenham, England.
Bernhardt, H. 1967. Aeration of Wahnbach Reservoir without changing the temperature profile. *J Am Water Works Assoc* 9:943–964.
Bernhardt, H. 1974. *Ten Years Experience of Reservoir Aeration,* Seventh Conf. on Water Poll. Res., Paris.
Bianucci, G., and Bianucci, E.R. 1979. Oxygenation of a polluted lake in northern Italy. *Effluent Water Treat J* 19:117–128.
Björk, S. 1974. *European Lake Rehabilitation Activities,* Inst. of Limnol. Rept. Univ. Lund, Sweden.
Fast, A.W. 1971a. *The Effects of Artificial Aeration on Lake Ecology,* Water Pollut. Cont. Res. Ser. 16010 Exe 12/71, USEPA.
Fast, A.W. 1971b. Effects of artificial destratification on zooplankton depth distribution. *Trans Am Fish Soc* 100:355–358.
Fast, A.W. 1973. Effects of artificial hypolimnion aeration on rainbow trout (*Salmo gairdneri* Richardson) depth distribution. *Trans Am Fish Soc* 102:715–722.
Fast, A.W. 1977. Artificial aeration and oxidation of lakes as a restoration technique. In J. Cairns, Jr., K.L. Dickson, and F.E. Herricks, eds., *Recovery and Restoration of Damaged Ecosystems,* Charlottesville, VA: Univ. Press of Virginia.
Fast, A.W. 1979. Artificial aeration as a lake restoration technique. *Proc. Natl. Conf. Lake Rest.,* USEPA.
Fast, A.W., Dorr, V.A., and Rosen, R.J. 1975. A submerged hypolimnion aerator. *Water Resour Res* 11:287–293.
Fast, A.W., and Lorenzen, M.W. 1976. Synoptic survey of hypolimnetic aeration. *J Environ Eng Div ASCE* 102:1161–1173.
Fast, A.W., Lorenzen, M.W., Glenn, J.H. 1976. Comparative study with costs of hypolimnetic aeration. *J Environ Eng Div ASCE* 102:1175–1187.
Fast, A.W., Moss, B., and Wetzel, R.G. 1973. Effects of artificial aeration on the chemistry and algae of two Michigan lakes. *Water Resour Res* 9:624–647.
Garrell, M.H., Confer, J.C., Kirchner, D., and Fast, A.W. 1977. Effects of hypolimnetic aeration on nitrogen and phosphorus in a eutrophic lake. *Water Resour Res* 13:343–347.

Hess, L. 1977. *Lake Destratification Investigations. Job 1-3: Lake Aeration June 1, 1972 to June 30, 1977*, Final report, West Virginia Dept. Nat. Res., D-J Proj. F-19-R.
Holton, H., and Holton, G. 1978. *Sammerstilling av Undersøkelsesresultates 1972-1977*, Norsk Inst. for Vannfors., Report No. 0-5/70 (Norwegian Institute for Water Research, Oslo).
Holton, H., Brettum, P., Holton, G., and Kjellberg, G. 1981. *Kolbotnvatn med tillop; Sammerstilling av undersøkelsesresultates 1978-1979*, Norsk Inst. for Vannfors., Report No. 0-78007.
LaBaugh, J.W. 1980. Water chemistry changes during artificial aeration of Spruce Knob Lake, West Virginia. *Hydrobiologia* 20:201-216.
Lorenzen, M.W., and Fast, A.W. 1977. *A Guide to Aeration/Circulation Techniques for Lake Management*. Ecol. Res. Ser. EPA-600/3-77-004, USEPA.
McQueen, D.J., and Lean, D.R.S. 1984. Aeration of anoxic hypolimnetic water: Effects on nitrogen and phosphorus concentrations. *Verh Int Ver Limnol* 22:267-276.
Overholtz, W.J., Fast, A.W., Tubb, R.A., and Miller, R. 1977. Hypolimnion oxygenation and its effects on the depth distribution of rainbow trout (*Salmo gairdneri*) and gizzard shad (*Dorosoma cepedianum*). *Trans Am Fish Soc* 106:371-375.
Pastorak, R.A., Ginn, T.C., and Lorenzen, M.W. 1980. *Evaluation of Aeration/Circulation as a Lake Restoration Technique*, Report No. TC-3947, USEPA.
Pastorak, R.A., Lorenzen, M.W., and Ginn, T.C. 1982. *Environmental Aspects of Artificial Aeration and Oxygenation of Reservoirs: A Review of Theory, Techniques, and Experiences*, Tech. Report No. E-82-3, U.S. Army Corps of Engr.
Shapiro, J. 1979. The need for more biology in lake restoration. *Proc. Natl. Conf. on Lake Restoration*, USEPA.
Smith, S.A., Kramer, D.R., and Wirth, T.L. 1975. *Aeration as a Lake Management Technique*, Wisc. Dept. Nat. Res., Tech. Bull. No. 87.
Steinberg, C., and Arzet, K. 1984. Impact of hypolimnetic aeration on abiotic and biotic conditions in a small kettle lake. *Environ Tech Letters* 5:151-162.
Taggart, C.T., and McQueen, D.J. 1981. Hypolimnetic aeration of a small eutrophic kettle lake: Physical and chemical changes. *Arch Hydrobiol* 91:150-180.
Welch, E.B. 1980. *Ecological Effects of Wastewater*, Cambridge Univ. Press.
Whipple, W. Jr., Hunter, J.V., Trama, F.B., and Tuffey, T.J. 1975. *Oxidation of Lake and Impoundment Hypolimnia*, Water Resources Res. Inst., Proj. No. B-050-N.J., final report.

III

Procedures to Control Plant Biomass

10

Artificial Circulation

Artificial circulation of lakes has been employed as a management technique since at least the early 1950s. Initially it was used to prevent fish kills during winter in shallow, ice-covered lakes (Halsey, 1968). Nearly all of the reported applications of the technique to control eutrophication effects and water quality occurred later than the mid 1960s. It has been the most commonly used technique of those discussed in this book.

Circulation has been achieved by pumps, jets, and bubbled air. Complete circulation is usually the objective, and in the majority of cases examined either stratification was prevented or destratification resulted. Unlike hypolimnetic aeration, the temperature of the whole lake is raised with complete circulation; the greatest increase in temperature occurs at depths that were previously part of the hypolimnion.

The principal improvements in water quality caused by complete circulation are aeration and chemical oxidation of substances in the entire water column (Pastorak et al., 1980; 1982). Similar to hypolimnetic aeration, its main benefit is enlarging the suitable habitat for aerobic animals. Initially it was thought that circulation would reduce internal loading of P from profundal sediments, through increased oxidation. It also was believed that circulation would reduce algal biomass by increasing the mixed depth, thereby reducing available light, and by subjecting mixing algal cells to rapid changes in hydrostatic pressure (Fast, 1979). Although there are reasons for expecting reduced internal P loading and decreased phytoplankton biomass, there are equally reasonable explanations as to why completely opposite effects could occur. Thus it is not surprising that, in most instances, phytoplankton biomass and P content either showed no change or increased following circulation.

THEORETICAL EFFECTS OF CIRCULATION

Dissolved Oxygen

The principal, and probably the most reliable, effect of circulation, is to raise the dissolved oxygen content throughout the lake. If the lake is destratified the dissolved oxygen content in what was the hypolimnion will increase, and that in the epilimnion will decrease. This can occur from simple dilution. Additional reasons why the surface water may decrease in dissolved oxygen, however, are the transfer of oxygen-demanding substances toward the surface, and a decrease in photosynthesis in the photic zone, due to increased mixing (Haynes, 1973; Ridley et al., 1966; Thomas, 1966). Dissolved oxygen will continue to increase as circulation is maintained, largely because water undersaturated with oxygen is brought into contact with the air. Where the vertical transport of water is achieved by entraining water through releasing compressed air at some depth, little oxygen increase is achieved through direct diffusion from bubbles (King, 1970; Smith et al., 1975).

Physical Control of Phytoplankton Biomass

Circulation, theoretically, can result in reduced phytoplankton biomass. The principal cause would be light limitation, brought about by providing a greater depth of mixing. Limitation would occur by mixing plankton cells deep enough in the water column so that the total net received while in the shallower photic zone is insufficient for net photosynthesis (photosynthesis in excess of respiration) and thus any substantial growth or increase in cell mass. This is known as the "critical depth" concept, first formulated to predict the timing of the spring diatom bloom in the ocean (Sverdrup, 1953). By knowing light at the surface, compensation depth, and the extinction coefficient, the critical depth could be calculated as the point above which net production was possible; when that calculated depth exceeded the mixed-layer depth, a bloom could occur. Of course, this model is dependent upon some relationship between light intensity and gross photosynthesis, assuming a constant rate of respiration.

Talling (1971) has applied the same concept in lakes. He found that the combination of low surface light intensity and deep mixing prevented net photosynthesis during winter in relatively deeper lakes (> 30 m) of the English Lake District, but not in the shallower lakes (10 m). Normally, lakes are shallow enough to allow some net photosynthesis even in winter, but decreasing mixing depth, as stratification develops and surface light intensity increases in the spring, usually accounts for the large increase in net photosynthesis and the spring diatom bloom in deeper lakes.

The concept of physical control of phytoplankton growth has been extended to the effects of artificial circulation (Lorenzen and Mitchell, 1975; Murphy, 1962; Oskam, 1978). Forsberg and Shapiro (1980) and Shapiro et al. (1982) integrated the effects of nutrients with those of physical factors. By increasing the depth of mixing, the lake can potentially be returned to a winter condition where light is limiting, assuming depth and light attenuation are sufficient. In most cases this would not be great enough to prevent net biomass production completely, which is not the intention anyway. This is clearly shown in Figure 10-1, which is based on results from Kezar Lake (Lorenzen and Mitchell, 1975). Increasing mixing depth through increased circulation would be expected to substantially reduce algal biomass. Note that biomass is plotted as mass per area (g m^{-2}) in Figure 10-1, and in that unit the quantity would decrease by only 38 percent for a mixing-depth increase of 2 m to 6 m. Biomass concentration (g m^{-3}), however, would decrease by 80 percent, which would also include dilution.

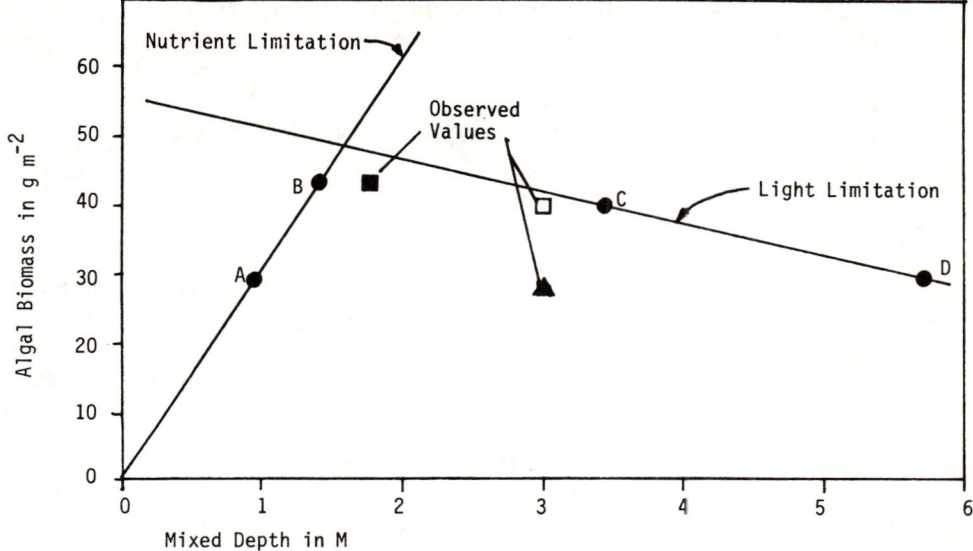

Figure 10-1 Theoretical and observed peak biomass of algae in Kezar Lake (from Lorenzen and Mitchell, 1975; Pastorak, et al., 1980, 1982; see text for explanation of points A–D). Solid circles: theoretical values; solid square: 1968, stratified; triangle: 1969, destratified; open square: 1970, destratified.

Lorenzen and Mitchell's model (Figure 10-1) also shows that nutrients may initially be limiting in the epilimnion, so that a slight increase in mixed depth may entrain water with higher nutrient content from below and biomass may increase (point A to point B in Figure 10-1). At some point, however, light will limit and productivity and biomass will decrease (point C to point D). This model predicts only the potential productivity without nutrient limitation and includes no losses from sinking, grazing, parasitism, or washout. Actual values may therefore fall below the line in Figure 10-1, as was the case for Kezar Lake.

Pastorak et al. (1980; 1982) suggest that little change in biomass may occur in oligotrophic lakes following circulation, because the slope of the ascending line in Figure 10-1 (nutrient limitation) would be less for such lakes. Because that line represents the maximum nutrient-limited biomass, any displacement of the biomass vertically by circulation would bring about a smaller change in biomass concentration in oligotrophic than in eutrophic lakes.

Oskam (1973; 1978) developed a model to express the effect of mixing-depth change on productivity and maximum biomass. Because net productivity (ΣP_{net}, mg C m^{-2} day^{-1}) is the difference between gross productivity and respiration in the mixed layer, the following equation should hold:

$$P_{net} = C\, P_{max} \left(\frac{F(i)\lambda}{\epsilon_W + C\epsilon_C} - 24\, r\, Z_m \right), \tag{10.1}$$

where C = chlorophyll a, in mg m^{-3}
P_{max} = maximum photosynthetic rate, in mg C mg chl^{-1} hr^{-1}
$F(i)$ = dimensionless function of light intensity
λ = daylight hours
ϵ_W = extinction of water, m^{-1}
ϵ_c = specific extinction coefficient per unit algae, m^2 mg chl^{-1}
Z_m = depth of mixing, m
24 = 24 hours day^{-1}
r = respiration/P_{max}.

According to this equation, as the depth of mixing increases, assuming uniform distribution of algae, net productivity decreases. The mixing depth can be increased by artificial circulation.

From equation (10-1), and setting P_{net} = 0, maximum biomass (mg chl m^{-3}) can be estimated as a function of mixing depth:

$$C_{max} = \frac{1}{\epsilon_C} \left[\frac{F(i)\lambda}{24\, rZ_m} - \epsilon_W \right]. \tag{10.2}$$

The maximum biomass possible is plotted for four different water extinction coefficients (Figure 10-2), assuming that ϵ_c = 0.02, $F(i)$ = 2.7, λ = 12, and r = 0.05. Further important assumptions are that nutrients are not limiting and there are no significant losses other than respiration. Accordingly, the maximum biomass concentration attainable in a lake with mixing depth = 5 m would be 220 mg m^{-3} chl a. If either nutrient limitation, grazing, sinking, or washout was significant, then the maximum would be correspondingly less. These relationships especially show the sensitivity of potential maximum biomass to mixing depth in shallow lakes.

Forsberg and Shapiro (1980) and Shapiro et al. (1982) developed an expanded model to include nutrient limitation and losses. Their equation for maximum biomass in the mixed layer is

$$C = \frac{\ln(Io/Iz')\, P_{max}^{sat} - D\theta Z_m \epsilon_q}{\epsilon_c D\theta Z_m + [\ln(Io/Iz')\, P_{max}^{sat}\, K_q]/TP}, \tag{10.3}$$

where C = chl a concentration, mg m^{-3}
Io = incident radiation
Iz' = radiation at a depth one-half the photosynthesis saturated light intensity (I_k; see Talling, 1971).
P_{max}^{sat} = maximum specific rate of photosynthesis under saturated nutrient concentration, mg C mg chl^{-1} day^{-1}
D = loss rate through sinking, grazing, parasitism, washout, etc., day^{-1}
θ = c/chl a ratio
Z_m = depth of mixing, m
ϵ_W = extinction coefficient of water, m^{-1}
ϵ_c = extinction coefficient of chl a, m^2 mg chl a^{-1}

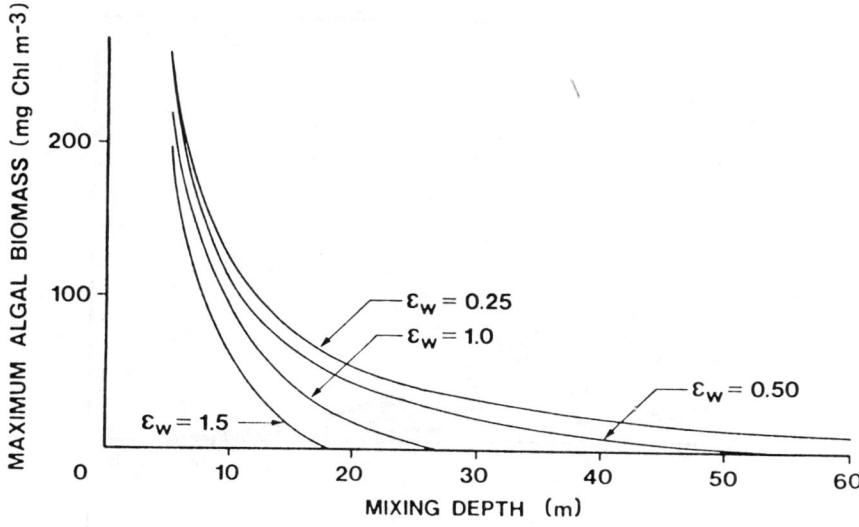

Figure 10-2 Relation of maximum chlorophyll concentration to mixing depth for different levels of nonalgal attenuation of light (from Oskam, 1978).

K_q = subsistence quota of TP, mg TP mg chl^{-1}
TP = total phosphorus concentration, mg m^{-3}.

The basis for the nutrient effect in equation 10.3 is an expression of cell nutrient quota, which is approximated by the ratio of TP to chl a:

$$P_{max} = P_{max}^{sat}\left(1 - \frac{K_q'}{TP/\text{chl }a}\right), \quad (10.4)$$

where P_{max} = maximum specific daily rate of photosynthesis at saturating nutrient level
K_q' = minimum ratio of TP/chl a required for photosynthesis to occur

The relationships, from this model, between maximum biomass (chl) per unit volume and per unit area in the mixed layer, and the depth of mixing, are shown in Figure 10-3. Clearly, the concentration of limiting nutrient determines the maximum biomass at any depth of mixing. This is an important point and should be part of predictions of improvements following circulation, because of the great potential for increasing the nutrient available to algae following destratification. Of course, if nutrient content is relatively high already and increases do not occur with mixing, then biomass concentration will decrease, with the greatest decreases occurring at mixing depths less than 10 m.

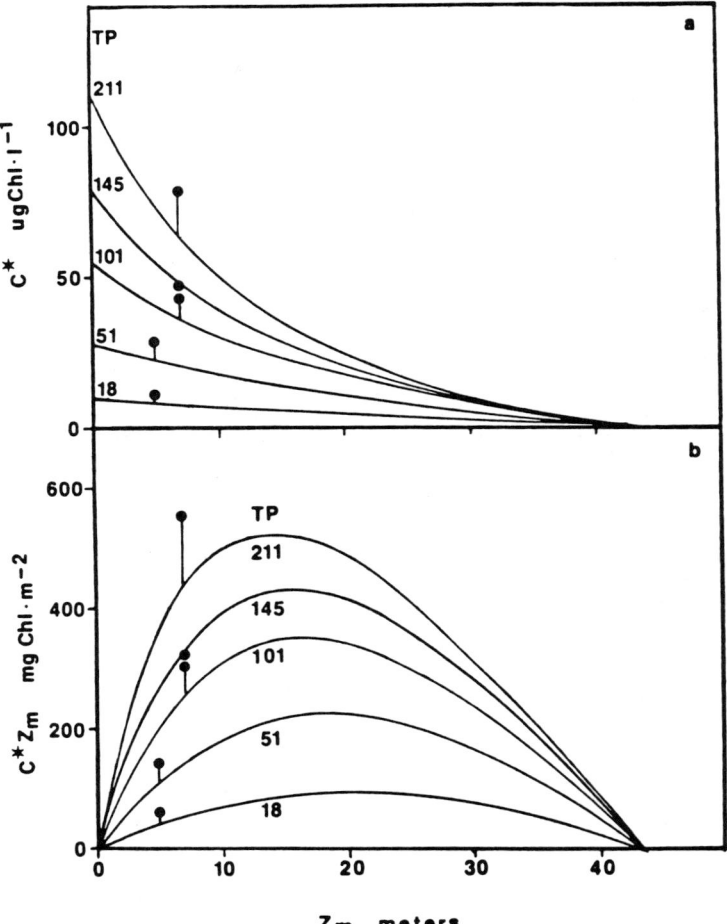

Figure 10-3 The effect of changes in the mixed depth and TP on (a) the maximum concentration of chlorophyll *a* and (b) the maximum aerial standing crop of chlorophyll *a* in the mixed layer of Twin Lake, as predicted by the model (closed circles indicate observed results; short lines connecting circles with the simulation lines indicate the deviation between predicted and observed results) (from Shapiro et al., 1982).

As Pastorak et al. (1982) indicate, there are several problems with application of this model. The most serious would appear to be difficulties in estimating loss rates and the effects of shifts in species composition and the nutrient history on the growth-rate response of algae. Nevertheless, rather good agreement between the model predictions and experimental results were observed and are shown in Figure 10-3 (Forsberg and Shapiro, 1980). The low level of complexity in this model makes it appealing as a tool to guide the application of the circulation technique.

Nutrients

Internal loading of phosphorus can theoretically be decreased through increased circulation. This would result in situations where anaerobic release of P from hypolimnetic sediments represented the dominant mechanism (Mortimer, 1942a, b). By aerating the sediment-water interface of lakes where iron is controlling P solubility, P should be absorbed from solution by ferric-hydroxy complexes (Stumm and Leckie, 1971). Thus P would be prevented from migrating from the high concentration in sediment interstitial water to overlying water. In some lakes, however, iron may not be controlling P. Calcium may control solubility in hard-water lakes, or the release rate may be largely a function of aerobic decomposition (Kamp-Nielson, 1975), and may actually increase as temperature at the sediment-water interface is raised in the circulation process. In the latter situation internal loading could actually increase following circulation. This is especially possible because sediment-water exchange rates are dependent upon water exchange at the interface, which would be increased by mixing (Lee, 1970). Because internal loading of P is recognized to be very great in unstratified, shallow lakes in which the sediment-water interface is usually aerobic (Jacoby et al., 1982; Kamp-Nielsen, 1975), reduced internal P loading probably cannot be expected to be a benefit of artificial circulation in such waters. Fast (1979) concurs with this evaluation of artificial circulation. Even if internal loading and total lake concentration of P decrease following circulation, the concentration available for growth in the photic zone may increase (Brosnan, 1983).

Other potential changes in chemical content resulting from circulation and aeration are conversion of ammonium to nitrate and the complexation and sedimentation of trace metals such as manganese and iron. Ammonium decrease can largely be attributed to increased nitrification brought about by aerobic conditions, which are necessary for that process (Brezonik et al., 1969; Toetz, 1979). This effect will be greater as the duration and completeness of hypolimnetic deoxygenation increases. The decrease in trace metals like manganese and iron should also tend to be greater as the oxygen deficit prior to reaeration increases. Because these metals diffuse from the sediment in their reduced, soluble forms, aeration will bring about their oxidation and subsequent complexation and precipitation. This can be an important benefit in lakes used for drinking-water supplies.

Effects on Phytoplankton Composition

There are at least three hypotheses to explain why a shift from dominance by bloom-forming blue-green algae to one by diatoms or green algae may result from circulation. These involve changes in (1) CO_2 and pH; (2) sinking rate; and (3) grazing by zooplankton, all of which could result from increased circulation.

Blue-green-algae-dominated cultures have been shown by King (1970; 1972), and Shapiro et al. (1973; 1975) and Shapiro (1984) to shift to ones dominated by green algae in response to decreased pH and associated increases in free CO_2 concentration. This has been explained by both authors as a greater capacity for blue-greens to absorb CO_2 at low concentrations, compared to green algae, although Shapiro et al.

(1982) have suggested that the rapid die-off of blue-greens following pH decrease is due to lysing of the blue-greens by viruses that are favored by low pH. King introduced the CO_2 hypothesis based on comparisons of algal populations and chemical conditions existing in sewage lagoons, and suggested that the potential for lakes to promote blue-green dominance increases as alkalinity (buffering capacity) decreases at any given P loading. Shapiro was able to shift dominance from blue-greens to greens with either HCl or CO_2 addition, but the shift was more complete if nutrient additions were also included. Increased circulation can cause CO_2 to increase and pH to decrease in the euphotic zone, by vertical transport of bottom water, in which CO_2 content is high due to respiration in the absence of photosynthesis. For circulation to promote the shift from blue-greens, the surface waters should not be nutrient-limited, because high content of N and P also exists in bottom water, which could increase blue-greens already present.

Because blue-green algae have pseudovacuoles, which can cause them to be buoyant, the increased stability brought about by thermal stratification and calm weather will favor blue-greens over nonbuoyant algae. Increased circulation can favor non-buoyant algae which tend to sink rapidly (especially diatoms) under stable conditions. Knoechel and Kalff (1975) were able to show that the specific growth rate of a blue-green (*Anabena*) was not different from that of a spring-dominating diatom (*Tabellaria*), which was succeeded by *Anabena* as thermal stratification developed in summer. This change was explained as a physical effect based on sinking-rate difference, rather than one based on growth-rate differences related to nutrient changes. Artificial circulation, preventing stratification, may have allowed the diatom to persist. Taylor (1966) did show that artificial destratification in a Thames River reservoir in late July promoted a second bloom of *Asterionella*, which had previously bloomed in the spring and had subsequently declined. Such a decline in *Asterionella* has been attributed to the combination of nutrient limitations and sinking losses (Lehman and Sandgren, 1978). One likely explanation for the second bloom in the Thames reservoir was a reinoculation of high Si content in the lighted zone, but conditions must have likewise been more favorable for the large diatom, because of a probable decreased sinking rate. Subsequent destratification in the fall, when blue-green algae were very abundant, did not result in a third *Asterionella* bloom. Destratification did not seem to affect blue-greens in this case. Bernhardt (1967) and Fast et al. (1973) have also observed increases in *Asterionella* following increased circulation.

At the same time that diatoms are sometimes favored by increased circulation, the advantage that blue-greens have in stable water, through their pseudovacuole regulation, is negated. Their vacuoles can adjust buoyancy, allowing their movement between the more lighted surface waters and the more nutrient-rich intermediate depths (Reynolds, 1975; Walsby and Reynolds, 1975) if the water column is stable. Increased circulation, however, can prevent that pattern. Bernhardt (1967) observed dispersal of an *Oscillatoria* population, located in the metalimnion, following circulation.

Forsberg and Shapiro (1980) and Shapiro et al. (1982) reported on artificial circulation experiments carried out in 1-m-diameter by 7-m-deep plastic bags in two lakes. To a great extent, results from these in situ mixing experiments verified the above hypotheses regarding changes in pH, CO_2, nutrients, and species composition.

At the slowest mixing rates, TP and chl a increased, with blue-green algae dominating the plankton. At intermediate and high rates of mixing, diatoms and greens tended to dominate the plankton, with biomass and TP increasing at the intermediate rate of mixing, but decreasing at the high rate. Green algae did not dominate unless pH was low and nutrients were high. Overall, abundance of algae was more related to nutrient content than to light availability, which was controlled by mixing up to 7 m.

These potential benefits of artificial circulation are illustrated together in Figure 10-4. Phosphorus and trace element inactivation due to aeration may result in reduced internal loading and, consequently, less algae. Algal abundance may also decrease because mixed depth increases and because silt stirred up by circulation could cause light to limit photosynthesis. This would be most likely in nutrient-rich lakes. Epilimnetic CO_2 may increase and pH decline as a result of mixing bottom waters enriched with CO_2. The lowered pH would stimulate cyanophage activity, lysing blue-greens, while increased free CO_2 concentration could provide green algae with a growth advantage. Thus the more edible size of phytoplankton, plus the enlarged aerobic, dimly-lit habitat, serving as a refuge from zooplanktivorous fishes, could result in greater loss rates of phytoplankton through grazing (Figure 10-4).

Not included in this diagram is the potential shift in phytoplankton composition from blue-greens to diatoms because of the increased mixing depth. As mentioned earlier, one explanation of the diatom-to-blue-green succession has been ascribed to increased diatom sinking as stability of the water column increases during summer. Succession was reversed by artificial circulation in a Thames River reservoir.

DEVICES AND AIR QUANTITIES

Introduction of compressed air through a diffuser or perforated pipe located at depth employs the "air-lift" method of circulating lakes, in which water is upwelled by the rising plume of air bubbles (Pastorak et al., 1980; 1982). Although techniques using pumps and water jets have been used to circulate lakes, the air-lift method, through diffusion of compressed air, is apparently the least expensive and is easiest to operate (Lorenzen and Fast, 1977). If the lake is already stratified, mixing is usually achieved only above the depth of air injection. If the lake is not stratified, however, injection near the surface can prevent stratification (Pastorak et al., 1980; 1982).

Injection of compressed air at maximum depth usually affords the greatest rate of mixing, because water flow is a function of depth of release and air flow rate. Lorenzen and Fast have concluded that an air flow rate per lake surface area of 9.2 m^3 min^{-1} km^{-2} should provide adequate surface reaeration and other benefits of circulation. In only 42 percent of the cases cited in Table 10-1 did the air flow rate approach or exceed that critical value. The flow rate is substantiated by the cases in Table 10-1 where before-and-after temperature data were provided (see Pastorak et al., 1982). Figure 10-5 is a plot of the degree of destratification (% reduction in Δt in the water column) related to air-flow rate per unit area. Except for three observations, flow rates approaching or in excess of 9.2 m^3 min^{-1} km^{-2} produced complete mixing, or 100 percent decrease in Δt. In two of the three lakes to the right of the line in Figure 10-5, the final Δt was $< 3°C$.

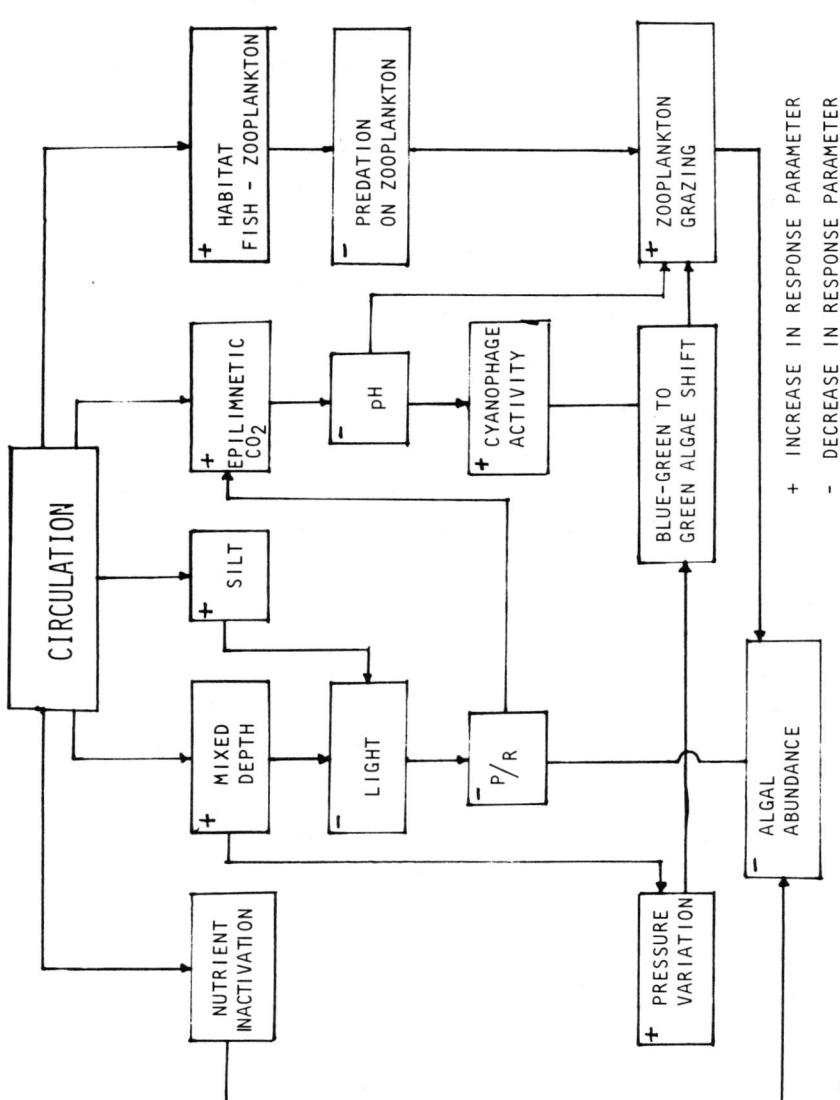

Figure 10-4 Potential beneficial effects of artificial circulation on phytoplankton (from Pastorak et al., 1980; modified from Shapiro, 1979).

Table 10-1 Lakes Receiving Treatment by Artificial Circulation, with Associated Characteristics (modified from Pastorak et al., 1980, 1982).

Lake	Max. Depth	Mean Depth	Device Depth	Volume ($10^6 m^3$)	Area (ha)	Q Air ($m^3 min^{-1}$)	Q Air ($m^{-3} \times 10^6$)	Q Air (km^{-2})	References
Clines Pond, Oregon	4.9	2.5	4.9	0.003	0.13	0.028[a]	10.2	21.6	Malueg et al., 1973
Parvin, Colorado	10.0	4.4	10.0	0.849	19.0	2.1[a]	2.5	11.18	Lackey, 1972
Section 4, Michigan	19.1	9.8	18.3	0.110	1.1	2.21[a]	20.0	200.0	Fast, 1971a
Boltz, Kentucky	18.9	9.4	18.9	3.614	39.0	3.17[a]	0.88	8.17	Symons et al., 1967; 1970; Robinson et al., 1969
University, No. Carolina	9.1	3.2	9.1	2.591	80.9	0.40[a]	0.15	0.49	Weiss and Breedlove, 1973
Kezar, New Hampshire	8.2	2.8	8.2	2.008	73.0	2.83[a]	1.41	3.88	Anon, 1971; Haynes, 1973
Indian Brook, New York	8.4	4.1	2.2	0.302	7.3	4.53[a]	15.0	62.06	Riddick, 1957
Prompton, Pennsylvania	10.7	3.7	10.7	0.193	112.0	4.53[a]	1.08	4.04	McCullough, 1974
Cox Hollow, Wisconsin	8.8	3.8	8.8	1.480	38.8	2.04–4.08[a]	1.38–2.76	5.26–10.52	Wirth and Dunst, 1967; Wirth et al., 1970
Stewart, Ohio	7.5	3.4	7.0	0.090	2.6	0.25[b]	2.83	9.80	Barnes and Griswold, 1975
Wahnbach, West Germany 1961–62 1964	43.0	19.2	43.0	41.618	214.0	2.01[b] 5.95[b]	0.048 0.143	0.94 2.78	Bernhardt, 1967
Starodworskie, Poland	23.0		23.0		7.0	0.27[a]		3.81	Lossow et al., 1975
Roberts, New Mexico	9.1	4.4	9.1	1.233	28.3	3.54[a] 2.26[a]	2.87 1.84	12.51 8.00	U.S. EPA, 1970; McNally, 1971
Falmouth, Kentucky	12.8	6.1	12.8	5.674	91.0	3.26 ?	0.58	3.58	Symons et al., 1967; 1970; Robinson et al., 1969
Test II, United Kingdom	10.7	9.4	10.7	2.405	25.4	2.01[a]	0.84	7.92	Knoppert et al., 1970
Test I, United Kingdom	10.7	9.4	10.7	2.097	22.7	2.01[a]	0.96	8.86	Knoppert et al., 1970

236 Procedures to Control Plant Biomass

Table 10-1 (continued)

Lake	Max.	Depth Mean	Device	Volume ($10^6 m^3$)	Area (ha)	$Q Air$ ($m^3 min^{-1}$)	$Q Air$ ($m^{-3} \times 10^6$)	$Q Air$ (km^{-2})	References
Mirror, Wisconsin	13.1	7.6	12.8	0.40	5.3	0.45^a	1.13	8.55	Smith et al., 1975; Brynildson and Serns, 1977
Växjosjön, Sweden	6.5	3.5	6.0	3.1	87.0	7.2^a	2.32	8.28	Bengtsson and Gelin, 1975
Corbett, British Columbia	19.5	7.0	19.5	1,689	24.2	4.5^a	2.66	18.52	Halsey, 1968; Halsey and Galbraith, 1971
Buchanan, Ontario	13.0	4.9	13.0	0.42	8.9	0.28^a	0.67	3.17	Brown et al., 1971
Maarsseveen, United Kingdom	29.9	14.0	19.0 / 29.9	8.018	60.7	2.49^a	0.31	4.10	Knoppert et al., 1970
Casitas, California	82.0	26.8	39.0 / 55.0	308.0	1,100.0	17.84^b	0.06	1.62	Barnett, 1975
Hyrum, Utah	23.0	11.9	15.2	23.1	190.0	2.83^b	0.17	1.49	Drury et al., 1975
Waco, Texas	23.0	10.7	23.0	128.0	2,942.0	3.11^b	0.02	0.10	Biederman and Fulton, 1971
Catharine, Illinois	11.8	5.0	8.5	3.034	59.5	0.76^c	0.25	1.27	Kothandaraman et al., 1973
El Capitan, California 1965 1966	62.0	9.8 / 9.4	21.3 / 28.3	17.99 / 21.05	183.9 / 222.0	6.09^b / 6.09^b	0.34 / 0.29	3.31 / 2.74	Fast, 1968
Calhoun, Minnesota	27.4	10.6	23.0	18.01	170.4	2.83^b / 3.54^b	0.16–0.20	1.66–2.08	Shapiro and Pfannkuch, 1973
Eufaula, Oklahoma	27.0	16.2	27.0	703.1	414.8×10^2	33.98^c	0.05	0.06	Leach et al., 1980
Pfaffikersee, Switzerland	35.0	18.0	28.8	56.5	325.0	6.0^b	0.11	1.85	Thomas, 1966 Ambuhl, 1967
Wahiawa, Hawaii	26.0	8.0	2.7	1.7	20.0	2.4^b	1.4	12.0	Devick, 1972
Trasksjön, Sweden	4.0	3.0	4	0.365	12.1	a			Karlgren and Lingren, 1963

Table 10-1 (continued)

Lake		Max.	Depth Mean	Device	Volume ($10^6 m^3$)	Area (ha)	$Q\ Air$ ($m^3\ min^{-1}$)	$Q\ Air$ ($m^{-3} \times 10^6$)	$Q\ Air$ (km^{-2})	References
Altoona, Georgia	1968 1969	46.0	9.4	42.7	453	4,800	21.6-[b] 27.7 27.7[b]	0.05- 0.86 0.06	0.45- 0.58 0.58	USAE, 1973; Raynes, 1975
Lafayette, California		24.0	9.1	18.0	5.243	53	1.68[c]	0.32	3.17	Loverty and Nielsen, 1970
Hot Hole, New Hampshire		13.3	5.7	13.3	0.733	12.9	0.59[a]	0.80	4.57	NHWSPCC, 1979
Heart, Ontario		10.4	2.7	10.0	0.392	14.5	0.23-[a] 0.92	0.58- 2.34	1.56- 6.33	Nicholls et al. 1980 Nicholls[d]
Clear, California		15.0	10.2	14.0	115.9	1,217	17[a]	6.82	114	Rusk[d]
Kremenchug, Poland		3.0	2.0	2.6	0.002	0.12	4.38[a]	1,750	3,500	Ryabov et al., 1972 Sirevko et al., 1972
Tarago, Australia		23.0	10.5	14.0	27.6	360	3.0-[c] 9.0 3.0-[c] 7.50	0.08 0.24 0.08- 0.20	0.83- 2.50 0.83- 2.08	Bowles et al., 1979
Silver, Ohio		12.0	4.22	10.0	1.68	38.44	3.37[b]	2.01	8.77	Brosnan, 1983
			Pumping or Jet				$Q\ Water$ ($m^3\ min^{-1}$)			
King George VI, United Kingdom		16.0	14.0	10.0	20.0		142.0 Water Jet[c]			Ridley et al., 1966
Queen Elizabeth II, United Kingdom		17.5		17.5			128.0 Water Jet[c]			Ridley et al., 1966
Ham's, Oklahoma		10.0	2.9	1.2	115.0		40.0 Axial-Flow Pump[a]			Steichen et al., 1974; Toetz, 1977a;b
Stewart Hollow, Ohio		7.6	4.6	7.6	0.148		3.2 Axial-Flow Pump[a]			Garton et al., 1978
Cladwell, Ohio		6.1	3.0	6.1	0.123		4.0 Axial-Flow Pump[b]			Irwin et al., 1966
Pine, Ohio		5.2	2.1	5.2	0.121		5.7 Axial-Flow Pump[b]			Irwin et al., 1966

Table 10-1 (continued)

Lake		Depth		Volume $(10^6 m^3)$	Area (ha)	$Q\ Water$ $(m^3\ min^{-1})$	$Q\ Air$ $(m^3\ min^{-1})$	$Q\ Air$ $(m^{-3} \times 10^6)$	$Q\ Air$ (km^{-2})	References
		Max.	Mean	Device						
				Pumping or Jet						
Vesuvius, Ohio		9.1	3.6	9.1	1.554	42.5	Axial-Flow Pump[c]			Irwin et al., 1966
Arbuckle, Oklahoma	1975	24.7	9.5	6.0	89.3	951.0	Axial-Flow Pump[c]			Toetz, 1977 a; b; 1979
	1977			2.0	$\times 10^2$					
West Lost, Michigan		12.8	6.2	11.9	0.089	1.4	Pump[c]			Hooper et al., 1953

[a]Flow rate produced destratification.
[b]Partly Mixed.
[c]Flow rate inadequate to destratify.
[d]R.A. Pastorak, personal communication.

Artificial Circulation 239

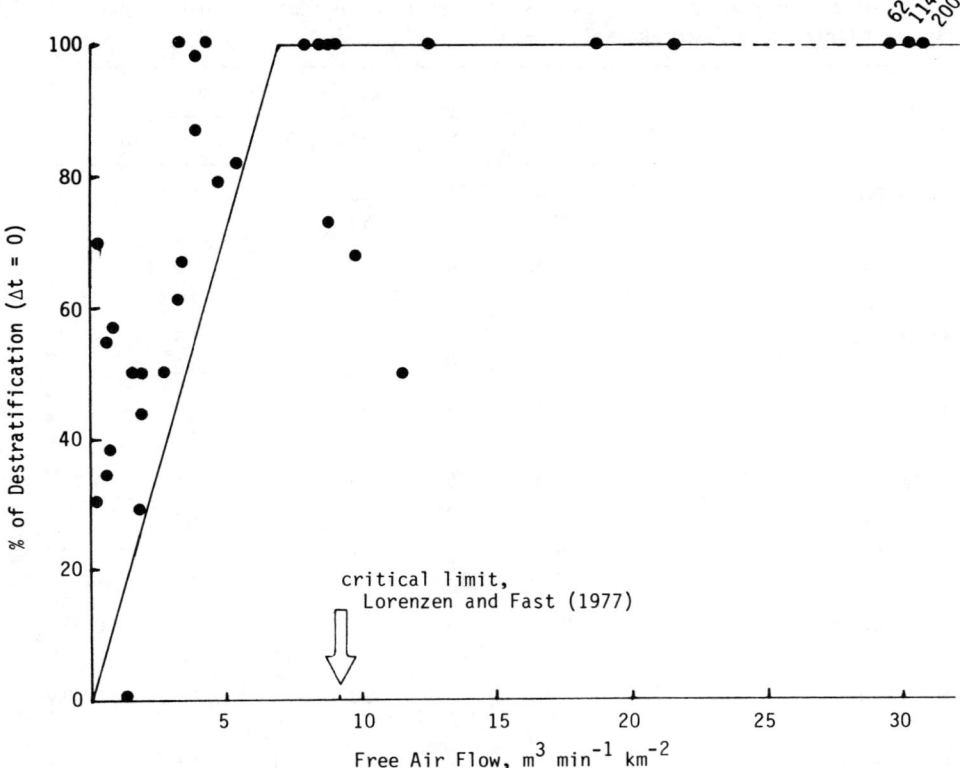

Figure 10-5 Percent destratification, based on surface-to-bottom temperature differences (Δt) before and after circulation, and free air flow (data from Pastorak et al., 1982).

In 30 of the 45 cases cited for the air-lift technique, where temperature data were available, the air-flow rates presented were apparently adequate to either cause destratification or prevent stratification (Table 10-2). The criterion for satisfactory mixing was a temperature difference between surface and bottom, after circulation, less than or equal to 3°C (Pastorak et al., 1982).

Mechanical mixing devices have been used less frequently than compressed air (Table 10-1). Apparently there has been no attempt to define a critical flow rate for pumping to destratify a lake, though destratification was complete ($\Delta t < 3°C$) for five of the 11 cases cited in Table 10-1.

The usual situation during artificial circulation is that surface waters in a stratified lake cool slightly during the destratification process, and the bottom waters heat up as much as 15° to 20°C, to approach a temperature similar to the surface water (Pastorak et al., 1982). If the air-flow rate is inadequate, according to Lorenzen and Fast's (1977) critical limit—which was true for about 58 percent of the cases listed in Table 10-1—then microstratification may develop at the surface (Fast, 1973). That would provide light conditions that are highly desirable for phytoplankton

Table 10-2 Summary of Lake Responses to Artificial Circulation, Diffused-Air Systems Only (from Pastorak et al., 1982)

Parameter	N		Lake Responses				χ^2
			+	−	0	?	
Δt After[a]	45	No.	15	30			5.00[b]
		%	33	67			
Secchi depth	19	No.	4	10	2	3	6.50[b]
		%	21	53	11	16	
Dissolved oxygen	41	No.	33	1	2	5	55.2[d]
		%	80	1	5	12	
Phosphate	17	No.	3	5	7	2	1.60
		%	18	29	41	12	
Total P	20	No.	5	6	8	1	0.74
		%	25	30	40	5	
Nitrate	20	No.	7	8	3	2	2.33
		%	35	40	15	10	
Ammonium	20	No.	3	13	3	1	10.5[c]
		%	15	65	15	5	
Iron/manganese	22	No.		20	2		33.1[d]
		%		91	9		
Epilimnetic pH	21	No.	1	9	8	3	6.33[b]
		%	5	43	38	14	
Algal density	33	No.	6	14	8	5	3.71
		%	18	42	24	15	
Biomass/chlorophyll	23	No.	5	6	6	6	0.12
		%	22	26	26	26	
Green algae	18	No.	7	4	7		1.00
		%	39	22	39		
Blue-green algae	25	No.	5	13	5	2	5.57
		%	20	52	20	8	
Ratio of green to blue-green algae	21	No.	11	3	6	1	4.90
		%	52	14	29	5	

[a] Temperature differential between surface and bottom water during artificial mixing. + means $\Delta t > 3°C$; − means $\Delta t \leq 3°C$.
[b] $p < 0.05$ Goodness-of-fit test to uniform frequency distribution for +, −, 0 responses only.
[c] $p < 0.01$
[d] $p < 0.001$

production, because the ratio of effective mixing depth to "critical depth" would be rather small. On the other hand, if the air-flow rate is too high, sediment can be suspended in the water column.

The basis for the "rule of thumb" air-flow rate of 9.2 m^3 min^{-1} km^{-2} is a relationship among air-flow rate, depth, and flow rate of upwelled water above an orifice (Lorenzen and Fast, 1977; Pastorak et al., 1982):

$$Q_w(X) = 35.6 \, C \, (X + 0.8) \, \frac{-V_0 \ln\left(1 - \frac{X}{h + 10.3}\right)^{1/2}}{\mu_b}, \quad (10.5)$$

where $Q_w(X)$ = water flow rate, m^3 s^{-1}
  C = $2V_0 + 0.05$ m^3 s^{-1}
  X = height above orifice, m
  V_0 = air flow, m^3 s^{-1} at 1 atm
  h = depth of orifice, m
  μ_b = $25 V_0 + 0.7$ m^3 s^{-1}.

Using this estimate of water flow rate, the effect of various air-flow rates on hypothetical lake and reservoir morphometry was studied (Chen and Orlob, 1975). As will be seen from an examination of the results of 38 air lift cases over a range of lake and reservoir areas, volumes, and depths, an air-flow rate equal to or greater than the 9.2 m^3 min^{-1} km^{-2} level consistently achieves destratification.

The diffuser should be a pipe with multiple orifices, usually located in the deepest point in the lake, but suspended sufficiently well off the bottom (1-2 m) to minimize sediment entrainment. Orifice spacing should be about 0.1 times the depth of air release, because the rising water plume will spread horizontally at 0.05 m per m rise (Lorenzen and Fast, 1977).

Another approach to designing an air-lift system to destratify lakes and reservoirs was given by Davis (1980). This approach requires that:

1. temperature or density profile be known or assumed,
2. theoretical energy required to overcome existing stability and added heat input be calculated,
3. free air-flow rate at the compressor be calculated,
4. length and internal diameter of perforated pipe (0.8 mm diameter holes) be calculated, and
5. anchor weight be determined.

Stability is calculated first, as the difference between the unmixed, existing density gradient, and the mixed condition:

$$S = g \sum_{i=1}^{n} p_{im} V_i h_i - \sum_{i=1}^{n} p_{is} V_i h_i, \quad (10.6)$$

where S = stability, joules $(kg\ m^2\ s^{-2})$
 g = acceleration due to gravity, ms^{-2}
 ρ_i = density of layer i, $kg\ m^{-3}$
 V_i = volume of layer i, m^3
 h_i = height of centroid of layer i, m
 m = mixed
 s = stratified

The energy required for destratification is calculated by

$$E = S + R - W, \tag{10.7}$$

where S = stability, R = heat input, and W = wind energy, all in joules. Wind is neglected, as a conservative approach, so that mixing is possible without wind. R can be approximated as 5 J m^{-2} day^{-1}.

The required air-flow rate (Q) in m^3 s^{-1} is given by

$$Q = \frac{0.196\,E}{T \ln\left(1 + \frac{D}{10.4}\right)}, \tag{10.8}$$

where E = energy input required; 20 times the theoretical level is recommended
 T = time to achieve destratification, sec
 D = depth of diffuser, m.

The volume of water entrained by the air bubbles from a line source (V_e) is recommended to be 2.5 times the volume of the lake or reservoir and is calculated according to

$$V_e = 0.486\,LT \left(\frac{gQ}{L}\right)^{1/3} \left(1 + \frac{D}{10.4}\right)^{-1/3} \ln\left(1 + \frac{D}{10.4}\right), \tag{10.9}$$

where L = length of perforated pipe, given by

$$L = 3.73 \left\{ \frac{V^3 \left(1 + \frac{D}{10.4}\right)}{T^3 Q \left[\ln\left(1 + \frac{D}{10.4}\right)\right]^3} \right\}^{1/2}, \tag{10.10}$$

where V = reservoir volume, m^3.

Pastorak et al. (1982) compared the calculated flow rates required by the two procedures, using an example from Davis (1980) for a body of water with a volume of 20×10^6 m^3, a maximum depth of 20 m, and an area of 1.2×10^6 m^2. The flow rate recommended by the Davis procedure would be 3.5 m^3 min^{-1} km^{-2}, or 70 1 s^{-1}. By the Lorenzen and Fast (1977) procedure the rate would be 6 m^3 min^{-1} km^{-2}, or

120 l s^{-1}, nearly twice the Davis rate. The "rule of thumb" rate of 9.2 m^3 min^{-1} km^{-2} is the midpoint of a range, 6.1-12.3 m^3 min^{-1} km^{-2}, the lower end of which was used here because deeper lakes generally require less air to mix than do shallow lakes (Pastorak, personal communiation).

According to equation 10-10, the pipe length would be 216 m, based on 70 l s^{-1} and 182 m based on 120 l s^{-1} for destratification to occur in five days. The air pressure at the compressor can be estimated by summing the hydrostatic pressure represented by the water depth over the pipe, the mean excess pressure above the hydrostatic pressure at the end of the pipe (related to pipe length), and the friction loss in the pipe (related to pipe diameter). For the example lake, Davis (1980) selected a high-density polyethylene pipe of diameter 50.8 mm, perforated with 1-mm-diameter holes spaced at 0.3 m, and a pressure of 5.3 bar at the compressor, all to provide the 70 l s^{-1} free air flow.

COSTS

Cost information for artificial circulation on a project basis is scarce. However, Lorenzen and Fast (1977) cite an annual cost of $113,000 (1983 dollars) for two air compressors producing air at 1200 ft^3 (34.3 m^3) min^{-1} at standard conditions. This price includes pipes and air diffusers. At the recommended rate of 9.2 m^3 min^{-1} km^{-2}, this represents a cost of $303 ha^{-1}, which is modest relative to other restoration techniques.

EFFECTS OF CIRCULATION ON TROPHIC INDICATORS

The three indicators that have improved most consistently following artificial circulation are dissolved oxygen, ammonium, epilimnetic pH, and the trace nutrients iron and manganese. Table 10-2 shows an analysis of the results presented by Pastorak et al. (1982). Dissolved oxygen increased and trace metal decreased in a very high percentage of cases studied, while favorable changes in ammonium and pH were less frequent. All four variable changes were statistically significant. These changes are a result of increased aeration of the water column, caused by increased contact with the atmosphere.

The increased aeration that results in the complexation and precipitation of Fe and Mn should also be effective in decreasing P content, for the same reason. The results for P, however, are much less impressive (Table 10-2). The cases where P increased or did not change following circulation were more frequent (65 %) than those where decreases occurred. For many of the cases examined, there may have been other sources of internal loading that could be more significant than release from pelagic sediments into anoxic overlying water. These could include aerobic release from littoral sediment, plant decomposition, or littoral release from photosynthetically caused high pH, although pH usually decreased with circulation. Also, external loading may have represented most of the input to the water column. If aerobic release through microbial decomposition is the principal mechanism for P internal loading in unstratified lakes (Kamp-Nielsen, 1974; 1975), then destratification, along with increased water exchange at the mud-water interface and increased temperature, may have resulted in greater release of P than occurred before circulation.

Transparency worsened more often than it improved (53 % versus 21 %) in cases examined following circulation (Table 10-2). Transparency may decrease following treatment, if (1) the photic zone is initially nutrient-limited such that phytoplankton content increases following circulation; (2) circulation is too weak, resulting in microstratification and more favorable light climate for productivity; or (3) circulation is so intense that particulate matter becomes resuspended. These complicating effects were probably responsible for transparency increase being unimpressive in most cases.

Phytoplankton content decreased following circulation in less than half the cases examined (Table 10-2). A decrease in blue-green algae, on the other hand, occurred in the majority of cases. The ratio of green to blue-green algae also increased proportionately. In eight of the 15 cases in which mixing was complete, the amount of blue-greens decreased, while in three cases they increased and in four cases there was no change. In seven cases where mixing was incomplete, five showed blue-green increases and two showed decreases. There is a strong indication that increased circulation reduces phytoplankton and especially blue-green algae in waters where light can be made limiting and where air-flow rate is adequate to prevent microstratification. The results of pH and CO_2 change following mixing lend some support to a contribution by those variables to decreased blue-greens, increased ratios of greens to blue-greens, or both (Pastorak et al., 1982). Where pH declined to a value less than 7.5, a shift of blue-greens usually occurred. Where no shift occurred, the change in pH was generally insignificant. Shapiro (1984) contends that the CO_2/pH mechanism is the principal one controlling the blue-green to green shift and that if mixing is fast enough to allow sufficient CO_2 transport from the atmosphere to drop the pH substantially, a shift may occur. Reynolds et al. (1984) have suggested, based on enclosure experiments, that intermittent mixing will reduce the total biomass as well as that of blue-greens because of an interference in the growth of fast-growing species and the population build-up of slower growing species.

Circulation can increase the aerobic environment, as noted earlier, which in turn can greatly influence the depth distribution of zooplankton. If allowed to distribute themselves to greater but more poorly lit depths, zooplankton should be able to avoid predation by fish (Zaret and Suffern, 1976). Shapiro et al. (1975) have shown the most positive results in this regard. *Daphnia* increased from five to eight times and was distributed to greater depths in an incomplete destratification of Lake Calhoun. Fast (1971b) also found that circulation resulted in most of the zooplankton occupying water below 10 m following circulation, while most occupied depths less than 10 m before circulation. It was noted by Pastorak et al. (1980; 1982) that, in either of the 13 cases examined, circulation increased the depth distribution of some or all zooplankters. In others zooplankton were distributed to depth before circulation. Abundance increased in ten of 15 cases examined. Such abundance increases could be due to reduced predation from planktivorous fishes, as found by Shapiro et al. (1975), Andersson et al. (1975), and Kitchell and Kitchell (1980).

Pastorak et al. (1980; 1982) have reviewed results on macroinvertebrates and fish following circulation, although the cases where those populations were studied are few. In most instances (6 to 8), the expanded aerobic habitat resulted in increased abundance of macroinvertebrates, and increased diversity resulted in seven of eight cases examined. In all cases fish expanded their depth distribution following circulation, but growth-rate increases were seldom observed, possibly because studies were not of long enough duration to detect changes; higher trophic levels usually respond more slowly to manipulation. Fish kills were averted in several cases, and where temperature remained satisfactory (temperatures $> 20°$ should be avoided) salmonids survived in eutrophic lakes.

UNDESIRABLE EFFECTS

There are several potential adverse effects of artificial circulation, some more likely to occur than others; 13 are illustrated in Figure 10-6. If nutrients are limiting productivity in the epilimnion, then circulation may increase particulate P, which could be mineralized to the usable form, or highly dissolved P itself may be transferred to the lighted zone. Water transparency could then become worse than before circulation, due to increased silt, as well as algal biomass. Increased algal abundance and photosynthesis would lower epilimnetic CO_2, raise pH, and prevent the succession from blue-greens to greens. Because blue-greens would tend to dominate as a result of the increased productivity and concomitant chemical changes, zooplankton would have a lesser effect on algal loss rates through grazing. The existing blue-green crop could then increase.

Figure 10-6 suggests that reduced algal sinking may cause increased algal abundance, but as indicated under the description of benefits, such an increase could be represented by diatoms. In that case it would be recognized as a benefit, because diatoms would tend to dominate the phytoplankton and would be less objectionable.

The diagram also assumes that the air-flow rate is adequate to attain complete mixing. If air flow is not adequate, then partial stratification may occur during periods of summer heating, and algal abundance (especially blue-greens) would increase, not because of increased nutrients, decreased grazing, or decreased sinking, but because of increased available light.

Temperature increase is also omitted from Figure 10-6. The increase of 15° to 20°C in the hypolimnetic waters as a result of complete circulation may in fact be the most adverse effect. This is especially true where cold-water fish species are involved.

An effect on fish from supersaturated N_2 has been suggested. See Chapter 9, Hypolimnetic Aeration, for comments on its potential significance.

SUMMARY AND RECOMMENDATIONS

Artificial circulation has been recommended as an inexpensive, efficient restoration technique (Pastorak et al., 1980). The technique should be most applicable in

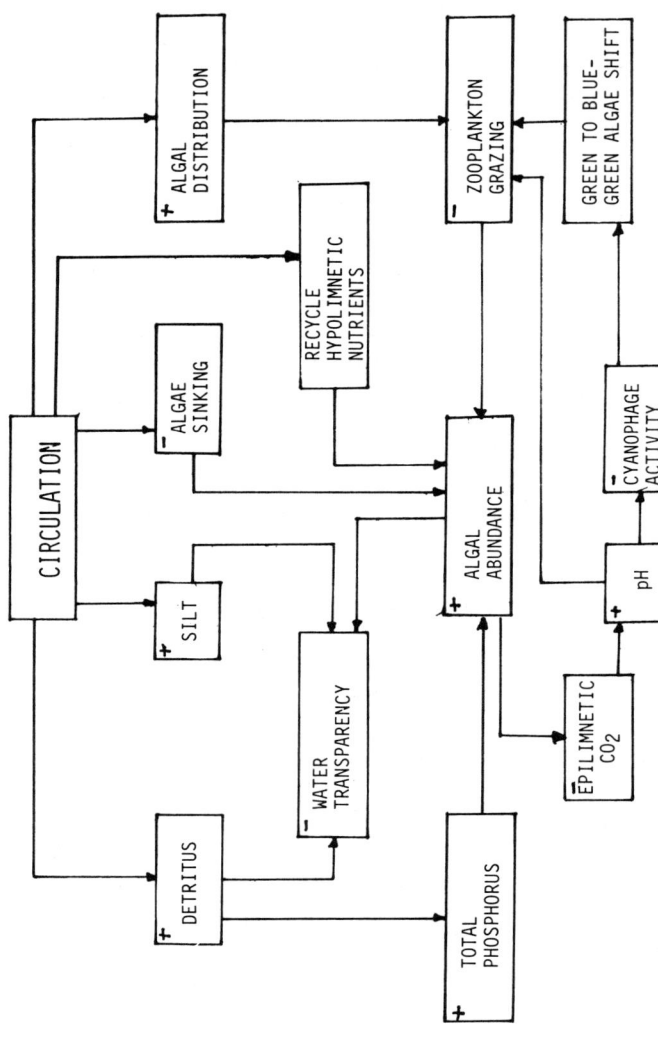

Figure 10-6 Potential adverse effects of artificial circulation, including the promotion of blue-green algal blooms (from Pastorak et al., 1980; modified from Shapiro, 1973).

lakes that are not nutrient-limited and where oxygen depletion is a threat to warm-water fish and the quality (metal content) of water supplies. The best record of improvement has been observed with dissolved oxygen, Fe and Mn content, ammonium, and pH. The principle of increasing the depth of mixing, thereby decreasing available light to plankton algae, however, may also have a good chance of working in lakes where nutrients usually are not limiting. Furthermore, such mixing may discourage blue-green algae while encouraging diatoms and greens. If the lake is marginally nutrient-limited, then more algae, and even more blue-greens (especially if pH is raised) may result, which is probably why algal abundance and blue-greens have decreased following circulation in only about half of the cases cited.

The depth at which compressed air is released is critical to the problem of preventing the persistence of anaerobic bottom water. Likewise the sizing of the system for flow rate, whether using compressed air or a pump, is critical to achieving complete mixing and preventing microstratification at the lake surface. To avoid these problems, placing air diffusers at the lake bottom and using a combination of compressed air and a surface pump for very deep lakes is recommended (Pastorak, 1980; 1982). The flow rate of free air per unit of lake surface should be about 9.2 m^3 min^{-1} km^{-2} and within the range 6.1-12.3 m^3 min^{-1} km^{-2}. Nutrient transport to the surface lighted zone should be minimized and sedimentation maximized if circulation is begun prior to stratification or done gradually when started after stratification.

Circulation would probably be best used alone as a restoration technique and not combined with other methods designed to reduce phosphorus content. This is because the benefits for algal control would be best achieved in nonnutrient-limited situations, and because increased circulation may encourage internal loading of P and counteract other efforts to lower P. The exception may be diversion in which a large internal loading persists.

REFERENCES

Ambuhl, H. 1967. Discussion of impoundment destratification by mechanical pumping, by W.H. Irwin, J.M. Symons, and G.G. Robeck. *J San Eng Div ASCE* 93:141-143.

Am Water Works Assoc. 1971. Artificial destratification in reservoirs. Committee Report, 63:597-604.

Andersson, G., Berggren, H. Cronberg, G., and Gelin, C. 1978. Effects of planktivorous and benthivorous fish on organisms and water chemistry in eutrophic lakes. *Hydrobiologia* 59:9-15.

Anon. 1971. *Algae Control by Mixing*. New Hampshire Water Supply and Pollut. Cont. Comm.

Barnes, M.D., and Griswold, B.L. 1975. Effects of artificial nutrient circulation on lake productivity and fish growth. *Proc. Conf. on Lake Reaeration Research*, ASCE, Gatlinburg, TN.

Barnett, R.H. 1975. Case study of reaeration of Casitas reservoir. *Proc. Conf. on Lake Reaeration Research*, ASCE, Gatlinburg, TN.

Bengtsson, L., and Gelin, C. 1975. Artificial aeration and suction dredging methods for controlling water quality. *Proc. Symposium on Effects of Storage on Water Quality*, Water Resour. Center, Medmenham, England.

Bernhardt, H. 1967. Aeration of Wahnbach Reservoir without changing the temperature profile. *J Am Water Works Assoc* 59:943–964.

Bierderman, W.J., and Fulton, E.E. 1971. Destratification using air. *J Am Water Works Assoc* 63:462–466.

Brezonik, P.L., Delfino, J., and Lee, G.F. 1969. Chemistry of N and Mn in Cox Hollow Lake, Wisconsin, following destratification. *J San Eng Div ASCE* 95:929–940.

Brosnan, T. 1983. Physical, chemical and biological effects of artificial circulation on Silver Lake, Summit Co., Ohio. M.S. Thesis, Kent State Univ.

Brown, D.J., Brydges, T.G., Ellerington, W., Evans, J.J., Michalski, M.F.P., Hitchin, G.G. Palmer, M.D., and Veal, D.D. 1971. *Progress Report on the Destratification of Buchanan Lake*, Ont. Water Res. Comm., Aid for Lakes Prog.

Brynildson, O.M., and Serns, S.L. 1977. *Effects of Destratification and Aeration of a Lake on the Distribution of Planktonic Crustacea, Yellow Perch and Trout*, Wisc. Dept. Natur. Resour., Tech. Bull. No. 99.

Chen, C.W., and Orlob, G.T. 1975. Ecological simulation for aquatic environments. In *Systems Analysis and Simulation in Ecology, Vol. III*, New York: Academic, 475–588.

Davis, J.M. 1980. Destratification of reservoirs – A design approach for perforated-pipe compressed-air systems. *Water Serv* 84:497–504.

Drury, D.D., Porcella, D.B., and Gearheart, R.A. 1975. *The Effects of Artificial Destratification on the Water Quality and Microbial Populations of Hyrum Reservoir*, Utah. Wat. Res. Lab. Proj. EW 011-1.

Fast, A.W. 1968. *Artificial Destratification of El Capitan Reservoir by Aeration. Part I: Effects on Chemical and Physical Parameters.* Calif. State Dept. Fish and Game, Fish. Bull. No. 141.

Fast, A.W. 1971a. *The Effects of Artificial Aeration on Lake Ecology*, Water Pollut. Contr. Res. Ser. 16010 Exc. 12/71/ USEPA.

Fast, A.W. 1971b. Effects of artificial destratification on zooplankton depth distribution. *Trans Am Fish Soc* 100:355–358.

Fast, A.W. 1973. Effects of artificial destratification on primary production and zoobenthos of El Capitan Reservoir, California. *Water Resour Res* 9:607–623.

Fast, A.W. 1979. Artificial aeration as a lake restoration technique. In *Lake Restoration*, EPA 440/5-79-001, 121–132.

Fast, A.W., Moss, B., and Wetzel, R.G. 1973. Effects of artificial aeration on the chemistry and algae of two Michigan lakes. *Water Resour Res* 9:624–647.

Forsberg, B.R., and Shapiro, J. 1980. Predicting the algal response to destratification. In *Restoration of Lakes and Inland Waters*, EPA 440/5-81-010, 134–139.

Garton, J.E., Strecker, R.G., and Summerfelt, R.C. 1978. Performance of an axial-flow pump for lake destratification. In W.A. Rogers, ed., *Proc. 13th Annual Conf. SE Assoc. Fish and Wildl. Agencies*, 336–346.

Halsey, T.G. 1968. Autumnal and overwinter limnology of three small eutrophic lakes with particular reference to experimental circulation and trout mortality. *J Fish Res Board Can* 25:81–99.

Halsey, T.G., and Galbraith, D.M. 1971. *Evaluation of Two Artificial Circulation Systems Used to Prevent Trout Winter-Kill in Small Lakes*, B.C. Fish and Wildl. Br., Fish Mgmt. Publ. No. 16.

Haynes, R.C. 1973. Some ecological effects of artificial circulation on a small eutrophic lake with particular emphasis on phytoplankton. I. Kezar Lake experiment. *Hydrobiologia* 43:463-504.

Hooper, F.F., Ball, R.C., and Tanner, H.A. 1953. An experiment in the artificial circulation of a small Michigan lake. *Trans Am Fish So* 82:222-241.

Irwin, W.H., Symons, J.M., and Robeck, G.G. 1966. Impoundment destratification by mechanical pumping. *J San Eng Div ASCE* 92:21-40.

Jacoby, J.M., Lynch, D.D., Welch, E.B. and Perkins, M.A. 1982. Internal phosphorus loading in a shallow eutrophic lake. *Water Res* 16:911-919.

Kamp-Nielsen, L. 1974. Mud-water exchange of phosphate and other ions in undisturbed sediment cores and factors affecting the exchange rates. *Arch Hydrobiol* 73:218-237.

Kamp-Nielsen, L. 1975. Seasonal variation in sediment-water exchange of nutrient ions in Lake Esrom. *Verh Int Ver Limnol* 19:1057-1065.

Karlgren, L. and Lindgren, O. 1963. Luftningastudier; Traskjön (Aeration studies at Troskjön). Sartryck ur Vattenygien 3:67-69.

King, D.L. 1970. The role of carbon in eutrophication. *J Water Pollut Control Fed* 42:2035-2051.

King, D.L. 1972. Carbon limitation in sewage lagoons. In *Nutrients and Eutrophication*, Special Symposium, Vol. 1, Am. Soc. Limnol. and Oceanogr., 98-110.

Kitchell, J.A., and Kitchell, J.F. 1980. Size-selective predation, light transmission, and oxygen stratification: Evidence from the recent sediments of manipulated lakes. *Limnol Oceanogr* 25:389-402.

Knoechel, R., and Kalff, J. 1975. Algal sedimentation: The cause of a diatom-bluegreen succession. *Verh Int Ver Limnol* 19:745-754.

Knoppert, P.L., Rook, J.J., Hofker, T., and Oskam, G. 1970. Destratification experiments at Rotterdam. *J Am Water Works Assoc* 62:448-454.

Kothandaraman, V., Roseboom, D., and Evans, R.L. 1973. Pilot lake restoration investigations: Aeration and destratification in Lake Catherine, Il. *State Water Sup*

Lakey, R.T. 1972. Response of physical and chemical parameters to eliminating thermal stratification in a reservoir. *Water Res Bull* 8:589-599.

Leach, L.E., Duffer, W.R., and Harlin, C.C. Jr. 1980. *Induced Hypolimnion Aeration for Water Quality Improvement of Power Releases. Water Pollut Cont. Res. Ser.* 16080, USEPA.

Lee, G.F. 1970. *Factors Affecting the Transfer of Materials Between Water and Sediments*, Univ. of Wisc. Water Res. Cent.

Lehman, J.T., and Sandgren, C.D. 1978. Documenting a seasonal change from phosphorus to nitrogen limitation in a small temperate lake, and its impact on the population dynamics of *Asterionella*. *Verh Int Verh Limnol* 20:375-380.

Lorenzen, M.W., and Mitchell, R. 1975. An evaluation of artifical destratification for control of algal blooms. *J Amer Water Works Assoc* 67:373-376.

Lorenzen, M.W., and Fast, A.W. 1977. *A Guide to Aeration/Circulation Techniques for Lake Management*, Ecol. Res. Ser. EPA-600/3-77-004.

Lossow, K., Sikorowa, A., Drozd, H., Wuckowa, A., Nejranowska, H., Sobierajska, M., Widuto, J., and Zmyslowska, I. 1975. Results of research on the influence of aeration on the physico-chemical systems and biological complexes in the Starodworskie Lake obtained hitherto. *Pol Arch Hydrobiol* 22:195-216.

Malueg, K.W., Tilstra, J.R. Schultz, D.W., and Powers, C.F. 1973. Effect of Induced Aeration on Stratification and Eutrophication Processes in an Oregon Farm Pond. *Geophys Monogr Ser* 17:578-587.

McCullough, J.R. 1974. Aeration revitalizes reservoir. *Water Sewage Works* 121:84–85.
McNally, W.J. 1971. *Destratification of Lakes,* Federal Aid to Fisheries, Proj. Comp. Rep., State of New Mexico, F-22-R-11, J of C-8.
Mortimer, C.H. 1942a. The exchange of dissolved substances between mud and water in lakes. (Parts 1 and 2). *J Ecol* 29:280–329.
Mortimer, C.H. 1942b. The exchange of dissolved substances between mud and water in lakes. (Parts 3, 4, summary and references). *J Ecol* 30:147–301.
Murphy, G.I. 1962. Effects of mixing depth and turbidity on the productivity of freshwater impoundments. *Trans Am Fish Soc* 91:69–76.
Oskam, G. 1973. A kinetic model of phytoplankton growth and its use in algal control by reservoir mixing. *Geophys Monogr Ser* 17:629–631.
Oskam, G. 1978. Light and zooplankton as algae regulating factors in eutrophic Biesbosch reservoirs. *Verh Int Verin Limnol* 20:1612–1618.
Pastorak, R.A., Ginn, T.C., and Lorenzen, M.W. 1980. *Evaluation of Aeration/Circulation as a Lake Restoration Technique,* Bellevue, WA: Tetra Tech Inc for USEPA.
Pastorak, R.A., Lorenzen, M.W., and Ginn, T.C. 1982. *Environmental Aspects of Artificial Aeration and Oxygenation of Reservoirs: A Review of Theory, Techniques, and Experiences,* Technical Rept. No. E-82-3, U.S. Army Corps of Engrs.
Reynolds, C.S. 1975. Interrelations of photosynthetic behavior and buoyancy regulation in a natural population of a blue-green alga. *Freshwater Biol* 5:323–338.
Reynolds, C.S., Wiseman, S.W., and Clarke, M.J.O. 1984. Growth and loss rate responses of phytoplankton to intermittent artificial mixing and their potential application to the control of planktonic algal biomass. *J Applied Ecol* 21:11–39.
Riddick, T.M. 1957. Forced circulation of reservoir waters yields multiple benefits at Ossining, New York. *Water Sewage Works* 104:231–237.
Ridley, J.E. 1970. The biology and management of eutrophic reservoirs. *Water Treat Exam* 19:374–399.
Ridley, J.E., Cooley, P., and Steel, J.A.P. 1966. Control of thermal stratification in Thames Valley reservoirs. *Proc Soc Water Treat Exam* 15:225–244.
Robinson, E.L., Irwin, W.H., and Symons, J.M. 1969. Influence of artificial destratification on plankton populations in impoundments. *Trans Ky Acad Sci* 30:1–18.
Shapiro, J. 1973. Blue green algae: Why they become abundant. *Science* 197:382–384.
Shapiro, J. 1979. The need for more biology in lake restoration. In: *Lake Restoration.* EPA-440/5-79-001. p. 161–167.
Shapiro, J., Forsberg, B., Lamarra, V., Lindmark, G., Lynch, M., Smeltzer, E., and Zoto, G. 1982. *Experiments and Experiences in Biomanipulation – Studies of Biological Ways to Reduce Algal Abundance and Eliminate Blue Greens,* USEPA-600/3-82-096.
Shapiro, J., Lamarra, V., and Lynch, M. 1975. Biomanipulation: an ecosystem approach to lake restoration. In P.L. Brezonik and J.L. Fox, eds., *Proc. Symp. Water Quality Management Through Biological Control,* Univ. Florida and USEPA, 85–95.
Shapiro, J., and Pfannkuch, H.O. 1973. *The Minneapolis Chain of Lakes: A Study of Urban Drainage and its Effects,* Inter. Report No. 9 Limnol. Res. Center, Univ. Minn.

Shapiro, J. 1984. Blue green dominance in lakes: The role and management significance of pH and CO_2. *Int Rev ges Hydrobiol* 69:765-780.
Smith, S.A., Knauer, D.R., and Wirth, T.L. 1975. *Aeration as a Lake Management Technique*, Wisc. Dept. Nat. Res., Tech. Bull. No. 87.
Steichen, J.M., Garton, J.E., and Rice, C.E. 1974. The effects of lake destratification on water quality parameters. *Ann Meeting Amer Soc Agr Eng.*
Stumm, W., and Leckie, J.O. 1971. Phosphate exchange with sediments: Its role in the productivity of surface waters, In *Proc 5th Int Conf Water Pollution Res* III-26/1-16.
Sverdrup, H.U. 1953. On conditions for the vernal blooming of phytoplankton. *J Cons Int Explor Mer* 18:287-295.
Symons, J.M., Irwin, W.H., Robinson, E.L., and Robeck, G.G. 1967. Impoundment destratification for raw water quality control using either mechanical- or diffused-air pumping. *J. Amer Water Works* 59:1268-1291.
Symons, J.M., Carswell, J.K., and Robeck, G.G. 1970. Mixing of water supply reservoirs for quality control. *J Am Water Works Assoc* 62:322-334.
Talling, J.F. 1971. The underwater light climate as a controlling factor in the production ecology of fresh water phytoplankton. *Mitt Int Ver Limnol* 19:214-243.
Taylor, E.W. 1966. *Forty-Second Report on the Results of the Bacteriological Examinations of the London Waters for the Years 1965-66*, Met. Water Bd., New River Head, London.
Thomas, E.A. 1966. Der Pfaffikersee vor, wahrand, und nach kunstlicher Durchmischung. *Verh Int Ver Limnol* 16:144-152.
Toetz, D.W. 1977a. *Biological and Water Quality Effects of Whole Lake Mixing*, Water Resour. Res. Inat., Tech. Rep. A-068-OKLA.
Toetz, D.W. 1977b. Effects of lake mixing with an axial flow pump on water chemistry and phytoplankton. *Hydrobiologia* 55:129-138.
Toetz, D.W. 1979. Biological and water quality effects of artificial mixing of Arbuckle Lake, Oklahoma, during 1977. *Hydrobiologia* 63:255-262.
U.S. Environmental Protection Agency. 1970. *Induced Aeration of Small Mountain Lakes*, Water Pollut. Cont. Res. Ser., 16080-11/70, USEPA.
Walsby, A.E., and Reynolds, C.S. 1975. Water blooms. *Biol Rev* 50:437-481.
Weiss, C.M., and Breedlove, B.W. 1973. *Water Quality Changes in an Impoundment as a Consequence of Artificial Destratification*, North Carolina Water Resour. Res. Inst., Rept. No. 80.
Welch, E.B. 1980. *Ecological Effects of Waste Water*, Cambridge Univ Press. London and New York.
Wirth, T.L., and Dunst, R.C. 1967. *Limnological Changes Resulting from Artificial Destratification and Aeration of an Impoundment*. Wisc Conserv. Dept., Fish. Res. Rep. No. 22.
Wirth, T.L., Dunst, R.C., Uttormark, P.D., and Hilsenhoff, W. 1970. *Manipulation of Reservoir Waters for Improved Quality and Fish Population Response*. Wisc. Dept. Nat. Res., Rep. No. 62.
Zaret, T.M., and Suffern, J.S. 1976. Vertical migration in zooplankton as a predator avoidance mechanism. *Limnol Oceanogr* 21:804-813.

11

Water-Level Drawdown

Water-level drawdown is a well established, multipurpose technique of reservoir and pond management. It is used to control certain aquatic plants, to manage fish populations, to repair structures such as dams or docks, and to carry out other improvement procedures such as dredging or installation of sediment covers (Cooke, 1980). This chapter describes the use of this technique through case studies of water-level drawdown in four different North American climates. A table listing the responses of 74 species of aquatic plants to whole-year, winter, or summer drawdown is presented as a guideline for future users of the technique, and positive and negative factors are summarized.

CASE STUDIES

The object of water-level drawdown to control or eliminate nuisance plants is to expose the plant to freezing-dessication or to heat-dessication to destroy both the thallus and the rhizomes or roots. Exposure to heat or cold may also be detrimental to seeds and other reproductive structures. As will be seen from the case studies, success is dependent upon many factors, including luck with regard to weather. In some regions (e.g., Louisiana), water-level fluctuations have been the principal means of plant control (Richardson, 1975), whereas in others (e.g., the Pacific Northwest) the extremes of climate are usually too narrow to provide the harsh conditions necessary.

Tennessee Valley Authority (TVA) Reservoirs

Hall et al. (1946) were among the first to describe the effects of flooding and dewatering on aquatic plants. They found that black willow (*Salix nigra*), a woody species that quickly colonizes silt deposits and newly formed islands, requires dewatering for establishment. Other woody species respond similarly. Seeds of plants such as buttonball (*Cephalanthus occidentalis*), trumpet vine (*Campsis radicans*), green ash, (*Fraxinus lanceolata*), tupelo gum (*Nyssa aquatica*), and bald cypress (*Taxodium destichum*) all must be dewatered to become established. Similarly, herbaceous weeds will not develop on sites that remain inundated until June or later. The alligator weed (*Alternanthera philoxeroides*) is a nuisance plant in some TVA reservoirs. Subfreezing temperatures are lethal to above-ground parts, but below-ground roots show little or no injury and fragments of the plant overwinter in the water and will recolonize a site after spring reflooding. The water primrose (*Jussiaea diffusa*), however, which forms floating mats, is destroyed by dewatering and freezing. In the case of the primrose, as well as two other nuisance plants (bladderwort, *Utricularia gibba* and milfoil, *Myriophyllum scabratum*), the plants may survive if the soil remains moist during a winter drawdown.

The detailed work of Hall et al. (1946) illustrates one of the main conclusions about water-level drawdown as a plant management technique. The method is species specific, in that some are destroyed, some enhanced, and others unaffected, and when dewatering and freezing are used the plant must be exposed to subfreezing, dry conditions for several days to several weeks.

The TVA reservoirs have become seriously infested with Eurasian water milfoil (*Myriophyllum spicatum*). It is particularly troublesome in reservoirs with differences of only 0.6 to 1.0 meters between minimum and maximum water levels (Goldsby et al., 1978). The herbicide 2,4-D has been used for control; it is effective in protected areas such as coves, but less so along shorelines, where dilution occurs. Between 1962 and 1969 the TVA treated 14,422 ha (35,636 acres) with 2,4-D, at an average cost of $110 ha^{-1} ($45 acre^{-1}) and obtained temporary relief. Drawdown was found to be the most effective method of control. A 1.8-meter (6-foot) winter drawdown at Watts Bar and Chickamauga Reservoirs killed all milfoil plants on well-drained shorelines. In some areas land forms of milfoil developed, which later reverted to the aquatic form when inundated (Smith, 1971).

Goldsby et al. (1978) describe the management of Eurasian water milfoil with 2,4-D and winter drawdown at Melton Hill Reservoir. The reservoir was drawn down from December to mid-February 1971-72. The area colonized in 1972 was less than 1971, especially in shallow water. Plants in deeper waters did not increase in biomass. From 1973 to 1976, 2,4-D was used, but the costs increased steadily from $84 ha^{-1} ($34 acre^{-1}) in 1971 to $227 ha^{-1} ($92 acre^{-1}, or $168 acre in 1983 dollars) in 1975. The herbicide was effective, bringing about a 68-percent reduction in areal coverage in 1973. compared to the 1972 coverage after drawdown, but reinfestation occurred at a rapid rate unless herbicides were reapplied or water-level drawdown was used. It was found that semimonthly winter drawdowns were very effective in destroying root crowns of plants exposed to freezing. Even a harsh winter will not reduce infestations unless the hydrosoil is completely dewatered. Goldsby et al. (1978) concluded that a combination of maintenance 2,4-D applications and high-frequency, short-duration winter drawdowns is most effective and economical for control of *M. spicatum* in Melton Hill Reservoir.

Manning and Johnson (1975) also describe the successful use of water-level fluctuation and herbicide application, in this case for the control of *Hydrilla* in Louisiana.

Louisiana Impoundments

Water-level manipulation is the primary method of reservoir management in Louisiana for fisheries, aquatic plant control, and renovation of structures such as docks. Chemical controls have proven to be too costly, and harvesting was found to be not only expensive but promoted infestation of plants through spread of fragments. Water hyacinth (*Eichhornia crassipes*) is a significant nuisance, but it is controlled by drying and freezing. As with milfoil, plants left in a few centimeters of water will survive. Unfortunately, seed germination of water hyacinth requires drying, so that dewatering promotes germination, but after one or two years of drying and freezing there is a significant reduction in viable seeds (Richardson, 1975).

Lantz et al. (1964) and Lantz (1974) describe the effects of drawdown on the vegetation of several Louisiana reservoirs. These case studies are instructive, since different periods of the year were used.

Anacoco Reservoir was drawn down about 1.5 meters (5 feet) from midsummer until mid-October and refilled by mid-February. The drawdown reduced the area of water from 1052 ha (2600 acres) to 526 ha (1300 acres). About 40 percent of the reservoir had been closed to fishing due to *Potamogeton* sp. and *Najas guadalupensis*, but after the drawdown and refill only about five percent of the area was closed. The drawdown eliminated water shield (*Brasenia Schreberi*), restricted the spread of parrot feather (*Myriophyllum brasiliense*) and water lily (*Nymphaea odorata*), and enhanced the infestation by *Chara vulgaris*.

Bussey Reservoir, located in northeastern Louisiana, was drawn down in October and refilled in May. In the summer prior to drawdown, 280 ha (700 acres) were infested with *Potamogeton* sp. and *Najas guadalupensis*. In the two summers following refill, only 16 ha (40 acres) were infested. The treatment was considered 90 percent effective.

Lafourche Reservoir, also in Northeastern Louisiana, had a partial drawdown in winter and further water removal in the summer to test the effect on *Ceratophyllum demersum* (coontail), which infested about 80 percent of the reservoir. In Louisiana this plant has holdfasts, and it was believed that dewatering would control it. In the summer following refill, over 60 percent of the reservoir was clear of coontail.

Lantz (1974) summarizes the effects of drawdowns of several other Louisiana reservoirs. As indicated in Table 11-1, some plants such as water hyacinth and alligator weed became established, whereas others were unchanged or reduced in abundance (e.g., *Potamogeton diversifolius, Myriophyllum brasiliense*). Lantz concludes that drawdown is a successful control method for many plants, but cannot be used for eradication. He suggests that the lake manager stagger the fluctuation years to prevent plants from adapting to it. The recommended schedule is two to three years of drawdown followed by two years without water-level fluctuation. Nichols (1975b) also recommends staggered drawdowns for Wisconsin reservoirs.

Presumably this use of intervals without water fluctuation will allow susceptible species to regain dominance over drawdown-resistant species. Subsequent drawdown would then free the reservoir once again from susceptible nuisance plants.

Florida

Hestand and Carter (1975) describe the effects of a fall-winter drawdown (September 1972 to February 1973) on nuisance aquatic vegetation in a central Florida reservoir,

Table 11-1 Responses of 74 Aquatic Plants to Water-Level Drawdown. A = whole-year drawdown, W = winter drawdown, S = summer drawdown. Numbers refer to references, listed in footnotes.

Species	Increased			Decreased			No Change		
	A	W	S	A	W	S	A	W	S
Alternanthera philoxeroides	10	9	15						
Bidens sp.			13						
Brasenia Schreberi					1	13			
					11	14			
					22	15			
						24			
Cabomba caroliniana			15		11	17			
					22	24			
Carex spp.			13						
Cephalanthus occidentalis			15						
Ceratophyllum demersum	26	19		14	1	13		20	15
					2	17			
					9				
					11				
					16				
Chara vulgaris		16	17			15		28	14
Cyperus spp.	10								

Table 11-1 (continued)

	Increased			Decreased			No Change		
Species	A	W	S	A	W	S	A	W	S
Eichhornia crassipes		9	15	10	11 22				
Eleocharis baldwinii			15			17			
Eleocharis acicularis			13 17		1 21				
Elodea canadensis		20			1 6 19			2 28	
Elodea densa					9 16	12 17			
Elodea sp.					11				
Glyceria borealis		20							
Hydrilla verticillata		3 9							
Hydrochloa caroliniensis				10					
Hydrotrida caroliniana					21				
Jussiaea diffusa					7				

Table 11-1 (continued)

Species	Increased			Decreased			No Change		
	A	W	S	A	W	S	A	W	S
Leersia oryzoides		20	13						
Lemna minor	26								
Lemna sp.					1				
Limnobium spongia						24			
Myriophyllum brasiliense						15		28	14
Myriophyllum exalbescens					2				
Myriophyllum heterophyllum						24		28	15
Myriophyllum spicatum		4			5 23				
Myriophyllum sp.					1 11				
Megalodonta Beckii		1							
Najas flexilis	26	1 6 20	13					25	15

Table 11-1 (continued)

Species	Increased			Decreased			No Change		
	A	W	S	A	W	S	A	W	S
Najas guadalupensis				10	9				17
					14				
					16				
Nelumbo lutea						15		22	7
Nuphar advena					21				
Nuphar luteum						24			
Nuphar macrophyllum		9							
Nuphar polysepalum									12
Nuphar variegatum					19				13
					20				
Nuphar sp.					1				
Nymphaea odorata			24			14			12
						15			
Nymphaea tuberosa					19				
Panicum sp.	10								
Polygonum coccineum		20	8		1				

Water-Level Drawdown 259

Table 11-1 (continued)

Species	Increased			Decreased			No Change		
	A	W	S	A	W	S	A	W	S
Polygonum natans								20	
Pontederia cordata		20					10		
Potamogeton americanus		19			1				
Potamogeton amplifolius					2				
Potamogeton crispus								6	
Potamogeton diversifolius		1					15		
		18							
Potamogeton epihydrous		18						1	
		20							
Potamogeton foliosus		18						6	
Potamogeton gramineus		18						6	
Potamogeton natans		1				13			
Potamogeton pectinatus		26						6	
Potamogeton Richardsonii		20						1	

Table 11-1 (continued)

Species	Increased			Decreased			No Change		
	A	W	S	A	W	S	A	W	S
Potamogeton Robbinsii					1				
Potamogeton zosteriformis		19			2			1	
					19			2	
Potamogeton spp.	8								14
Ranunculus tricophyllus								1	
Sagittaria graminea							10		
Sagittaria latifolia							1		
Sagittaria sp.		19							
Salix interior		20							
Scirpus americanus		1							
Scirpus californicus				10					
Scirpus validus		20	27						
Sium suave		20							
Sparganium chlorocarpum						27		1	13

Water-Level Drawdown   261

Table 11-1 (continued)

Species	Increased			Decreased			No Change		
	A	W	S	A	W	S	A	W	S
Spirodela polyrhiza					1				
Typha latifolia		20				27	10	1	
Utricularia purpurea						24			
Utricularia vulgaris					1				
Utricularia sp.					21				17
Valisneria americana		1		10			2		

1. Beard, 1973
2. Dunst and Nichols, 1979
3. Fox et al. 1977
4. Geiger, 1983
5. Goldsby et al., 1978
6. Gorman, 1979
7. Hall et al., 1946
8. Harris and Marshall, 1963
9. Hestand and Carter, 1975
10. Holcomb and Wegener, 1971
11. Hulsey, 1958
12. Jacoby et al., 1983
13. Kadlec, 1962
14. Lantz et al., 1964
15. Lantz, 1974
16. Manning and Johnson, 1975
17. Manning and Sanders, 1975 (summer-fall drawdown)
18. Nichols, 1974
19. Nichols, 1975a
20. Nichols, 1975a
21. Pierce et al., 1963
22. Richardson, 1975
23. Smith, 1971
24. Tarver, 1980
25. Tazik et al., 1982
26. van der Valk and Davis, 1978
27. van der Valk and Davis, 1980
28. Wile and Hitchin, 1977

Lake Ocklawaha (Rodman Reservoir). The dominant plants before drawdown were *Ceratophyllum demersum, Elodea* (= *Egeria*) *densa, Hydrilla verticillata, Eichhornia crassipes* (water hyacinth), and *Pestia stratioles* (water lettuce). Water level was lowered 1.5 meters (5 feet) and the study sites were dry or had very shallow water.

By the end of the second growing season following refill, *Ceratophyllum* percent coverage had been reduced by 47 percent and *Elodea* by 56 percent. Hydrilla and water hyacinth, however, spread after refill, with a lake-wide increase of 6,440 percent for hydrilla and 3,300 percent for water hyacinth. Failure to control these two species was due in part to a mild winter without frost, which allowed them to spread to many new areas. Lantz et al. (1964) and Lantz (1974) have shown that water hyacinth can be controlled by drying and freezing in northern Louisiana reservoirs, although drying appears to enhance seed germination. In the milder climate of central Florida conditions appropriate for control of these species through winter drawdown may seldom occur.

Wisconsin

Beard (1973) has described the successful winter water-level drawdown of 173-ha (180-acre) Murphy Flowage, Wisconsin, to open the flowage to recreation. Between mid-October and mid-November in 1967 and 1968 the water level was lowered 1.5 meters (5 feet) and maintained at that level until March, when it was brought back to full volume. In 1967, 30 ha (75 acres) were closed to fishing from late spring from late spring through summer by *Potamogeton robbinsii, P. amplifolius, Ceratophyllum demersum, Myriophyllum* spp., and *Nuphar* spp. The first winter drawdown released 26 of the 30 ha for fishing, and none of these species returned as dominant in 1969. *Megalodonta beckii* (water marigold), *Najas flexilis*, and *P. diversifolius* all increased following drawdown, and *P. natans* was unchanged. Even with these resistant plants there were still 24 of the original 30 ha open to fishing in 1969. The results are illustrated in Fig. 11-1 and Fig. 11-2.

Beard attributed the success to the effect of freezing and drying on vegetative reproductive structures. The reduction of the *Nuphar* population was thought to be due to deep frost and upheaval of bottom sediments. The three resistant species were beginning to come back, but a flood destroyed the flowage in 1970 so that a longer-term evaluation was impossible.

The primary negative effect of the drawdown was the appearance of a "pea soup" phytoplankton bloom during August of 1968. Also, one shallow (6 meters) well went dry after the flowage was completely down.

Among the beneficial effects was the increase in fishing success for largemouth bass in summer, 1968. This was attributed to the increased area open to fishing and the decrease in crayfish after drawdown, both of which could have made the bass more vulnerable to fishing (Snow, 1971).

Oregon

An unsuccessful attempt to control *Myriophyllum spicatum* in a Portland, Oregon, reservoir through winter water-level drawdown was reported by Geiger (1983). The

264 Procedures to Control Plant Biomass

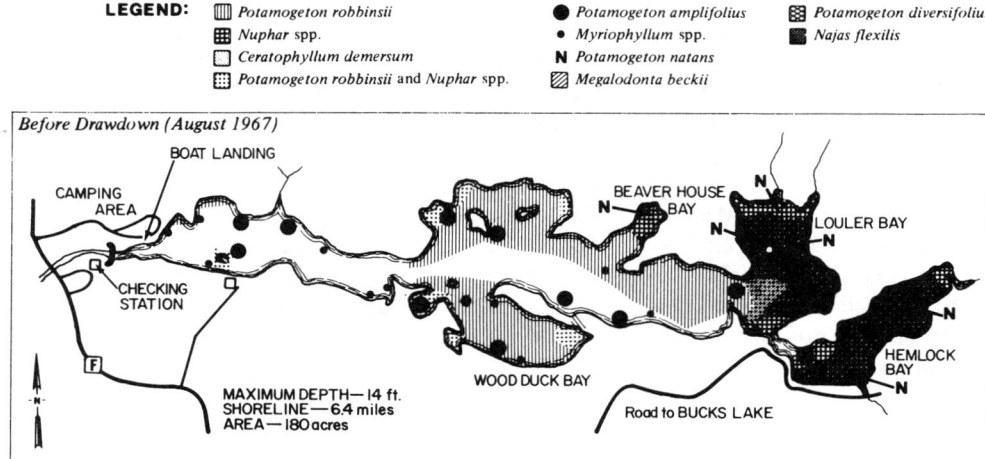

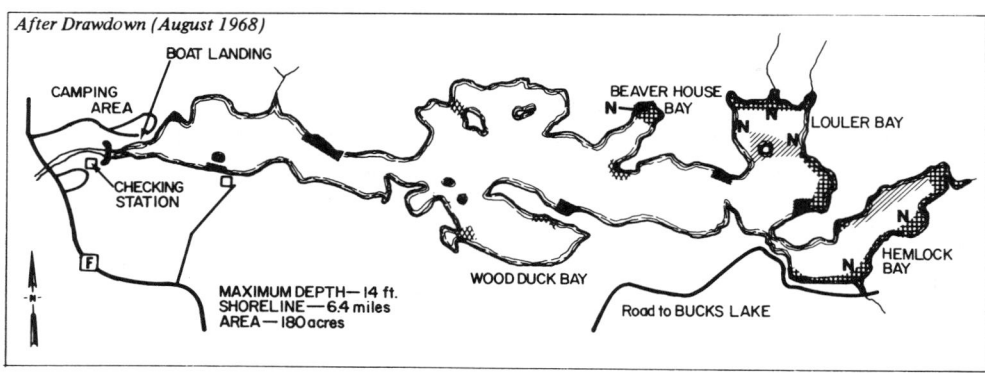

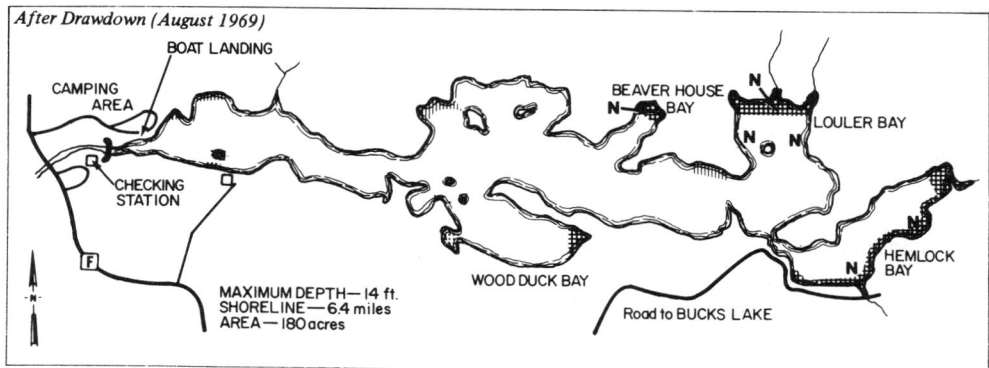

Figure 11-1 Abundance of aquatic plants before and two years after an over-winter drawdown at Murphy Flowage, Wisconsin. Ranking was based on the percentage within the 210 quadrants, covering the entire flowage (from Beard, 1973).

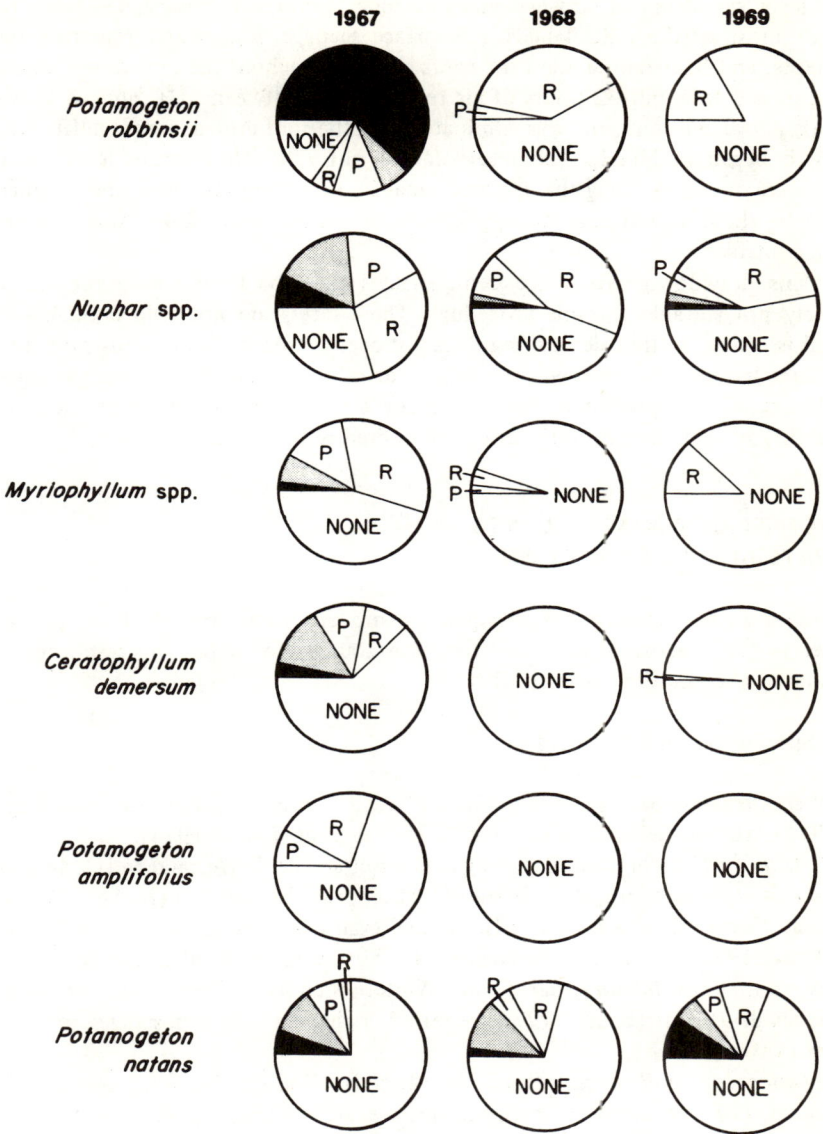

Figure 11-2 Distribution of the major species of aquatic plants in Murphy Flowage, Wisconsin, before and after over-winter drawdowns. (The distribution includes only the areas in which the species were abundant, common, and present.) (from Beard, 1973).

water level was dropped between mid-December and mid-February, 1981-82, to the base of most of the milfoil beds. Subsurface seepage, high water retention by the sediments, and high rainfall kept the roots moist throughout the drawdown. The roots were exposed to about 32 hours of air temperatures between -1°C and -4°C. Milfoil biomass above the sediment was eliminated, but the root crowns were unaffected, and regrowth began in March. The reservoir did not refill to previous levels, and live plants were common along the exposed areas in July, demonstrating the hardiness of milfoil to these conditions. An application of the herbicide 2,4-D was required to obtain control.

This drawdown experience demonstrates that the Pacific maritime climate is probably not suitable for this procedure. The winters are not cold enough and the rainfall is heavy, so that dewatering does not occur. This study also illustrates another significant hazard of drawdown, the failure to refill. Long-term weather for a specific location cannot be predicted, but past records should be consulted to estimate the probability of sufficient run-off during the desired refill period.

RESPONSES OF AQUATIC PLANTS TO WATER-LEVEL DRAWDOWN

The responses to drawdown of 74 aquatic plants are listed in Table 11-1, along with the season of the drawdown and literature sources for data. The responses of 19 common species are summarized below.

Species that usually increase:

1. *Alternanthera philoxeroides* (alligator weed): Annual (Holcomb and Wegener, 1971), winter (Hestand and Carter, 1975), summer (Lantz, 1974).
2. *Hydrilla verticillata* (hydrilla): Winter (Fox et al., 1977; Hestand and Carter, 1975).
3. *Leersia oryzoides* (cutgrass): Winter (Nichols, 1975b), summer (Kadlec, 1962).
4. *Najas flexilis* (bushy pondweed): Annual (van der Valk and Davis, 1978), winter (Beard, 1973; Gorman, 1979; Nichols, 1975b), summer (Kadlec, 1962).
5. *Polygonum coccineum* (smartwood): Winter (Nichols, 1975b), summer (Harris and Marshall, 1963). (Beard, 1973, reported a decrease in this species in a winter drawdown.)
6. *Potamogeton epihydrous* (leafy pondweed): Winter (Nichols, 1974; 1975b). (Beard, 1973, reported no change in this species in a winter drawdown.)
7. *Scirpus validus* (softstem bulrush): Winter (Nichols, 1975b), summer (van der Valk and Davis, 1980).

Species that usually decrease:

1. *Brasenia Schreberi* (water shield): Winter (Beard, 1973; Hulsey, 1958; Richardson, 1975), summer (Kadlec, 1962; Lantz et al., 1964; Lantz, 1974; Tarver, 1980).
2. *Cabomba caroliniana* (fanwort): Winter (Hulsey, 1958; Richardson, 1975), summer (Manning and Sanders, 1975; Tarver, 1980).

3. *Ceratophyllum demersum* (coontail): Annual (Lantz et al., 1964), winter (Beard, 1973; Dunst and Nichols, 1979; Hestand and Carter, 1975; Hulsey, 1958; Manning and Johnson, 1975), summer (Kadlec, 1962; Manning and Sanders, 1975). (Increases or no change in this species were reported by Lantz, 1974; Nichols, 1975a; Nichols, 1975b; and van der Valk and Davis, 1978).
4. *Elodea* (= *Egeria*) *densa* (Brazilian elodea): Winter (Hestand and Carter, 1975; Manning and Johnson, 1975), summer (Jacoby et al., 1983; Manning and Sanders, 1975).
5. *Myriophyllum* spp. (milfoil): Winter (Beard, 1973; Dunst and Nichols, 1979; Goldsby et al., 1978; Hulsey, 1958; Smith, 1971), summer (Lantz, 1974; Tarver, 1980). (Increases and no change in milfoil have occasionally been reported; see Table 11-1 for species and references.)
6. *Najas guadalupensis* (southern naiad): Annual (Holcomb and Wegener, 1971), winter (Hestand and Carter, 1975; Lantz et al., 1964; Manning and Johnson, 1975). (Manning and Sanders, 1975, reported no change in this species in a summer-fall drawdown.)
7. *Nuphar* spp. (yellow water lily): Winter (Beard, 1973; Nichols, 1975a; 1975b; Pierce et al., 1963), summer (Tarver, 1980). (Increases and no change in *Nuphar* have occasionally been reported; see Table 11-1 for species and references.)
8. *Nymphaea odorata* (water lily): Summer (Lantz et al., 1964; Lantz, 1974). (Jacoby et al., 1983, reported no change in this species in a summer drawdown; Tarver, 1980, reported an increase in a summer drawdown.)
9. *Potamogeton robbinsii* (Robbins's pondweed): Winter (Beard, 1973; Dunst and Nichols, 1979; Nichols, 1975a).

Species that do not change, or whose response is variable:

1. *Eichhornia crassipes* (water hyacinth): Hestand and Carter (1975), Holcomb and Wegener (1971), Hulsey (1958), Lantz (1974), Richardson (1975).
2. *Elodea canadensis* (elodea): Beard (1973), Dunst and Nichols (1979), Gorman (1979), Nichols (1975a; 1975b), Wile and Hitchin (1977).
3. *Typha latifolia* (cattail): Beard (1973), Holcomb and Wegener (1971), Nichols (1975b), van der Valk and Davis (1980).

Cutgrass and smartweed are among the several species that grow well in moist soils and shallow water and would thus be expected to proliferate in some situations of drawdown. Alligator weed and hydrilla are serious nuisances in southern waters, and they are rarely controlled by this procedure. Milfoil and water hyacinth have been controlled by it, particularly *Myriophyllum spicatum* (Eurasian water milfoil). P.R. Newroth (personal communication) has found that partial winter drawdown (1.2 meters) of several Okanangan Valley lakes (British Columbia) exposed *M. spicatum* to ice scouring and thus to some control. This plant, however, as shown by experience in TVA reservoirs and in Oregon, is capable of withstanding low temperatures if the thallus remains moist or if the exposed hydrosoil is not frozen for several weeks.

Clearly, this technique is highly species specific. In lakes with a mixture of species, the normal case in most situations, exposure of littoral communities to dry and hot or dry and cold conditions may eliminate or curtail one plant nuisance and favor the development of a resistant species. Some of the susceptible plants, such as milfoil, are normally so successful that few other species coexist. In these cases, several years of winter water-level drawdown, followed by no drawdown for one to two years, may allow the manager to prevent establishment of resistant species, by allowing reestablishment of the milfoil. The drawdown cycle would then be repeated.

Winter drawdowns seem to be more successful than summer, although the number of summer drawdowns reported in the literature is far too small to evaluate. The advantages of winter drawdown, in addition to its effectiveness on some target plants, are that there will be no invasion of moist lake soils by semiterrestrial plants (e.g., cutgrass) nor proliferation of aquatic emergents (e.g., cattails) and little interference with recreation. Also, runoff is often highest in the spring, so that refill should occur. The decision to employ a summer or a winter drawdown for the purpose of plant control will depend upon the susceptibility of the target species and upon uses of the reservoir and any other management objectives.

POSITIVE AND NEGATIVE FACTORS

Positive Factors

Control of susceptible nuisance aquatic plants is only one of several ways that lake-level drawdown can be used to improve or restore lakes. Ideally, if this procedure is to be implemented for plant control, the lake manager or consultant should examine the possibility of carrying out every other lake improvement procedure that drawdown makes possible.

Water-level drawdown can be an important tool in fish management. It has been used to increase fish standing crop (Lantz et al., 1964), to shift species composition (Hulsey, 1958), and to enhance spawning of bass and crappie (Richardson, 1975). When water levels are low, "structure" or cover for fish can be added to the lake bottom to enhance fishing success. Not all drawdowns have positive effects on fish communities. Gaboury and Patalas (1984) report that summer drawdown of Cross Lake, Manitoba, reduced the amount of fish habitat, and that the standing crops of lake whitefish (*Coregonus clupeaformis*), walleye (*Stizostedion virtrisum vitreum*), northern pike (*Esox lucius*), and cisco (*Coregonus artedii*) are now lower. A severe fish kill occurred after a winter drawdown. Prolonged winter drawdowns reduced whitefish and cisco hatching success, whereas low spring water levels denied walleye and pike access to spawning areas. More details on the significance of fish management to lake restoration are presented in Chapter 13, Biological Controls.

The littoral zone has been implicated as a significant source of internal nutrient loading to the open water, via macrophyte senescence and tissue sloughing (e.g., Landers, 1982; Rich and Wetzel, 1978) and by release from oxic sediments (e.g., Lennox, 1984). Jacoby et al. (1982; 1983) examined the effects of a summer drawdown of shallow ($\bar{Z}=2$ m) Long Lake, Washington, on the biomass of the principal nuisance plant, *Elodea densa*, and on internal phosphorus loading. It was believed that macrophyte decay during the winter provided phosphorus for the spring algal bloom, and that ensuing high rates of photosynthesis brought about elevated pH, providing conditions for release of phosphorus from lake sediments and stimulation of further blue-green algal blooms. Drawdown in summer, 1979, to interrupt this cycle, reduced the 1980 standing crop of *Elodea densa* by 84 percent, although *Nuphar polysepalum* and *Nymphaea odorata* were unaffected because their extensive root systems were able to provide water. Internal phosphorus release was not evident, water-column total phosphorus concentration was 50 percent lower, pH was lower, and dense blue-green blooms did not occur in summer, 1980. The results of this experiment strongly support the view that macrophyte control methods such as drawdown or harvesting, in which plants are not allowed to decay at season's end, have the effect of curtailing internal nutrient loading and subsequent algal blooms. Further discussion of this point is found in Welch et al. (1982) and Carpenter (1983).

Loose, flocculent sediments are common in eutrophic systems and can represent a significant source of turbidity, be uncomfortable to swimmers, or release nutrients to the water column. Water-level drawdown can be an effective means of consolidating some types of lake sediments. Fox et al. (1977) examined in the laboratory the effect of drying on muck-type (organic and nutrient-rich, high in water content, flocculent, poorly defined sediment-water interface) and peat-type (fibrous, organic, low water content) sediments from Lake Apopka, Florida. The muck-type sediments consolidated 40-50 percent after exposure to rain and sun for 170 days. Peat consolidated about 7 percent under identical conditions. The percent water loss may be sufficient to make the sediments firm to walk on, and consolidated sediments appear to remain firm after reflooding (Kadlec, 1962). Jacoby et al. (1982), however, report only a slight consolidation of Long Lake, Washington sediments after a summer drawdown.

Sediment removal could be combined with sediment consolidation to bring about considerable deepening of selected areas of a lake. Sediments could be removed by bulldozer instead of the more expensive hydraulic dredge operation. Since consolidated sediments have much lower water content and little water is removed with them during a bulldozer operation, runoff from disposal sites is minimized and land reuse at the disposal site could be immediate. When sediments are exposed can also be a good time to remove stumps and debris and to build artificial reefs for fishermen.

Some controlled-level systems are maintained to attract ducks. Kadlec (1962) describes a Michigan waterfowl impoundment that was drawn down in summer for the purpose of stimulating the growth of plants attractive to ducks. Cover plants such as the emergents *Typha* (cattail) and *Scirpus* (bulrush) must have bare mudflats as a seedbed, a condition not met in stable-water-level systems; drawdown provides conditions for the germination of seeds. Van der Valk and Davis (1978) found up to 20,000 seeds m^{-2} in the upper 5 cm of exposed sediments in an Iowa marsh, a seed bank that should bring about establishment of a large community of emergent and annual species. Kadlec's (1962) work supports the belief that fluctuating water levels are essential to maintaining vegetation to support a substantial waterfowl community.

Water-level drawdown presents the lake manager with many other possibilities for lake improvement. Sediment coverings, one of the most effective controls of rooted plant growth (see Chapter 14), may be more easily and cheaply installed on dry, consolidated sediments than by the use of SCUBA equipment. Also, repair or construction of docks, placement of rip-rap on banks, maintenance of dams, and removal of stumps and litter can be carried out more effectively after water-level drawdown.

Negative Factors

Algal blooms after reflooding have been described by Hulsey (1958), Beard (1973), and others. The causes of these blooms are poorly understood, since there is little documentation regarding algal species and factors such as internal and external loading. The experience at Long Lake, Washington (Jacoby et al., 1983), where internal loading and algal blooms were controlled by elimination of plants through drawdown, suggests that high external loading could have been involved in these cases where algal blooms following drawdown were reported.

Several authors have suggested that exposure of lake sediments to the air will bring about oxidation of organic matter and subsequent nutrient release to the water column at reflooding (Cook and Powers, 1958; Hulsey, 1958; Kadlec, 1962; Harris and Marshall, 1963; Wegener and Williams, 1974). There is little experimental evidence to support this position. Fox et al. (1977) conducted a series of laboratory experiments with muck-type sediments from Lake Apopka, Florida. Their results strongly support the belief that no significant decomposition of organic matter in sediments occurs during drying. Measurements of microbial activity in, and nutrient release from the reflooded sediments in laboratory containers, and algal growth in the refill water overlying dried sediments support their conclusion that algal blooms following water-level drawdown are not due to nutrient release from the dried sediments. Plotkin (1979), however, found that the phosphorus content of interstitial waters of the muck-type sediments of Long Lake, Washington, increased following two, three, and four months of drying. Further studies on this issue are needed.

Rooted aquatic plants and the organic sediments of the littoral zone form a habitat for invertebrate species. These animals may exhibit great changes in density and diversity following water-level drawdown. Kadlec (1962) observed an absence of molluscs in the year following a summer drawdown in which loose, flocculent sediments became solid enough to walk on. Insect populations were greatly reduced as well, and presumably these hardened sediments would retard recolonization by certain species. Bennett (1954) believed that the sharp decrease in fishing in an Illinois lake following an autumn water-level drawdown was due to a decline in abundance of invertebrates. Hunt and Jones (1972) reported that drawdown caused a decrease in invertebrate species diversity, and Paterson and Fernando (1969) found that 150 days of exposure (southern Ontario), during which sediments froze to a depth of 20 cm, destroyed a large portion of the benthic fauna. In situations where drawdown and exposure are not severe, the density of invertebrates and rooted plants may increase, along with improved fishing (Wegener et al. 1974). McAfee (1980) suggests that short-term drawdowns during cool periods will protect burrowing invertebrates. This, however, will not accomplish control of nuisance macrophytes, and may not bring about desired changes in fish species composition.

Failure to promptly refill following water-level drawdown is one of the most serious problems with this lake improvement method and may be due to a failure to close the dam at the proper time or to drought. Great care should be exercised here, especially after a winter drawdown, so that lake users can be assured of access to the lake during recreation season.

Lake managers need to be aware of the potential for low dissolved oxygen and an associated fish kill, particularly if the incoming water is rich in nutrients and organic matter and the remaining pool is small in volume. Once the water level is down, there are few possibilities for enriching it with oxygen.

Fish kills due to low dissolved oxygen have been a matter of concern with this lake improvement technique, but the reports are contradictory. Beard (1973) found no fish mortality despite a 70-percent winter drawdown in a eutrophic flowage, and E.B. Welch (personal communication) found that dissolved oxygen in Long Lake, Washington, did not fall below 5 mg l^{-1} during a summer drawdown of 2 m (Z_{max} = 3.5 m). Nichols (1975a) reported low dissolved oxygen (but no fish kill) in Mondeaux Flowage, Wisconsin, during a winter drawdown. In contrast, Geagan (1960) observed a summer fish kill in Chicot Lake, Louisiana, during a summer drawdown, and Gaboury and Patalas (1984) observed a fish kill in a Manitoba lake following winter drawdown and a loss of dissolved oxygen. Richardson (1975) has pointed out that dissolved oxygen problems can occur when drawdown causes turnover of a thermally stratified lake, so that water low in oxygen is suddenly introduced into surface waters.

Shaw (1983) has described dissolved oxygen problems during a winter drawdown of a highly enriched Wisconsin reservoir. As drawdown proceeded, organic sediments from spring runoff and summer algal production became resuspended in the river-like areas of the upper reservoir. These sediments were not only high in organic matter and anaerobic, but also contained significant amounts of H_2S and reduced iron. The high COD and BOD extracted any remaining oxygen from the water. This condition moved downstream, removing oxygen from the lower reaches of the reservoir. Shaw recommends delay of drawdown until mid-January and a release limit of 25 percent of reservoir volume. These steps, however, might not provide sufficient exposure to cold for control of macrophytes.

The possibility that drawdown will produce an oxygen depletion in the remaining pool should be assessed. The installation of an aeration or artificial circulation device (Chapters 9 and 10) may be necessary.

SUMMARY

Water-level drawdown has been used successfully in many instances to produce at least short-term control of certain species of aquatic plants. The lake manager must be aware that this procedure is highly species specific, and some plants are unaffected or may even thrive on it, particularly if competitive plants are eliminated. At present, there is insufficient information to determine whether a summer or a winter drawdown is most effective in plant control. It has been done most often in the winter, in part to avoid inconvenience and unsightliness to lake users. It must be remembered, however, that long periods of drying and freezing are needed (three weeks or more) to kill plants such as Eurasian water milfoil. As pointed out in Cooke (1980), this technique is probably inappropriate in moist, mild climates and in lakes where seepage in winter will keep lake sediments moist.

This procedure is unquestionably among the least expensive lake management techniques. Its use can also reduce the cost of other procedures such as sediment removal or application of sediment covers. Construction of a dam, such as might occur with a new real estate lake, should therefore be planned so that water level can be easily regulated.

A great deal of additional research is needed in the areas of species responses to drawdown, the release of nutrients from reflooded sediments, and the comparative merits of dry-hot versus dry-cold exposure. There is also contradictory evidence regarding the impact of this procedure on fish and other animal populations. Drawdown of enriched or shallow water impoundments will increase the risk of oxygen depletion.

REFERENCES

Beard, T.D. 1973. *Overwinter Drawdown. Impact on the Aquatic Vegetation in Murphy Flowage, Wisconsin*, Tech. Bull. No. 61, Madison, WI: Department of Natural Resources.

Bennett. G.W. 1954. The effects of a late summer drawdown on the fish population of Ridge Lake, Coles County, Illinois. *Trans. 19th North American Wildlife Conf.* 259-270.

Carpenter, S.R. 1983. Submersed macrophyte community structure and internal loading: Relationship to lake ecosystem productivity and succession. In *Lake Restoration, Protection and Management*, EPA-440/5-83-001, 105-111.

Cook, A.H., and Powers, C.F. 1958. Early biochemical changes in the soils and waters of artificially created marshes in New York. *NY Fish and Game J* 5:9-65.

Cooke, G.D. 1980. Lake level drawdown as a macrophyte control technique. *Water Res Bull* 16:317-322.

Dunst, R., and Nichols, S.A. 1979. Macrophyte control in a lake management program. In J.E. Breck, R.T. Prentki, and O.L. Loucks, eds., *Aquatic Plants, Lake Management, and Ecosystem Consequences of Lake Harvesting*, Madison, WI: Center for Biotic Systems and Institute for Environmental Studies, Univ Wisconsin, 411-418.

Fox, J.L., Brezonick, P.L., and Keirn, M.A. 1977. *Lake Drawdown as a Method of Improving Water Quality*, EPA-600/3-77-005.

Gaboury, M.N., and Patalas, J.W. 1984. Influence of water-level drawdown on the fish populations of Cross Lake, Manitoba. *Can J Fish Aquatic Sci* 41:118-125.

Geagan, D. 1960. A report of a fish kill in Chicot Lake, Louisiana during a water-level drawdown. *Proc La Acad Sci* 23:39-44.

Geiger, N.S. 1983. Winter drawdown for the control of Eurasian watermilfoil in an Oregon oxbow lake (Blue Lake, Multnomah, County). In *Lake Restoration, Protection and Management*, EPA-440/583-001, 193-197.

Goldsby, T.L., Bates, A.L., and Stanley, R.A. 1978. Effect of water level fluctuation and herbicide on Eurasian watermilfoil in Melton Hill Reservoir. *J Aquatic Plant Manage* 16:34-38.

Gorman, M.E. 1979. Effects of an overwinter drawdown and incomplete refill on autotroph distribution and water chemistry in a permanent recreational pond, M.S. thesis, Kent State University, Kent, Ohio.

Hall, T.F., Penfound, W.T., and Hess, A.D. 1946. Water level relationships of plants in the Tennessee Valley with particular reference to malaria control. *J Tenn Acad Sci* 21:18-59.

Harris, S.W., and Marshall, W.H. 1963. Ecology of water-level manipulations on a northern marsh. *Ecology* 44:331-343.

Hestand, R.S., and Carter, C.C. 1975. Succession of aquatic vegetation in Lake Ocklawaha two growing seasons following a winter drawdown. *Hyacinth Control J* 13:43-47.

Holcomb, D., and Wegener, W. 1971. Hydrophytic changes related to lake fluctuation as measured by point transects. *Proc Southeast Assoc Game and Fish Comm* 25:570-583.

Hulsey, A.H. 1958. A proposal for the management of reservoirs for fisheries. *Proc Southeast Assoc Game Fish Comm* 12:132-143.

Hunt, P.C., and Jones, J.W. 1972. The effect of water level fluctuations on a littoral fauna. *J Fish Biol* 4:385-394.

Jacoby, J.M., Lynch, D.D., Welch, E.B., and Perkins, M.A. 1982. Internal phosphorus loading in a shallow eutrophic lake. *Water Res* 16:911-919.

Jacoby, J.M., Welch, E.B., and Michaud, J.T. 1983. Control of internal phosphorus loading in a shallow lake by drawdown and alum. In *Lake Restoration, Protection and Management*, EPA-440/5-83-001, 112-118.

Kadlec, J.A. 1962. Effects of a drawdown on a waterfowl impoundment. *Ecology* 43:267-281.

Landers, D.H. 1982. Effects of naturally senescing aquatic macrophytes on nutrient chemistry and chlorophyll *a* of surrounding waters. *Limnol Oceanogr* 27:428-439.

Lantz, K.E. 1974. *Natural and Controlled Water Level Fluctuation in a Backwater Lake and Three Louisiana Impoundments*, Baton Rouge, LA: Louisiana Wildlife and Fisheries Comm.

Lantz, K.E., Davis, J.T., Hughes, J.S., and Schafer, H.E. 1964. Water level fluctuation – Its effects on vegetation control and fish population management. *Proc Southeast Assoc Game Fish Comm* 18:483-494.

Lennox, L.J. 1984. Lough Ennell: Laboratory studies on sediment phosphorus release under varying mixing, aerobic and anaerobic conditions. *Freshwater Biol* 14:183-187.

Manning, J.H., and Johnson, R.E. 1975. Water level fluctuation and herbicide application: An integrated control method for hydrilla in a Louisiana reservoir. *Hyacinth Control J* 13:11-17.

Manning, J.H., and Sanders, D.R. 1975. Effects of water fluctuation on vegetation in Black Lake, Louisiana. *Hyacinth Control J* 13:17-21.

McAfee, M. 1980. Effects of a water drawdown on the fauna in small cold water reservoirs. *Water Res Bull* 16:690-696.

Nichols, S.A. 1974. *Mechanical and Habitat Manipulation for Aquatic Plant Management. A Review of Techniques*, Tech. Bull. No. 77, Madison, WI: Wisconsin Department of Natural Resources.

Nichols, S.A. 1975a. The use of overwinter drawdown for aquatic vegetation management. *Water Res Bull* 11:1137-1148.

Nichols, S.A. 1975b. The impact of overwinter drawdown on the aquatic vegetation of the Chippewa Flowage, Wisconsin. *Trans Wisc Acad Sci* 63:176-186.

Paterson, C.G., and Fernando, C.H. 1969. The effect of winter drainage on reservoir benthic fauna. *Can J Zool* 47:589-595.

Pierce, P.C., Frey, J.E., and Yawn, H.M. 1963. An evaluation of fishery management techniques utilizing winter drawdowns. *Proc Southeast Assoc Game Fish Comm* 17:347-363.

Plotkin, S. 1979. Changes in selected sediment characteristics due to drawdown of a shallow eutrophic lake, M.S. thesis, University of Washington, Seattle, Washington.

Rich, P.H., and Wetzel, R.G. 1978. Detritus in the lake ecosystem. *Am Nat* 112:57-71.

Richardson, L.V. 1975. Water level manipulation: A tool for aquatic weed control. *Hyacinth Control J* 13:8-11.

Shaw, B.H. 1983. *Agricultural runoff and Reservoir Drawdown Effects on a 2760-Hectare Reservoir*, EPA-600/S3-82-003.

Smith, G.E. 1971. Resume of studies and control of Eurasian watermilfoil (*Myriophyllum spicatum* L.) in the Tennessee Valley from 1960-1969. *Hyacinth Control J* 9:23-25.

Snow, H.E. 1971. *Harvest and Feeding Habits of Largemouth Bass in Murphy Flowage, Wisconsin*, Tech. Bull. No. 50, Madison, WI: Department of Natural Resources.

Tarver, D.P. 1980. Water fluctuation and the aquatic flora of Lake Miccosukee. *J Aquatic Plant Manage* 18:19-23.

Tazik, P.P., Kodrich, W.R., and Moore, J.R. 1982. Effects of overwinter drawdown on bushy pondweed. *J Aquatic Plant Manage* 20:19-21.

van der Valk, A.G., and Davis, C.B. 1978. The role of seed banks in the vegetation dynamics of prairie glacial marshes. *Ecology* 59:322-335.

van der Valk, A.G., and Davis, C.B. 1980. The impact of a natural drawdown on the growth of four emergent species in a prairie glacial marsh. *Aquatic Bot* 9:301-322.

Wegener, W., Williams, V., and McCall, T.D. 1974. Aquatic macroinvertebrate responses to extreme drawdown. *Proc Southeast Assoc Game Fish Comm* 28:126-144.

Wegener, W., and Williams, V. 1974. Fish population responses to improved lake habitat utilizing an extreme drawdown. *Proc Southeast Assoc Game Fish Comm* 28:144-161.

Welch, E.B., Michaud, J.T., and Perkins, M.A. 1982. Alum control of internal phosphorus loading in a shallow lake. *Water Res Bull* 18:929-936.

Wile, I., and Hitchin, G. 1977. *An Evaluation of Overwinter Drawdown as an Aquatic Plant Control Method for the Kawartha Lakes*, Ontario Ministry of the Environment.

12

Harvesting

Aquatic plant harvesting has developed gradually over the past 20 years as a water resource management technology, simultaneously with increasing interest in the maintenance of multiple uses of aquatic systems. The most important aspects of harvesting approaches are reviewed in the following sections. Included are objectives and benefits of harvesting, engineering approaches and limitations, practical and social aspects of harvesting operations, costs and effectiveness of harvesting, and the environmental consequences of applications of this technology. Some information on other technologies is also provided, for comparison purposes.

OBJECTIVES AND BENEFITS OF AQUATIC PLANT CONTROL

The objective of mechanical control of aquatic vegetation, broadly interpreted, is to remove unwanted plants or plant parts to permit the desired use of the water or littoral area. This may be achieved by nonselective means (e.g., dragging, dredging, bottom tillage, dewatering, burial) or by selective cutting of plant shoots, which may be collected or harvested. Manual methods continue to be utilized worldwide to control nuisance aquatic plants, and may be economically and environmentally appropriate in many cases (Sculthorpe, 1967). Mechanization of plant removal has led to development of mowing machinery, tillage equipment, and suction and diver-operated dredging and hydraulic washing equipment (Anonymous, 1969; Armour et al., 1980; Bryan and Armour, 1982; Dunst et al., 1974; Maxnuk, 1979; Nichols, 1974).

Integrated control methods have become popular in aquatic plant management, in efforts to minimize environmentally negative side effects; to improve the speed, overall effectiveness, or duration of control; and to reduce costs. Information on some of the methods used in British Columbia as part of the integrated control program developed by the provincial Ministry of Environment is included later. In British Columbia harvesting has been performed to facilitate herbicide applications, by removal of excessive surfacing vegetation prior to applications of granular formulations of 2,4-D (Anonymous, 1980). The dense surface mats prevented the granules from reaching the plant roots. Also, shoot growth was stimulated by harvesting late in the growing season, prior to fall application of 2,4-D, so that herbicide uptake and translocation would be encouraged. In another example of integrated control, Shireman and Maceina (1981) have recommended either harvesting or herbicide application in combination with the stocking of grass carp as a means of vegetation control. It seems possible that harvesting-induced stress in rooted aquatic plants could be integrated with introduction of other biological control agents such as pathogens or insects. Harvesting prior to water level changes or drawdown might also facilitate overall control.

Aquatic plant harvesting has a number of advantages over other conventional control technologies, including:

- Mechanical harvesting is not inhibited, as herbicidal control may be, by local regulations, adverse public opinion, or requirements to ban use of the water for 10-14 days after treatment.
- Harvesting is fully controlled by the machine operator, and the size of treatment areas may be determined and limited readily.
- The main nuisance is removed immediately without addition of potentially deleterious substances, and the nutrients incorporated in this biomass are removed.
- Harvesting removes biomass that otherwise will release nutrients at senescence and will contribute to an oxygen depletion that will stimulate further nutrient release from reduced sediments.
- Harvesting prior to herbicide treatments may facilitate dispersal of the formulation to the target vegetation and enhance overall effectiveness by increasing the susceptibility of the plants to herbicides.
- Multiple use of the water body may continue with minor interference during harvesting.
- Harvesting activities pose little hazard to nontarget organisms other than those inadvertently removed with the cut vegetation.
- Mechanical harvesting may be less expensive than other types of physical control, and operating costs compare favorably with herbicide treatment in most circumstances.
- Harvested vegetation may be a beneficial product, sought after by farmers and gardeners.

Disadvantages of aquatic plant harvesting include:

- Cut vegetation must be collected and removed from the water; these steps may be energy- or labor-intensive and relatively costly.
- Effective harvesting in temperate waters is seasonal, since in most cases it must be deferred until growth has peaked and nuisance conditions are apparent.
- Only relatively small areas can be treated by individual machine units during the harvesting season, when nuisance conditions may require simultaneous treatments over large areas in a short time period.
- In locations where a short operating season prevails, a high capital outlay is needed for the required machine capacity, which is used only for brief time periods.
- Harvesting may encourage vegetative fragmentation of target and nontarget plants and may encourage more rapid growth or shifts in species composition by encouragement of opportunistic species.

- Harvesters may inadvertently kill large numbers of small fish.
- Operating depths are usually limited if cutting units incorporate collection and storage functions.
- Favorable weather is essential to safe and effective operations.
- Harvesting operations must rely, at the present, on mechanical systems that usually involve high capital cost, are technically specialized, and could break down or require extensive maintenance.
- Harvesting operations are limited by access to the site, by confined spaces that limit movement, by obstacles such as docks and breakwaters, and by physical factors such as submerged rocks, deadheads, and other bottom irregularities.
- Because of slowness, harvesting operations may create public dissatisfaction and disputes over which treatment areas have priority.
- Harvested vegetation may become a waste material that must be disposed of, adding to costs.

The selection of aquatic plant harvesting as a major tool in lake restoration or in aquatic plant control may depend on factors that include:

- site-specific physical or climatic characteristics and limitations to machine operations,
- the types, densities, and distribution of nuisance aquatic vegetation,
- public interest and degree of concern, financial considerations, and institutional frameworks.

As reported by Koegel et al. (1978), water depth limits some types of equipment, and in the Okanagan Valley of British Columbia some shallow beach areas could not be harvested effectively with commercial equipment. Also, aquatic plant harvesting may not be efficiently carried out unless the target vegetation is localized, preferably in sites near convenient shoreline disposal areas.

DESCRIPTIONS OF MECHANICAL CONTROL TECHNOLOGIES

The mechanical control technologies developed by the British Columbia Ministry of Environment include the major technologies for mechanical control of submersed aquatic plants. They are summarized here:

Aquatic plant fragment barriers: Floating barrier systems were developed and tested to retard downstream spread of buoyant Eurasian water milfoil fragments; the barriers could be deployed across lake outlet channels or around mechanical control operations (Stephenson et al., 1980).

Lake-bottom barriers: A variety of opaque materials have been extensively tested for localized control of aquatic vegetation, by application to shallow lake shorelines. Porous materials with negative buoyancy have proven most effective, and common burlap is the least expensive (Armour et al., 1979).

Hydraulic dredging: Several conventional suction dredges have been considered for Eurasian water milfoil control, and an auger-type dredge was tested; this technology is relatively expensive, and its use may not be justified solely for aquatic plant control (Bryan, 1978).

Diver-operated dredging: Small portable dredges with suction heads, operated by SCUBA divers, have been developed. They are most useful for intensive root removal in sparse, localized populations where fragmentation must be minimized; operations are slow and relatively costly (Armour et al., 1980; Maxnuk, 1979).

Rototilling: Barge-mounted equipment has been developed for root removal and clearing by a hydraulically operated rotovating head that can be operated in up to 4 m depth, preferably in spring or fall (Maxnuk, 1979; Bryan and Armour, 1982).

Shallow-water cultivation: An amphibious tracked vehicle was used to test a variety of agricultural cultivation implements. The best results were with a field cultivator mounted on a three-point hitch for shearing of Eurasian water milfoil roots on beach areas (maximum depth 1 m) during lowered lake levels in winter and spring.

Hydraulic washing: Preliminary testing of prototype equipment that used high-pressure water jets demonstrated that Eurasian water milfoil root masses could not be removed solely by use of this technology, although the method could be used to improve root removal after rotovating. The method is no longer used (Bryan and Armour, 1982).

Harvesting: Cutting and harvesting of summer growth of Eurasian water milfoil has been performed routinely, and modifications to facilitate this operation have been made to conventional harvesting equipment (Newroth, 1979; Armour et al., 1980).

Although this review concentrates on the processes and effects of vegetation cutting and harvesting, some circumstances justify consideration of other methods, such as those outlined. Extensive testing and development of the methods listed were integrated in a comprehensive program to control Eurasian water milfoil (*Myriophyllum spicatum*) in British Columbia, Canada (Newroth, 1979, 1980). The basic construction and operation of commercial aquatic plant harvesters (Figure 12-1) and prototype rototilling (Figure 12-2) and diver-operated dredging equipment (Figure 12-3) are illustrated. All of this equipment has been extensively utilized in ongoing Eurasian water milfoil control by the British Columbia Ministry of Environment, as described in a later section.

Aquatic plant harvesting may be employed during the growing season when submersed vegetation has grown densely to or near the water surface. Cutting only, without collection of plant fragments, is not normally considered unless the vegetation is sparse or wind and currents can be utilized to collect floating cuttings. Tables 12-1 and 12-2 list manufacturers of aquatic plant cutters and harvesting equipment available at the present time. Where available, prices current in 1982 have been listed. Canellos (1981) provided a detailed review of mechanical methods for aquatic plant control. Included in his review was additional information, including specifications of major control equipment, that might help to guide potential purchasers of such equipment.

Diver-operated dredges have been used to remove root systems for intensive control of sparse colonies of Eurasian water milfoil in lakes where prevention of

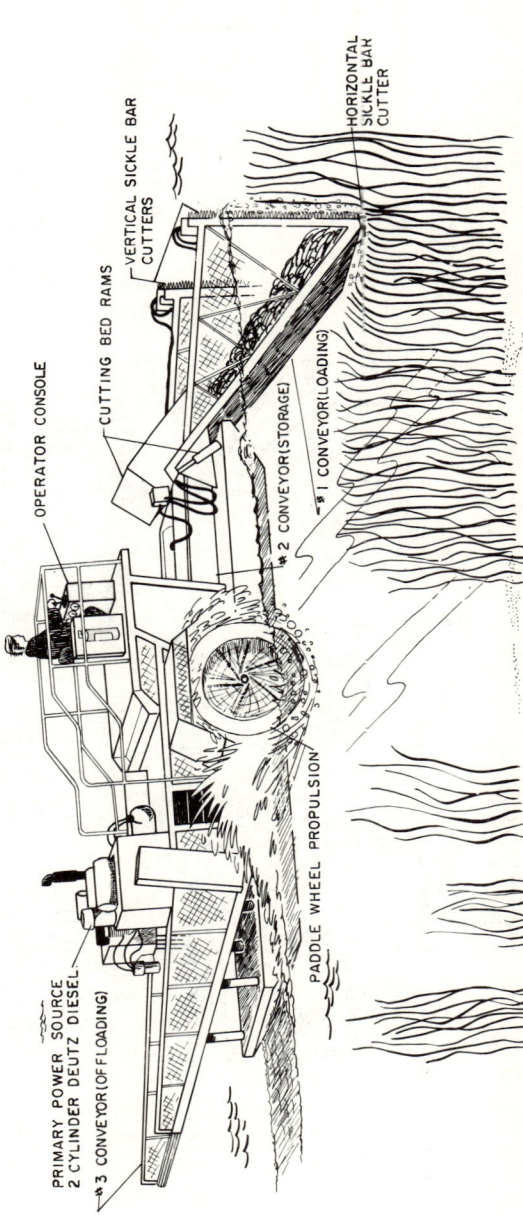

Figure 12-1 The Aquamarine Corporation's (Waukesha, Wisconsin) H650 harvester. Plants are cut with a cutter bar that is 1.6 m (5 feet) long, conveyed into a hold that can contain up to 18 cubic meters (650 ft^3) of plants and offloaded at the shore with another conveyor (Courtesy of the Aquamarine Corporation).

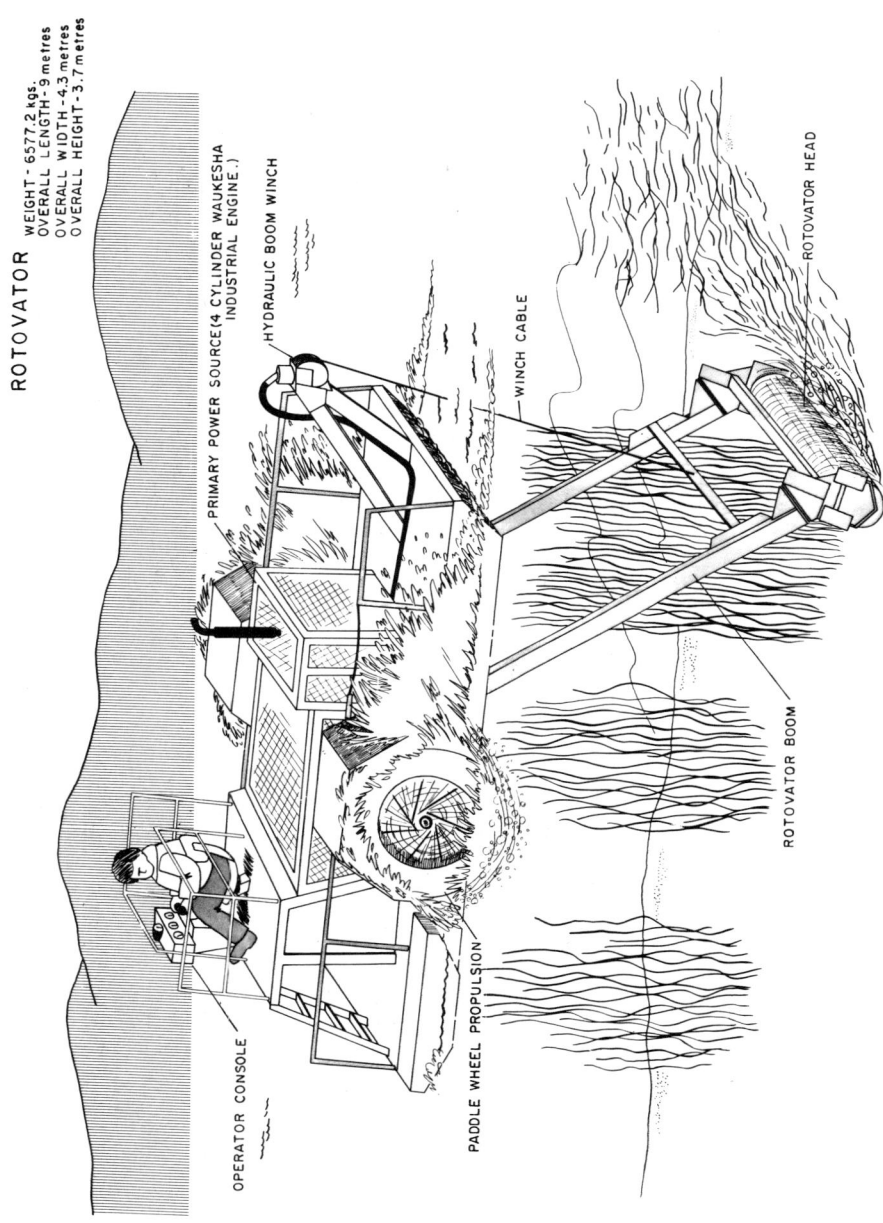

Figure 12-2 The Rotovator, for root tillage.

Harvesting 281

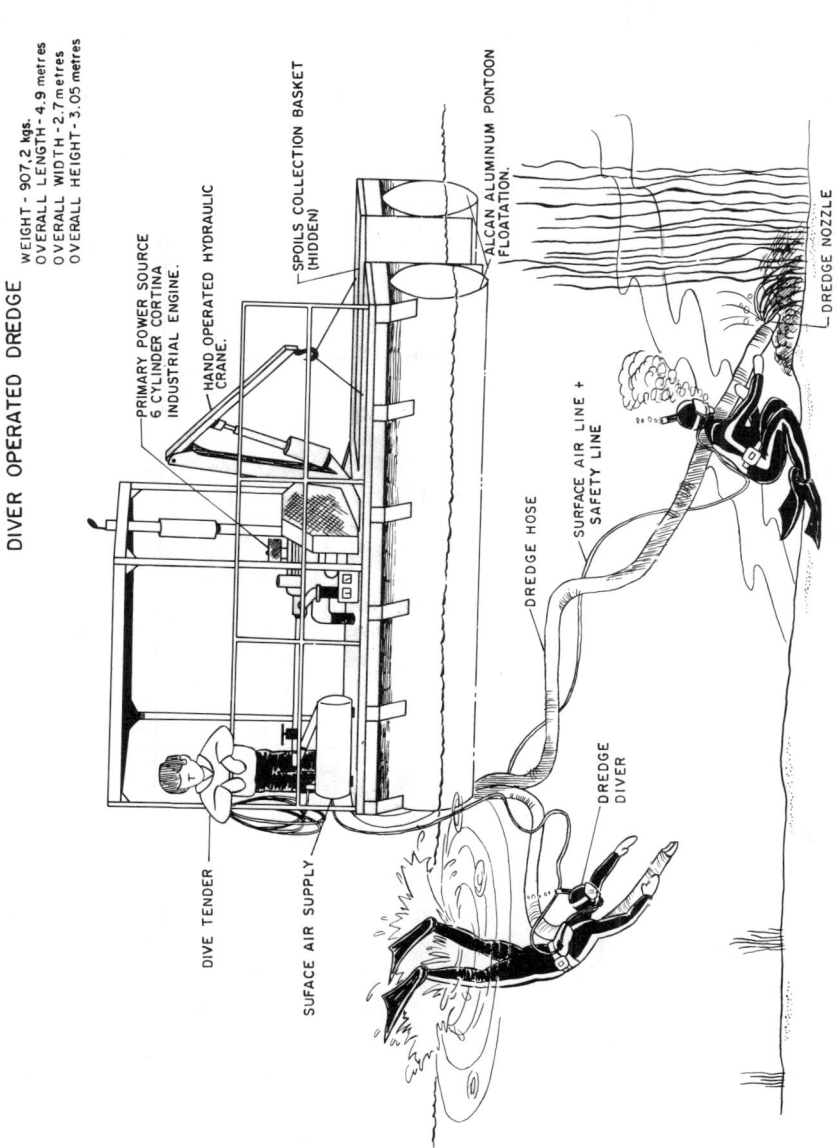

Figure 12-3 A diver-operated dredge for plant and root removal.

Table 12-1 Aquatic Plant Harvesters

Allied Aquatics International Ltd.	800 cu. ft. capacity	$223,000[a]
5029 Flournoy Lucas Road	1200 cu. ft. capacity	330,000[a]
Shreveport, Louisiana 71129	2500 cu. ft. capacity	450,000[a]
Aquamarine	H4 = 100 harvester	$ 19,700[b]
Division of Erectoweld Inc.	(100 cu. ft. capacity)	($ 25,825[c])
1604 South West Avenue	H5 = 200 harvester	30,675[b]
Waukesha, Wisconsin 53186	(200 cu. ft. capacity)	($ 40,200[c])
and	H7 = 400 harvester	51,975[b]
586 Third Line	(400 cu. ft. capacity)	($ 68,075[c])
Oakville, Ontario	H10 = 800 harvester	81,375[b]
Canada L6L4A7	(800 cu. ft. capacity) Cutting depths 4 to 8 ft. and cutting widths 4 to 10 ft. Trailers and shore conveyors also available.	($106,000[c])
Mariner Water Weed Harvesters	"Yankee Clipper" model (200	$ 16,500[a]
104 Locust Street	cu. ft. capacity) cuts 5-ft.-wide	
Palmyra, Wisconsin 53156	swath up to 5-ft. deep. Shore conveyor available	4,500[a]
Mudcat Division	Several models with various	Prices not
National Car Rental Co.	capacities up to 1200 cu. ft.;	available
P.O. Box 16247	5-ft. to 11-ft. swath widths;	
St. Louis Park, Minnesota 55416	5-ft. to 6-ft. maximum cutting depth.	

[a]Prices effective 1982, U.S. dollars.
[b]1985 prices in U.S. dollars.
[c]1985 prices in Canadian dollars.

further spread and long term control is desired. This approach may be integrated with other methods (e.g., following herbicide treatments of bottom tillage) to achieve maximum reductions of plant density. Diver dredging is best performed early in the growing season, when shoot growth is beginning, to reduce opportunities for plant fragmentation (Maxnuk, 1979; Armour et al., 1980).

A variety of root tillage methods have been tested. The purpose of these methods is to stress rooted aquatic plants and to prevent the development of surfacing colonies that could cause rapid fragment dispersal or nuisance conditions. Rototilling or tillage using agricultural plows is best applied during the nongrowing season, when shoot material is minimal. Root masses may be buried, stressed, or dislodged (and float to the surface), and containment using floating boom systems may be required, or root masses may be washed into shoreline areas and raked manually or gathered by shore-based equipment (Maxnuk, 1979).

Table 12-2 Aquatic Plant Cutters

Company	Description	Price
AIR-LEC Industries Ltd. 3300 Commercial Avenue Madison, Wisconsin 53714 USA	Aquatic weed mower (attaches to boat). Cuts swath 3.5 ft. wide and up to 3½ ft. deep. Rake available.	$2,270 323
Allied Aquatics 5029 Flournoy Lucas Road Shreveport, Louisiana 71129 USA	"Water Bug" model cuts swath 12 ft. wide and up to 4 ft. deep. Pusher Boat with rake.	$19,500 8,500
Arundo, Ltd. of Louisiana Suite 2, 783 Engineer's Road Belle Chasse, Louisiana 70037 USA	"TORTOISE" amphibian and "AQUA 690 BRAZIL" barge offer submerged weed cutting capability to width of 9 ft. and maximum depth of 13 ft. Rakes and other attachments available	Not available
Aquamarine Corporation 1604 South West Ave. Waukesha, Wisconsin 53186 USA	"Sawfish" cutter barge cuts 8-ft.-wide swath up to 3.5 feet deep. Rake available.	$29,680
Bouwmeester Machine fabriek en Constructiewerk plaats Noorderstraat 316- 9611 At Sappemeer, Holland	Weed boat with various mower and rake attachments.	start at $28,000 Cdn.
Hockney Company Silver Lake, Wisconsin 53170 USA	Portable model (7-ft. swath width, 4-ft. maximum depth of cut); complete model (10-ft. swath width, 5-ft. maximum depth of cut). Rakes available.	$2,350–$5,600
Wilder British Engineering Hithercroft Works Wallingford, Oxford OX10 9AR England	"Water Warrior" mower boat cuts swath 12 ft. wide and up to 5.5 ft. deep. Rake available.	start at $40,000 Cdn.

Prices effective 1982 and in U.S. Dollars unless otherwise stated.

Hydraulic washing methods have been employed to achieve higher degrees of root removal and are most successful in soft substrates or following tillage, where root systems have been sheared or dislodged (Bryan and Armour, 1982).

In British Columbia, several inventors have developed prototype equipment for Eurasian water milfoil control, using hydraulic devices to aid root removal. To date, none of these prototypes have demonstrated practicality or root removal rates that would justify the construction of production equipment.

The principles of aquatic plant harvesting are derived from agricultural practices in which vegetation is cut, transported, processed, and utilized. In the case of aquatic vegetation, the basic steps associated with harvesting are

1. cutting or separation of vegetation
2. collection of plant material
3. processing and storage
4. transportation to the shore
5. disposal or utilization

Koegel et al. (1978) summarized the main steps and options for handling cut vegetation, which they illustrated as shown in Figure 12-4. The pathways shown here indicate most of the options practical with commercially available equipment. Other options would allow cut vegetation to drift downstream with no effort to collect fragments, or would dispose of cut and collected vegetation in enclosures or protected areas where it will remain and decompose.

Cutting of submerged vegetation usually requires shearing of stems both on the horizontal plane, to detach them from root systems, and on the vertical plane, to separate intertwined and upper branches, often matted with filamentous algae (see Figure 12-1). Reciprocating mower bars, powered by mechanical linkages or hydraulic motors, are normally used on conventional harvesters in a "U" shape, with a single mower or with two vertical and one horizontal bars. Rotary cutters have been developed and tested, but none is known to be in production at this time. On conventional equipment, in which plants are collected by the cutting unit (Figure 12-1), the depth of cut is normally limited to a maximum of two meters. Two reasons are advanced for this: the main nuisance vegetation is near the water surface, and hydrodynamic drag of the mower and conveyor systems increases with operating depth. On other types of systems cutting depth may be considerably deeper, because collection is a separate function and collectors and elevating conveyors are transported on another equipment unit.

As shown in Figure 12-4, cut vegetation can be collected immediately or left in the water for later collection. These dissimilar approaches are described by Livermore and Koegel (1979) in a detailed review of two basic harvesting procedures. Cutting, pickup, and transport to shore are executed by one machine in a single-stage harvesting procedure. Multiple harvesting procedures have been designed to perform the basic steps with separate units to perform cutting, collection and processing, and transportation, storage, and shore disposal functions.

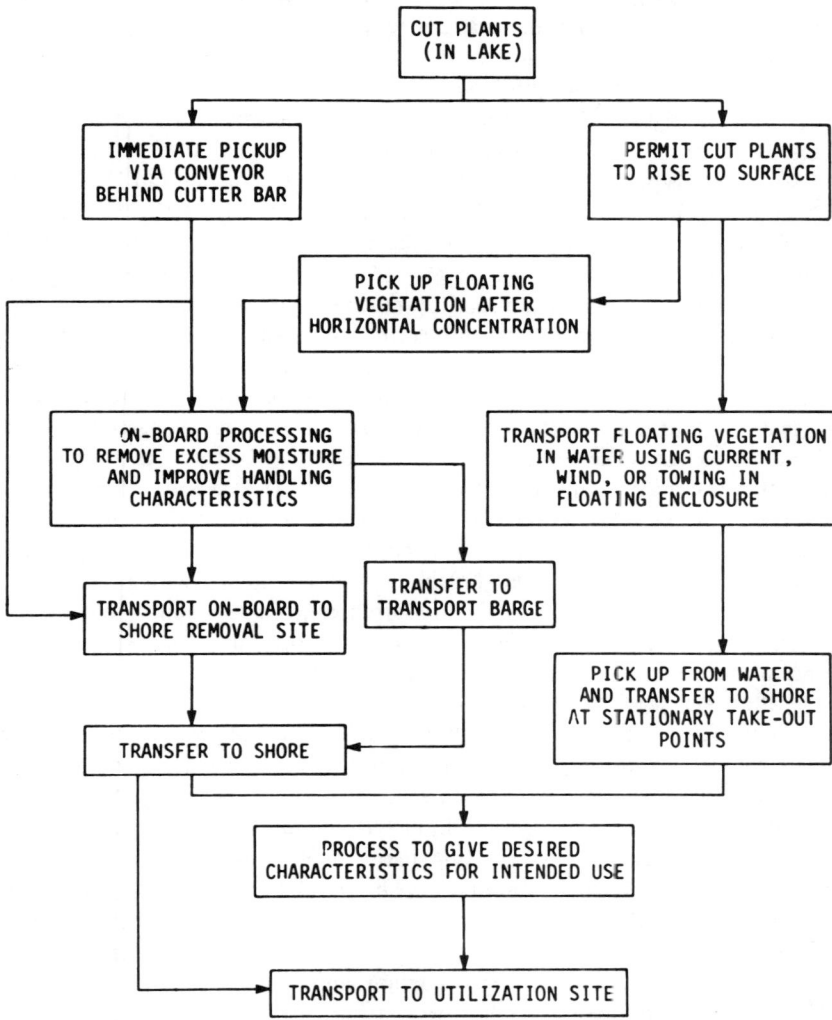

Figure 12-4 Options to be followed in the harvesting of aquatic plants (from Koegel et al., 1978).

Most control equipment is barge mounted, with mechanical or hydraulic systems that provide power for the cutters, conveyors, and propulsion (paddle wheels, Archimedean screws, air or water propellers). Cutting machines and harvesters usually have maximum cutting depths ranging from 1.5 to 2.5 meters, with swath widths of 2.0 to 3.7 meters. There is no minimum cutting depth, assuming there is enough water to float the harvester (18–45 cm, depending on barge size).

Single-stage harvesters have the reciprocating mowers mounted immediately ahead of inclined porous conveyors that collect plant cuttings and elevate them as the vessel proceeds. Collected vegetation may be stored on the barge or pontoon deck, or, in large harvesters, on a second horizontal or sloped conveyor that provides live-bed storage (example, Figure 12-1). This and additional conveyors are used to unload vegetation at shoreline disposal areas, or to transfer material to a second transport vessel. Use of this type of harvester was described by Culpepper and Decell (1978).

Because of the high plant moisture content (typically 95%-98%) and large volume of water entrained in harvested vegetation, various methods of vegetation processing have been studied (Koegel et al., 1973; Livermore and Koegel, 1979). By reduction of the volume and weight, and by cutting aquatic plants into short lengths, the handling characteristics may be significantly improved and savings in operational costs may be realized. If harvested vegetation can be utilized after transport to the shore, processing may reduce the subsequent handling costs and prepare the material for eventual use. Entrained surface moisture may be reduced by passing vegetation through rubber-covered rollers, but chopping and recutting (by forcing material through screens) makes the most significant reduction of volume (Koegel et al., 1973). Despite extensive research on this subject, no commercially manufactured harvesters offer processing equipment. This approach offers considerable promise, however, for operations in which limited equipment is available and where there are long transportation distances. Livermore and Koegel (1979) reported that Dane County, Wisconsin has incorporated agricultural forage choppers into two of their single-stage harvesters, but no information on improved efficiency has been established.

The multiple-stage harvesting procedures are characterized by separation of vegetation cutting from the fragment gathering, collection and processing, and disposal stages. The simplest variation is to permit cut and floating vegetation to be moved to a suitable collection area by natural processes (usually water currents). An example of this method is described by Koegel et al., (1978) in Buffalo Lake, an impoundment on the Fox River, Wisconsin. A floating barricade was used to prevent escape of floating vegetation. Water jets mounted on the barrier deflected vegetation to one end, where raking equipment and an elevating conveyor removed the vegetation from the water. This experimental project was continued for several years, and the combination of high-speed cutters and static shore collection equipment required a relatively low capital investment. Another low-capital project, also utilizing the natural buoyancy of plant cuttings, has been described in detail by Koegel and Livermore (1979).

Two other variations of multiple-stage harvesting have been tested extensively. Koegel et al. (1978) reported the development of experimental mechanical gathering systems mounted on collecting and storage barges. Previously cut and floating vegetation was concentrated between a pair of raking arms mounted in a "V" configuration at the front of the barge unit. The advantages of this approach are that the frontal area for vegetation collection may be several times wider than with conventional, single-stage harvesters, and the inclined conveyor system does not need to be immersed as deeply below the water surface. Because of the shallower draught of the collection barge, hydrodynamic resistance is substantially reduced and overall speed is increased, with less risk to equipment from underwater obstacles or shallow areas.

Neil (1978; 1979; 1980) described an innovative multiple-stage harvesting system utilizing four water-based units and shore-based disposal equipment. These units were:

1. a cutter barge with capability to cut plants to depths of 2.5 m with swath widths of 4.5 m
2. a pickup barge with rotating collector wheels that gather vegetation floating on the surface (over a width of 8.5 m) and direct it into an elevating conveyor, through a grinder or hammermill, and into a transport barge
3. two self-propelled transport barges that shuttle betweeen the pickup barge and shore; both provide storage for the vegetation (reduced to a 10-%-solids slurry) and use liquid manure pumps to discharge into a tank truck on shore.

Major advantages claimed for this method (Limnos Aquatic Plant Harvesting System) are improvements in speed of operation and operational flexibility. The capital costs of the basic system, however, are considerably higher than for a single-stage harvester (Neil, 1979), and the equipment is not available through commercial production at this time.

Smith (1981) described recent field tests of Limnos prototype equipment for harvesting *Hydrilla* in tests in Florida for the U.S. Army Corps of Engineers. The objectives of these tests were to

1. evaluate efficiency of the systems and each component in typical operations,
2. gather data for computer analyses of relationships between vegetation density and operational velocities of cutters and pickup equipment,
3. assess overall effectiveness and identify future research and development needs to improve harvesting operations.

The detailed review provided by Smith (1981) concluded that cutter productivity was about 1.7 ha/hr, and that the pickup unit gathered and processed vegetation at a rate of about 0.6 ha/hr at the same operating speeds. The combination processing rate varied, depending on the swath width, but rates up to about 28 metric tons/hr could be reached in dense weeds. Numerous mechanical improvements to the system and analyses of improved processing methods were recommended. Smith (1981) also recommended that barging and shore disposal steps could be eliminated if the processing steps could render plant fragments inviable and suitable for discharge directly back into the water, although nutrient and BOD loadings would be increased.

The possibility of disposal of harvested and processed *Hydrilla* directly into the water column has been investigated (Sabol, 1981). Langeland and Sutton (1980) investigated the capability of regrowth of fragments of *Hydrilla*, and showed that fragments retained viability and could regrow even if a single node remained. Sabol (1981) showed that *Hydrilla* processed by a hammermill might be expected to have regrowth capabilities. The impacts of regrowth from processed plant fragments and the release of organic materials from disrupted plant tissues on aquatic environments depend on site characteristics.

Utilization of harvested and processed aquatic vegetation was extensively reviewed by Little (1979), and the possibility that costs of harvesting might be recovered through economic utilization of this plant material has stimulated considerable research. The reader is directed to publications by Koegel et al. (1972, 1973), Bagnall et al., (1973), Bagnall (1980), Mitchell (1974), and Anonymous (1976), which describe major efforts to utilize aquatic vegetation. Potential areas for utilization fall into three categories:

1. feed or feed supplements for domestic animals or fish,
2. sources of energy (e.g., methane production),
3. soil conditioning or fertilizer additives.

The potential for successful and economical utilization depends on many variables, including the species of aquatic plants harvested, the volume of vegetation available, and the cost of processing and transporting the material. In most operational harvesting projects in North America, harvested vegetation (Eurasian water milfoil, *Hydrilla*, and water hyacinth) has not found a consistent, economical market, despite considerable experimental work (Anonymous, 1979; Riemer, 1984). At present the most promising way to utilize harvested vegetation appears to be as mulch or compost (Wile et al., 1978) made available to the public near the harvest areas at no additional cost to the harvesting operation. This may result in some financial savings to overall harvesting costs if transportation charges for disposal of harvested vegetation can be reduced.

In the case of Eurasian water milfoil, high moisture content and seasonal production of biomass in many localities where this species may become a nuisance limit interest in utilization. Also, this plant has a high mineral content (rendering it relatively unsuitable for feedstuffs) and low nutrient and fiber content. In British Columbia it appears unlikely that successful and economical utilization of harvested Eurasian water milfoil will develop (Anonymous, 1979).

CASE STUDIES

In Europe and Britain aquatic-plant mowing and root-cutting systems have primarily been used to clear canals and ditches (Robson, 1974). Although mechanical harvesting has been applied in other areas (e.g., New Zealand: Hughes, 1976), the main development and use of these systems has been in North America. A recent review of aquatic weed problems and their solutions has examined the control methods applied in each state in the United States. In summary, the largest application of mechanical treatments has been in the southeastern United States, where in Florida about 2,000 ha are treated annually by mechanical means (mostly harvesting). This should be compared to the estimated annual herbicide treatment of 40,000 to 80,000 ha in the southeastern states (Aurand, 1982). In comparison, the northeastern and north central states were estimated to use annual herbicide treatments of 6,000 to 12,000 ha per year, and mechanical treatments for the states of Minnesota and Wisconsin were estimated to total about 800 to 1,100 ha (Trudeau, 1982). In other parts of the United States, mechanical treatments were even less commonly used for aquatic plant control (Tyndall, 1982; Aurand, 1983), which serves to illustrate the relative importance of mechanical technologies as compared to the use of herbicides.

Grinwald (1968) reported the development of harvesting barge systems from simple cutters, beginning about 1947 and culminating in 1963 in manufacture of a production machine similar to conventional harvesters used today. Livermore and Wunderlich (1969), Livermore and Koegel (1979), and Canellos (1981) provide extensive reviews of available equipment and information on operational mechanical harvesting programs. The earlier publication described harvesting of water hyacinths and alligator weed in Louisiana, harvesting water chestnut and Eurasian water milfoil in Chesapeake Bay, and control of *Elodea, Ceratophyllum*, and water milfoil in New York and Wisconsin. Smith (1979) surveyed harvesting in California, Michigan, and New York and evaluated capital and operational costs.

Major harvesting projects recently documented include work performed in the Province of Ontario, Canada (Kawartha Lakes), the Province of British Columbia, Canada (Okanagan Valley lakes), the State of Wisconsin (Dane County), and the State of Ohio (Portage County, East and West Twin Lakes).

The objectives and details of these control programs are reviewed here as case studies because of their practical nature and to summarize examples of operational management. Although information about costs, rates of operation, and overall mechanical efficiency are available in many cases, more quantitative information on the overall effectiveness of harvesting operations is needed.

Harvesting in Ontario, Canada

Eurasian water milfoil spread through an extensive system of shallow lakes and navigable waterways in southern Ontario, and became a nuisance in the early 1970s, particularly in the Kawartha Lakes (Aiken et al., 1979). An extensive mechanical harvesting program was operated in the period 1973-1978, as described by Wile et al. (1977; 1979). The objectives of this program (funded and operated by the Provincial Ministry of Environment) were to evaluate mechanical harvesting as a practical solution to alleviate nusiance conditions and to observe and document environmental impacts (Wile, 1978).

Table 12-3 summarizes production levels, operating rates, and costs of harvesting in Chemung Lake, one of several lakes where these treatments were performed. Table 12-4 lists the equipment used.

Because extensive areas of nuisance vegetation covered substantial parts of these recreationally important lakes, the use of aquatic plant harvesters appeared particularly appropriate. The large volumes of Eurasian water milfoil that could be harvested in localized areas also encouraged extensive investigation of its utilization for compost and as supplementary feed for domestic animals (Wile et al., 1978). Although considerable information was obtained on possible values of harvested plants, no economically successful recycling of harvested plants has developed.

Table 12-3 Chemung Lake Harvesting Summary (from Wile et al., 1979)

	1973	1974	1975	1976	1977	1978
Total area harvested (ha)	265	261	297	357	328	356
Total biomass removed (metric tons, wet)	1,150	1,940	3,020	4,750	3,320	2,870
Total hours worked	1,125	2,365	2,663	3,327	2,794	2,200
Percentage downtime	12	26	30	29	31	19
Average hectare cut/hour (active)	0.15	0.15	0.16	0.15	0.17	0.20
Cost/hectare ($) (Cost/Acre)	346 (140)	282 (114)	331 (134)	343 (139)	390 (158)	376 (152)

All costs in Canadian dollars for the year shown.

Table 12-4 Harvesting Equipment Used in the Kawartha Lakes
(from Wile et al., 1977)

	Harvester	Transporter	Shore Conveyor
1973	1 Aquamarine harvester (8' cut)	1 Aquamarine transporter	1 Aquamarine shore conveyor
1974	2 Aquamarine harvesters (8' cut)	2 conventional outboard-powered-barges	1 tractor-mounted backhoe with clam
1975	2 Aquamarine harvesters (8' cut)	2 conventional outboard-powered-barges	1 tractor-mounted backhoe with clam
1976	2 Aquarmarine harvesters (8' cut)	2 conventional outboard-powered-barges 1 Aquamarine transporter	1 tractor-mounted backhoe with clam 1 Aquamarine shore conveyor
	1 Experimental harvester (16' cut)	(self-transporting)	Enlarged, high-capacity, continuous conveyor

The effectiveness of harvesting, in Chemung Lake, in maintaining a water column with few weeds was found to be dependent upon the time of year for harvesting, the number of harvests per season, and the depth of cut. Harvests in June and July were least effective in lowering rates of regrowth and plant densities. Two harvests and three harvests per season were found to be most effective in reducing stem numbers and stem height. Areas of the lake that had been harvested for three successive years had slower regrowth of plants. The results also indicated that regrowth is inhibited for longer periods if the plants are cut closer to the sediment. Cuts below the leafy portion of the stem were more effective in retarding regrowth than cuts above.

Wile et al., (1979) reported that the need for extensive harvesting in Chemung Lake was gradually reduced because of unexplained declines of Eurasian water milfoil about 1977, and the harvesting program was terminated after 1978.

Harvesting in British Columbia

In British Columbia, the introduction of Eurasian water milfoil to the Province and the recreationally valuable Okanagan Valley lakes, about 1970, stimulated the Provincial Ministry of Environment to evaluate a wide range of habitats into which this plant had spread. Nuisance growth was found by 1976. Extensive testing of herbicides, bottom barriers, and four major mechanical control techniques was integrated as part of a comprehensive management plan (Newroth, 1979, 1980). Table 12-5 summarizes the extent of treatments in the period 1975-1984, and Table 12-6 summarizes mechanical harvesting operations in the period 1978-1981.

Aquatic plant harvesting was first tested in the Okanagan Valley in 1972. Performance evaluations and operating characteristics of harvesters and impacts on target and nontarget organisms have been studied extensively. Armour et al. (1980) summarized the 1978 mechanical operations, when three Aquamarine machines were put into service, funded and operated by the Province. From 1980 until 1985 operations have been directed by regional agencies, and costs were shared 75 percent by the Province and 25 percent by the Region.

In contrast with mechanical harvesting performed (with similar equipment) in smaller lakes in Ontario, the British Columbia operations required considerably more time for transportation of equipment between the four or five lakes treated. Machine downtime was 34-44 percent (see Table 12-7), compared with 12-31 percent in the Ontario operation (see Table 12-3). Harvested plants also had to be transported over greater distances from many cutting areas to disposal sites, since the populations form narrow bands along the shallow recreational areas. The operation has been made more economical in recent years by reducing support staff and by discontinuing costly manual fragment cleanup on beaches (see Table 12-7). This was possible because the impact of further fragment spread induced by harvesting has become insignificant in these areas. Removal of weed fragments accumulated at boat launching ramps, however, has become more important, due to the risk of transport of viable plant material by boaters to uninfested lakes.

The Limnos harvesting system was developed in Ontario, but has not operated on a large scale there or in British Columbia. This is partially because little funding is available for harvesting in Ontario, and limited funding is available (insufficient for new capital equipment) for treatments of high-priority areas in British Columbia. Because nuisance populations of Eurasian water milfoil in the Okanagan Valley are distributed among numerous sites, few exceeding 30 ha, the benefits of use of a multistage system, as described by Neil (1978, 1979, 1980), would be difficult to realize. Because of many factors, including high labor and transportation costs, utilization of harvested biomass also has not proven economic in the Okanagan Valley or Ontario, so there has been no incentive to harvest unless this is necessitated by nuisance conditions.

Estimates of the operating costs and average operating rates of the main technologies applied for Eurasian water milfoil control in British Columbia are presented in Table 12-8.

Harvesting in Wisconsin

The conventional mechanical harvester appears to have originated in Wisconsin (Grinwald, 1968; Livermore and Wunderlich, 1969). Dunst and Nichols (1979)

Table 12-5 Littoral Area (ha) Treated for Eurasian Water Milfoil Control in the Okanagan Valley[a], 1975-1984

Method	1975	1976	1977	1978	1979	1980	1981	1982	1983	1984	Total
Harvesting[b]	15	56	44	64	116	120	153	122	114	110	858
Rototilling			17	34	1			7	16	19	150
Tillage						6	20	18	36	2	82
Diver-operated dredge			8	62	35	35				nil	140
Bottom barriers	0.04	1.25	0.002				0.04	0.5	1.0	1	3.83
Herbicide (2,4-D)		1	9	12	20	22	6			nil	70
Totals	15	58	78	172	172	183	179	148	167	132	1,304

[a]Figures are totals for treatments in Wood, Kalamalka, Okanagan, Skaha, Vaseux, and Osoyoos Lakes.
[b]Includes repetitive treatments in the same location during one season.

Table 12-6 Aquatic Plant Harvesting Summary – Okanagan Valley

	1978[a]	1979[a]	1980[b]	1981[b]
Operating cost ($)[c]	70,500	113,200	150,900	140,000
Area treated (ha)	64	116	120	138
Loads harvested		1,651	1,475	1,710
Estimated biomass (tons)		2,806	3,212	3,721
Cost per hectare ($)[c]	1,102	976	1,258	1,014
(Cost/acre)	(446)	(395)	(509)	(411)
Cost per ton ($)[c]		40	47	38

[a]Three Aquamarine H-650.
[b]Three Aquamarine H-650 and 1 Mudcat H-11-1100.
[c]Does not include capital cost depreciation. All values in Canadian dollars.

Table 12-7 Analysis of Harvesting Time (hours) – Okanagan Valley

	1978[a]	1979[a]	1980[b]
Total operating	1,433	2,540	2,764
Actual operating	818 (57%)	1,669 (66%)	1,547 (56%)
Downtime	615 (43%)	871 (34%)	1,217 (44%)
Downtime in detail			
Repair/maintenance	128 (21%)	524 (60%)	728 (60%)
Machine travel	107 (17%)	193 (22%)	313 (26%)
Inclement weather	113 (18%)	82 (9%)	176 (14%)
Fragment cleanup	243 (40%)	72 (8%)	
Unaccounted	24 (4%)		

[a]Three Aquamarine (H-650).
[b]Three Aquamarine (H-650) and one Mudcat (H-11-1100).

reported that 35–40 lakes in Wisconsin were harvested at a cost range of $370–490/ha ($150–198/acre); Dane County has utilized this method as a primary control method since 1965 (Koegel et al., 1977). Koegel et al. (1978) reported that this work was performed as a county public works function (1976 budget $144,000). Also, active multidisciplinary research on aquatic plant harvesting has been performed at the University of Wisconsin in Madison since about 1970.

Table 12-8 Estimated Operating Costs and Treatment Rates for Selected Eurasian Water Milfoil Control Methods

Method	Operating cost[a] (cost/hectare)	Average rate[b] (hectares/day)	Comments
2,4-D (BEE)	$700/ha @ 10 ha/day to $2,500/ha @ 1 ha/day (herbicide cost at 45 kg a.i. per ha = $200/ha; capital cost/applicator unit about $12,000)	30 ha/day	Treatment area limited to 1 ha requires the same preparation and monitoring effort as a 10-ha plot, therefore cost/ha is much larger for smaller treatments. These operating costs do not include supplying any alternate water, which can increase costs considerably. Allowance for minimal monitoring is included.
Hydraulic dredge	$5,000/ha (capital cost/dredge unit about $100,000)	0.1–0.2 ha/day	Operating cost obtained from a 2.9-ha test conducted by Aquatic Studies Branch in 1975. Rate is dependent on sediment depth removed and obstacles.
Diver-operated dredge	$2,500/ha to $19,000/ha (capital cost/dredge unit about $12,000)	0.20 ha/day extremely sparse 0.02 ha/day moderate 0.01 ha/day very dense	Treatment costs and rates are highly dependent on weed density.
Rototiller	$400/ha to $1,200/ha (capital cost/machine about $80,000)	0.2–0.5 ha/day	Cost and rate are dependent on the degree of difficulty in treating the site and number of passes necessary to remove most of the plant material.
Shallow water tillage	$400/ha (capital cost/machine about $60,000)	0.7 ha/day	Cost may increase in soft substrate areas or where obstacles are present.
Harvester	$1,200/ha (capital cost/machine about $90,000)	0.4 ha/day	Rate includes shore disposal of spoils.
Bottom barriers	Burlap $12,000/ha Polyethylene $13,000/ha Window screen $20,000/ha Texel $16,000/ha	0.05 ha/day	Polyethylene, window screen, and Texel may be reusable; their high cost could likely be amortized over several years.

[a] Does not include rental or depreciation of capital costs of machinery, expenses incurred in transport and launching of machines, or administration costs. Costs in 1982 Canadian dollars.
[b] Eight-hour work day; includes set-up time.

The shallow and eutrophic lakes of Wisconsin support extensive populations of nuisance plants such as Eurasian water milfoil, and there is considerable public demand for lake restoration and aquatic weed control. Nichols and Cottam (1972) studied the effectiveness of a harvesting procedure in Lake Mendota, Wisconsin, in which Eurasian water milfoil was cut as near to the sediment surface as possible. The effectiveness of single and multiple harvests in controlling biomass and in retarding regrowth in subsequent years was compared in shallow (1.0 m) and deeper plots (1.0-1.6 m). Harvesting was accomplished by cutting as near to the sediment surface as possible with a knife. The seasonal effects were clear. A single harvest reduced biomass to between one-fourth and one-tenth the original amount. Three harvests, one month apart, virtually eliminated plant material. The carryover to subsequent years was most noticeable in the deeper plots. Three harvests during the previous season were most effective, but two harvests followed closely in effectiveness. A single harvest had a carryover effect if executed in mid-June of the previous year. The carryover to subsequent seasons was probably due to the cutting of root crowns. Livermore and Koegel (1979), assessing the "state of the art," noted that the Wisconsin experience showed that harvesting was an effective management technique that may often be less expensive than herbicide treatments.

Harvesting in Ohio

Conyers and Cooke (1983) investigated the comparative effectiveness of harvesting and herbicide (Diquat and Cutrine) treatments in controlling macrophyte biomass in the shallow littoral (0.5-1.6 m) zone of mesotrophic East Twin Lake, Ohio. Three adjacent 810-m^2 plots were established, one serving as a control and the others for herbicide and harvesting treatments. Biomass (*Potamogeton crispus* in early summer, *Ceratophyllum demersum* in middle and late summer, and *Chara* and *Cladophora* in late summer) was sharply decreased by a single harvest, and biomass in this plot remained low all summer, perhaps due to the harvesting technique of cutting at the sediment-water interface. Biomass of the chemically treated plot was the same as the control, due to persistence of *Chara* and other macrophytes (Figure 12-5).

Table 12-9 lists the comparative costs of harvesting and herbicide treatments for both East and West Twin Lakes, assuming that half of the littoral area of the lakes was to be chemically treated. The harvester operating costs are for at least that much area. Herbicide applicator fee is not included. The harvesting cost is higher initially, because of the purchase cost of the machine. When that cost is amortized over the life of the harvester, however, costs are reduced substantially. For example, assuming that there will be no increase in the cost of herbicide chemicals over the next five years, the cost of their use would be 2.6 times greater than the use of the harvester over this period. It is likely that chemical costs will increase over the five-year program, making their use in comparison to the harvester even more costly. The Twin Lakes Association spent a maximum of $287/ha ($115/acre), using a harvester that required manual unloading and was therefore the least cost-efficient equipment available. Chemical treatment, as recommended by the herbicide manufacturer, would have been $657/ha ($266/acre).

Table 12-9 Cost summary for harvesting and chemical treatment of half of the littoral area (13.6 hectares or 33.5 acres) of East and West Twin Lakes, Ohio. Dose was that recommended by Applied Biochemists (see reference).[a] Herbicide operator fee is not included (from Conyers and Cooke, 1983).

Chemical treatment	$17,802[b]
Harvesting[c]	
Initial investment for harvester and trailer	15,000
1981 Operations	
Gasoline	350
Maintenance	230
Auxiliary equipment	80
Personnel	3,152
	$ 3,812
Total	$18,812

[a] Minimum treatment program of two Cutrine-Plus applications and one Diquat application per summer. Cutrine dose was 1.2 gal/acre-foot, Diquat was 1.4 gal/acre-foot, based on 134 acre-feet.
[b] 1981 prices, $22.60/gal for Cutrine, $56.15/gal for Diquat.
[c] Data provided by the Twin Lakes Association, Kent, Ohio

Note: The revised 1979 edition of "How to identify and control water weeds and algae" (Applied Biochemists, Inc., Mequon, Wisc.) recommends a dose of 2 gallons of Diquat per acre instead of the 1976 edition's recommendation of 1.4 gallons per acre-foot for control of *Ceratophyllum*. We were advised by Applied Biochemists to use the dose in the 1976 edition. If we had computed the herbicide cost per year with the dose in the 1979 edition, the per-year cost would have been $9,818.85 for both the Diquat and Cutrine-Plus, excluding herbicide operator fee and assuming no increase in herbicide cost in subsequent years. The harvester will pay for itself in the second year in either case.

SUMMARY OF COSTS

The costs of harvesting vary widely, depending upon area, density of vegetation, weather, machine reliability, and transport distance. Canellos (1981) has summarized operating conditions and operating costs for small-scale and large-scale harvesting equipment. Tables 12-3 and 12-6 summarize costs at Chemung Lake, Ontario, and the Okanagan Valley Lakes, British Columbia. Four North American lakes (Smith, 1979), Twin Lakes, Ohio, a Florida lake (Sassic, 1982), Lake Mendota, Wisconsin (Livermore and Koegel, 1979) and 35-40 other Wisconsin lakes have been added to these and the data summarized in Table 12-10. The Okanagan Valley program and Oneida Lake, New York (Smith, 1979) were far more expensive than any of the other operations. As suggested earlier, the Okanagan costs were influenced by the lake size and the large distances between cutting sites and disposal areas, as well as by the distribution of plants in narrow bands around the perimeters of these lakes. According to Smith (1979), the Oneida Lake operation could have been conducted with far greater efficiency. If these lakes are omitted, harvesting costs range from $148/ha to $490/ha ($60-$198/acre).

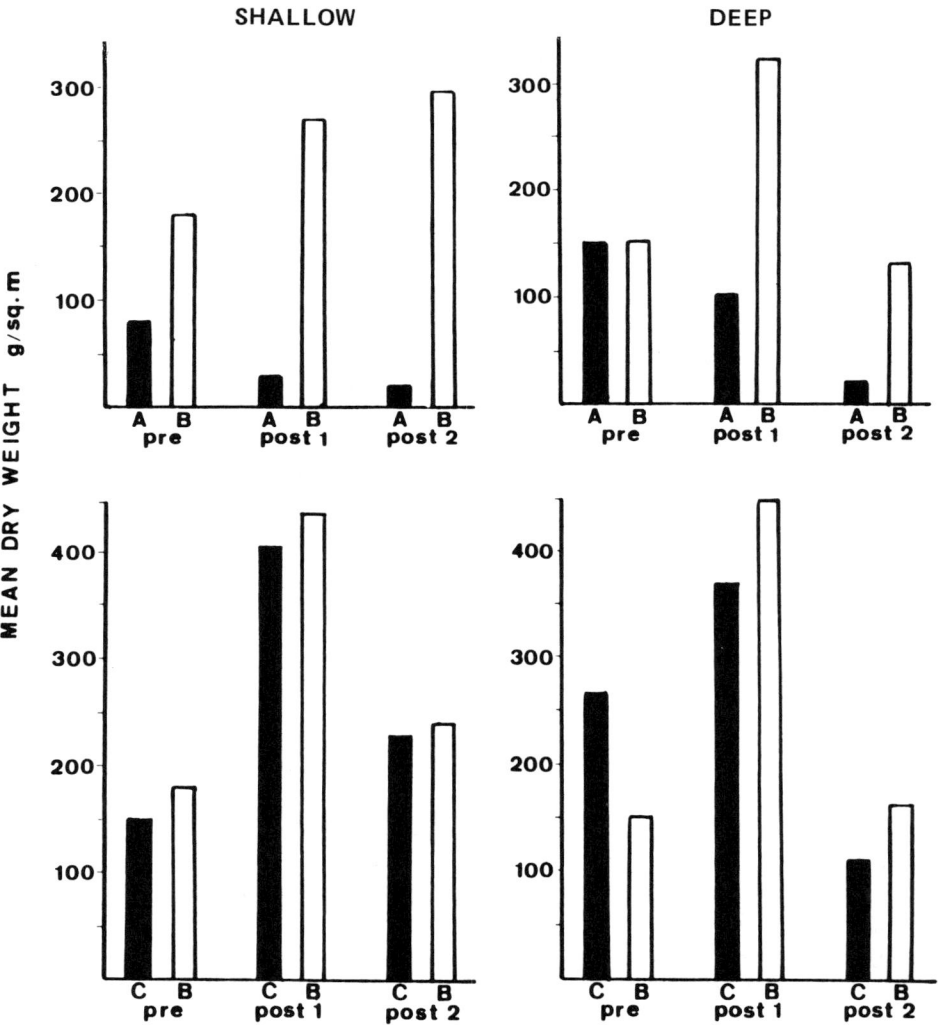

Figure 12-5 Pretreatment and posttreatment biomass of aquatic macrophytes (gm dry wt/m^2) in control, harvested, and chemically treated plots in East Twin Lake, Ohio. A = harvest plot, B = control plot, C = chemical (Diquat and Cutrine) plot. Shallow sampling area ranges from 0.5 to 1.0 meters deep; deep sampling areas ranged from 1.0 to 1.6 meters deep (from Conyers and Cooke, 1983; Conyers, 1983).

Table 12-10 Average Cost, Cost Range, and Number of Observations for Aquatic Plant Harvesting in North America (Values for Canadian Projects in Canadian Dollars)

	Chemung Lakes[a]	Okanagan Valley[b]	North America[c]	Wisconsin[d]	Dane Co., Wisconsin[e]	Ohio[f]	Florida[g]
Number of observations	6	4	4	35	1	1	1
Mean cost per							
hectare	$345	$1088	$551			$282	$186
acre	$140	$ 440	$223			$115	$ 75
Range of costs per							
hectare	$282–390	$976–1258	$148–1483	$370–490	$168–232		
acre	$114–158	$395– 509	$ 60– 600	$150–198	$ 68– 94		

[a] Wile et al. (1979)
[b] Table 12.6.
[c] Smith (1979)
[d] Dunst and Nichols (1979)
[e] Livermore and Koegel (1979)
[f] Conyers and Cooke (1983)
[g] Sassic (1982)

A frequent concern is fuel, maintenance, and repair costs. Smith (1979) found these to range from 5.8 percent to 25 percent of the total costs.

Harvesting is often considered to be no more expensive than herbicide treatment. Dunst and Nichols (1979) reported a cost range (n = 300 lakes) of $250-$1,077/ha ($100-$436/acre) for herbicide use in Wisconsin. Sassic (1982) estimated the cost range for herbicide use in Florida lakes to be $371-$741/ha. ($150-$300/acre). If the harvesting cost range of $148-$490/ha ($60-$198/acre) is widely applicable, then harvesting is less costly than herbicide treatment.

ENVIRONMENTAL EFFECTS

Because aquatic plant harvesting usually is limited to relatively small areas of water bodies, negative effects on water quality would be expected to be minor (Cottam and Nichols, 1970). Carpenter and Adams (1977) reviewed impacts of mechanical harvesting and separated their discussion into the following categories:

1. immediate physicochemical effects
2. protracted physicochemical effects
3. effect on biota
4. effect on ecosystem processes

They reported that harvesting can cause resuspension of epiphytes, detritus, and marl, and release of organic materials from sheared stems, noting that some of these materials would have minor effects because of rapid settling. Dilution would also minimize the impacts on water quality of exudation from cut plant stems, although further study is recommended. More long-term effects on nutrient cycling between the water column and lake sediments include littoral zone erosion, depression of photosynthesis (with decrease in pH), and possibly changes in oxygen levels (Burton et al., 1979). Mechanical harvesting may affect sediment accumulation or enrichment by removal of organic matter, part of which would normally decompose in situ. Particulate materials may be trapped in unharvested vegetation and, in water courses with moderate flows, aquatic plants may remove significant amounts of dissolved and particulate nutrients (Anderson, 1984).

In reviewing the effects of harvesting on biota, Carpenter and Adams (1977) stressed that many impacts could take relatively long periods of time to develop, and that repercussions could be complex. Variable regrowth of target macrophytes was reported, and they noted that harvesting alone was unlikely to cause immediate changes in lake flora (also noted by Johnson and Bagwell, 1979), but that, over a number of repetitive harvesting seasons, species more tolerant to harvesting might dominate. Neel et al. (1973) found that while one year of harvesting in Lake Sallie, Minnesota, greatly reduced weed concentration, phytoplankton productivity increased.

Nichols (1973) examined the question whether control of aquatic macrophytes by harvesting would lead to an increased algal problem. He found a significant increase in algal biomass the year after two successive years of harvesting in the shallow water, but in deep water the decline in macrophytes from harvesting was accompanied by a decrease in algal biomass. Similarly, Wile et al. (1979) reported that no significant changes in algal biomass were found during a seven-year harvesting operation. Zooplankton and other animals associated with aquatic plants may be profoundly affected by harvesting, primarily through elimination of food supplies and habitat and by shifts in the types and numbers of organisms frequenting the areas (Kern-Hansen, 1980).

The impacts of harvesting on fisheries are not well documented. Crowder and Cooper (1979) reported that aquatic plant harvesting may expose small fish (and invertebrates) to increased risk of predation, as well as reduce the basic nutrient source for these organisms. Exposure of stunted centrarchid populations to increased predation is thought to be beneficial to increased individual size and growth of both prey and predator. Wile (1978) reported that fish populations (except for yellow perch) remained stable throughout the extensive harvesting operation in Chemung Lake, Ontario. In the Okanagan Valley, B.C., preliminary monitoring of fish collected with aquatic vegetation by conventional single-stage harvesters showed about 50-100 juvenile fish were collected in each load (Newroth, unpublished). Many thousands of bass, perch, and sunfish were collected each season, but the overall impact of this depletion of fish populations has not been studied.

In far more detailed studies, Haller et al. (1980), Mikol (1984), and Storch and Winter (1983) have reported highly conflicting results with regard to the impact of harvesting on fish populations. Haller et al. examined the kinds and weights of fish removed by harvesting from a Florida lake that had a dense infestation of *Hydrilla*. The harvester selectively removed small fish, mainly young of the year, but large numbers were taken (85 kg ha^{-1}). The majority were sunfish (51 kg/ha of the total of 85 kg/ha^{-1}). Haller et al. concluded that harvesting could have a measurable impact, especially in small lakes harvested several times each year. The assessed value of the fish removed was estimated at $410,000; the fish removed represented 32 percent of fish numbers and 18 percent of fish biomass. In contrast, Mikol, and Storch and Winter report (for Saratoga and Chautauqua Lakes, New York, respectively) very low removal of fish standing crop (2.4-2.6% in Saratoga Lake) with mainly small, slower-moving species such as bullheads, sunfish, and perch found in the harvest. Storch and Winter report that very few muskie, walleye, or black bass were removed. One explanation for the differences in these results may be smaller standing crops of fish in the New York lakes and dense *Hydrilla* mats, which make escape more difficult in the Florida lake.

In all instances, smaller fish, which can reduce or eliminate populations of phytoplankton grazers such as large-bodied *Daphnia*, were removed. The significance of this idea is discussed further in Chapter 13, Biological Controls. It appears, given the current evidence, that there is little prospect for significant and lasting damage to the fish community from harvesting, and there may be benefits to the lake in the form of loss of fish that can be significant in nutrient regeneration and in increasing algal biomass through their control of herbivorous zooplankton.

Minor spillage of fuels, lubricants, and hydraulic fluids can occur in operations with mechanized equipment. In British Columbia there were frequent public complaints about plant fragments that accumulated on beach areas after harvesting operations. Harvesting operations in Vaseux Lake, B.C., were reported to have detached filamentous algae, which later clogged rotating screens in irrigation systems downstream.

CONTROL OF NUTRIENT LEVELS

Burton et al. (1979) and King and Burton (1980) reviewed harvesting as a technique to reverse nutrient enrichment of lakes, by biomass removal. It was concluded that, while harvesting is an excellent tool for reducing growths of nuisance weeds (especially *Myriophyllum*) and for reducing oxygen stress, it will restore few lakes to lower nutrient levels. Most eutrophic lakes have external and internal nutrient loadings well in excess of the amount that can be removed by harvesting. For harvesting to accomplish a reduction in lake nutrient levels, Burton et al. (1979) point out, macrophyte densities must be high, phosphorus input must be less than 1 gm P m^{-2} yr^{-1}, most of the lake surface must be covered by plants, and macrophytes must regrow every year. A number of investigators have reported that this latter condition might not be met since macrophyte biomass, especially Eurasian water milfoil, is diminished in the year after intensive harvesting. Also, cut stems may briefly continue to release phosphorus (Carpenter and Gasith, 1978). Perhaps most importantly, Burton et al. (1979) calculated that while maximum harvest in very dense weed beds might be 600-900 g dry wt m^{-2} yr^{-1}, the most likely range for the northern United States is 50-400 g dry wt m^{-2} yr^{-1} (Table 12-11).

Wile et al. (1979) reported that about one-quarter of the stored phosphorus was in root systems of *M. spicatum*; this is unavailable to conventional harvesting since only plant shoots are removed. This report also described the factors that would have the greatest effect on the value of harvesting for nutrient removal, size of the littoral zone, amount and type of biomass produced, intensity of harvesting, magnitude of external loading, and sediment nutrient supply. Wile et al. (1979) concluded that total biomass removal by aquatic plant harvesting was not feasible in the vast majority of circumstances because of the short harvesting season, equipment limitations, and the potential for ecosystem disruption. Nevertheless, in their detailed

Table 12-11 Harvest Removal Potential for Nitrogen and Phosphorus from Eutrophic Lakes in the United States (from Burton et al., 1979)

Plant Type	Typical harvest $g\ m^{-2}\ year^{-1}$ (dry weight)	Nitrogen removal* $g\ m^{-2}\ year^{-1}$	Phosphorus removal* $g\ m^{-2}\ year^{-1}$
Submerged macrophytes			
Northern U.S.	50- 400	1.5- 12.0	0.15- 1.20
Southern U.S.	150- 650	4.5- 19.5	0.45- 1.95
Floating species (water hyacinth)	500-2500	15.0- 75.0	1.50- 7.50
Emergent species	500-5000	15.0-150.0	1.50-15.00

*Based on nitrogen content of 3 percent and phosphorus content of 0.3 percent; removal is for area harvested, not the entire lake.

study of seven years of harvesting in Chemung Lake, Ontario, they estimate that a quantity equivalent to 92 percent of the net external phosphorus loading was removed during the 1975 harvesting operation. This was possible because the lake had a relatively low loading rate of P.

Welch et al. (1979) also reviewed the prospect of utilizing aquatic plant harvesting as a means of phosphorus control. They concluded that in the case of Long Lake, Washington, potential removal of 60 percent of the annual external loading could be achieved, provided that 41 percent of the surface area is harvested in each growing season. Conyers and Cooke (1983) calculated that 46-100 percent of the net annual P loading to East Twin Lake, Ohio, would have been removed during the 1972-1976 seasons if half the littoral area had been harvested each summer. Peterson et al. (1979) found, however, that continued plant harvesting over the summer of Lake Sallie, Minnesota, was ineffective in reducing the P content of the lake because of high external loading. Lie (1979) and Riemer (1984) also reviewed lake nutrient management by harvesting and reiterated the practical limitations described by other authors, indicating that extensive and continuous harvesting would be required throughout the growing season if long-term benefits to lake quality were to be expected.

Harvesting may have a far more important role in the rehabilitation of certain types of lakes than the simple removal of nuisance vegetation to satisfy the momentary demands of lake users. The vast majority of lakes are small and shallow, with morphometries that provide large areas for macrophyte colonization (Wetzel, 1983). These macrophyte beds are significant sources to the water column of dissolved (DOC) and particulate (POC) organic carbon (Carpenter et al., 1979; Wetzel and Manny, 1972). Otsuki and Wetzel (1974) reported that *Scirpus subterminalis* released 63 percent to 85 percent of the total DOC input, including allochthonous DOC, to the pelagic zone's organic carbon budget in Lawrence Lake, Michigan. Carpenter and Adams (1979) and Carpenter (1980) have noted that the entire macrophyte biomass of the littoral zone may be replaced one or more times during the summer. This DOC and POC release would increase bacterial metabolism in the littoral, accelerating nutrient recycling (Carpenter et al., 1979; Wetzel, 1983; Wetzel and Manny, 1972). Transport of DOC and POC from the littoral to the pelagic zone represents an important and often dominant subsidy that regulates the structure and function of the lake. Detritus is an energy source, and during benthic metabolism and subsequent oxygen depletion, it stimulates the release of nutrients from reduced sediments to the water column (Rich and Wetzel, 1978).

Tissue sloughing from macrophytes is an important source of phosphorus to the pelagic zone and may be one of the most important reasons for controlling macrophyte biomass by harvesting. Several researchers have noted that 60 percent or more of the total loading of phosphorus to planktonic algae during summer stratification is from macrophytes (Barko and Smart, 1980; Carpenter, 1981; Hill, 1979; Prentki, 1979). Erickson (1980) concluded, from experiments at Fish Lake, Wisconsin, that this transport from the littoral represented, over the duration of stratification, a 300-1500 percent replacement of the mean pelagic pool of phosphorus. Jacoby et al. (1982) provide an important illustration of the role of macrophytes in phosphorus release. The biomass of *Elodea densa*, the dominant macrophyte, was reduced by

84 percent after a water-level drawdown of Long Lake, Washington. A similar reduction could have been accomplished by harvesting. Internal loading of phosphorus was reduced by 80 percent, water column total phosphorus was 50 percent lower than previous summers, and no dense blue-green algal blooms occurred. In previous winters the decomposition of plants was believed to provide the initial supply of phosphorus for the spring algal bloom. This did not occur in the winter-spring after drawdown. Similarly, the work of Funk and his associates (1977; 1978; 1980) (on Liberty Lake, where alum was successfully applied to intercept P release from decaying macrophytes, and thus control algal blooms) supports the view that harvesting may improve the littoral and pelagic zones in some cases.

Thus there is evidence suggesting that substantial plant removal, as could be accomplished by harvesting in the typical small lake, may seriously interrupt phosphorus-DOC-POC release to the pelagic zone. This interruption could significantly lower the supplies of algal nutrients from the littoral zone and in vertical transport from the hypolimnion following oxidation of detritus and subsequent nutrient release. The nutrients removed in a macrophyte biomass harvest, while possibly important in lakes receiving relatively high external P loading, are not nearly as important as interference by harvesting in the more dynamic process of nutrient uptake by macrophyte roots, loss during senescence, and transport to the pelagic zone as a subsidy to this community. Macrophyte removal, especially in small lakes and in lakes with high internal loading, may therefore not only improve the lake for near-shore recreation activities, but also contribute to the control of nuisance algal blooms.

EFFECTIVENESS

Kimbell and Carpenter (1981) surveyed the literature and reported that growth of *Myriophyllum spicatum* was reduced by harvesting. Perkins and Sytsma (1981) qualified this generalization and noted that the results reported in the literature are somewhat variable. Rawls (1975) measured the benefits, from one year to the next, of harvesting Eurasian water milfoil, and he concluded that no control beyond the year of treatment could be expected and often two treatments per season would be required. This observation has been confirmed by research in the Okanagan Valley of British Columbia (Newroth, unpublished). Experiments to extend the depth of harvesting from 1.5 m to 2.5 m or 3.5 m showed limited benefits in treating nuisance populations of Eurasian water milfoil in British Columbia. In shallow water (less than 2.5 m deep), harvesting to the bottom resulted in seasonal control, while regrowth to the surface occurred within the growing season when plants were harvested to about 1.5 m depth. In deeper water there was no major difference in the effectiveness of shallow and deeper cutting. This pattern may be a characteristic of Eurasian water milfoil, or it may reflect inhibition of growth by cutting close to the root crown in the shallow populations. Variability in the published results of harvesting treatment may reflect differences in cutting depth, and in some cases the harvester may have reached the substrate, removing or disrupting the root crowns. Additional research is needed on how (e.g., root crown cutting versus stem shearing) to harvest.

Perkins and Sytsma (1981) concluded that single harvests of Eurasian water milfoil may be relatively ineffective in stressing plants (in a system where water transparency was high) and suggested that two cuts during the summer could provide more short-term benefits. A transitory reduction of carbohydrate accumulation levels in root systems was achieved by harvesting, and late fall harvesting was recommended if a potential stress on plants for the following season was to be expected. This result seems to be supported by work reviewed earlier by Wile et al. (1977) and Nichols and Cottam (1972), where two or three cuts, including a late season harvest, were found to be most effective in reducing stem density and in preventing regrowth.

The immediate reduction of nuisance macrophyte biomass resulting from harvesting is generally considered beneficial, but the duration of this perceived benefit depends on the target plant species, the depth, frequency, and seasonal timing of the harvest, and on ecosystem factors. Although the spread of nuisance plants through fragmentation occurs naturally in infested lakes, inefficient harvester pick-up of cut plants may speed the process in some cases. The responses of various component plant species in the harvested area have been poorly studied.

Declines in aquatic vegetation have been reported, most noticeably for Eurasian water milfoil, and aquatic plant harvesting has been implicated as a primary causal factor in some situations. Nichols (1974) reviewed the effectiveness and costs of 32 harvesting operations in the upper midwest and concluded that harvesting reduced biomass not only in the year of the harvest but possibly in subsequent years. S.A. Peterson (personal communication) noted reduced harvestable plant biomass with each successive year (3 years) of harvesting in Lake Sallie, Minnesota. Carpenter (1979), however, could not link harvesting directly to the decline of Eurasian water milfoil in Wisconsin lakes, because robust stands, remaining after the general decline in other areas, had been harvested regularly. In Buckhorn and Chemung Lakes in Ontario, a research program (Painter, personal communication) is endeavoring to determine clearly whether or not extensive harvesting in the 1970s contributed to general declines in Eurasian water milfoil growth. An ongoing experimental program in the Okanagan Valley of British Columbia has not yet shown significant reductions of stem biomass or plant vigor when areas have been harvested repeatedly.

Nicholson (1981a) presents evidence that harvesting has promoted the growth of *Myriophyllum spicatum* at the expense of native *Potamogeton* species in Chautauqua Lake, New York. The pond weeds are generally species that emphasize sexual reproduction, regenerate poorly from fragments, heal and regrow slowly after cutting, and are tall. This makes them susceptible to harvesting, along with other Chautauqua Lake species such as *Ranunculus trichophyllus* and *Vallisneria americana*. Nicholson points out that milfoil, on the other hand, can grow rapidly from fragments and heals and elongates rapidly after cutting. Anderson (1984), for example, found that milfoil regrew to preharvest density in 23 days in an Ohio reservoir in which plants were cut 5-10 cm from the mud surface.

Although harvesting activities do not usually extend into the roots of aquatic plants, conventional harvesters may operate in contact with soft bottom areas without damage to cutter systems. Reported declines of Eurasian water milfoil following harvesting may be due in part to deliberate or accidental harvesting at the root crown level. In effect, this treatment of plants in shallow areas may lead to major uprooting and stress. Nicholson (1981b) documented a significant annual reduction of shoots and of total biomass of plants by manual uprooting and removal of Eurasian water milfoil and other species. Rotovating and shallow tillage also have shown significant benefits from treatments of Eurasian water milfoil in British Columbia (Maxnuk, 1979; Newroth, unpublished). Conyers and Cooke (1983) and Conyers (1983) also reported a significant decrease in the biomass of a macrophyte community consisting of *Potamogeton crispus* (curly-leafed pondweed), *Ceratophyllum demersum* (coontail), and *Chara* sp. (stonewort) following a harvesting procedure in which plants were cut at the sediment-water interface. At the first harvest 80 gm dry wt m^{-2} of plants were removed. Seven weeks later, only 4.7 gm dry wt m^{-2} of macrophytes could be harvested. A control area had a normal seasonal succession of plants. Cooke (unpublished manuscript) has found that re-growth of European watermilfoil is significantly reduced if the harvester cuts 1-2 cm into the mud so that root crowns are torn out. A line of bubbles forms just in front of the cutter bar when this depth in the mud is achieved.

Cattails (*Typha* sp.) can be a significant nuisance in waterways and lakes. Sale and Wetzel (1983) describe an experiment in cattail control by harvesting. Plants cut below the water level exhibited rapid decline in oxygen in submersed parts, due to respiration, and the below-water biomass decayed. Three cuts below water during the growing season were sufficient to kill nearly all underwater biomass. This species is therefore particularly susceptible to proper harvesting techniques.

SUMMARY

Earlier sections of this chapter summarized available documentation of major aspects of mechanical harvesting as an aquatic-plant control technology. The scope of the chapter was limited to the equipment and operational programs reported in North America, where harvesting for management of freshwater aquatic plants is most prevalent. Also, most of the information reviewed here relates to control of submerged species that have caused major nuisances in temperate parts of the North American continent.

Aquatic-plant harvesting is distinguished from other major mechanical control technologies in that the main objective is semicosmetic removal of vegetation (often only the uppermost parts of rooted plants) to provide a relatively short-term improvement in the multiple-use capacity of a water body. Despite limitations and disadvantages, mechanical harvesting is widely used, often without integration with other control methods. This technology is selected because of the flexibility of mechanical harvesting and the absence of environmental drawbacks such as those attributed to herbicidal control.

Several basic options for aquatic-plant harvesting have been developed and tested in North America. The major developments of harvesting technologies originated in Wisconsin, where researchers at the University of Wisconsin in Madison have constructed and evaluated numerous harvester prototypes. Most of this work has been performed during the past decade, and despite considerable promise in methods for vegetation processing and improved technologies for gathering cut plants, little of this research has led to further development and marketing of improved equipment by harvesting equipment and manufacturers.

The single-stage harvester, which combines cutting, collection, storage, and shoreward transportation modes on one unit, continues to be most popular. Perhaps this relates to the relatively higher capital equipment cost of multiple-stage harvesting systems, in which these functions are performed by separate equipment units. No multiple-stage harvesting systems, in which the cutter unit is separate from the collection unit, are marketed at the present time. This type of system shows the best promise for economical operation to deal with very extensive aquatic plant populations, such as the major nuisances created by *Hydrilla* in the southeastern United States. The extra capital cost and manpower requirements of operating multiple-stage harvesting, however, appear to discourage the application of this approach at this time.

The prospect of utilizing harvested aquatic vegetation has encouraged considerable research in processing the material as it is collected and after it is transported to the shore. Unfortunately, despite research results that show promise for improved material handling by chopping and pressing harvested plants, and some valuable uses for the processed material, no general cost-effective benefits have been identified. The main problems appear to be high moisture content and the cost of transportation, and the fact that only relatively low volumes of harvested material can be accumulated at any given site. The seasonal nature of aquatic plant growth in many areas also limits the commercial utilization of harvested nuisance vegetation.

West (1980; 1981) stressed the need to develop analytical models and to improve procedures for evaluation of existing equipment, and recommended development of design specifications for new equipment. All of these areas will require ongoing commitments from senior government agencies and progressive industry if significant improvements in basic harvesting equipment are to develop.

Comparisons of several operations demonstrated the considerable ranges of harvesting costs, which reflect the variations of local conditions. Also the degree of machine downtime (which may be considerable) is a function of inherent characteristics of the harvesting site, the special needs of the operation (e.g., need for control of escaping fragments), and the propensity for machines to break down.

The environmental impacts of aquatic-plant harvesting are acknowledged by most authors to be relatively short-lived and inconsequential in most operational circumstances. This is primarily because of the localized, small scale of the treatments; uncertainties were described, however, in regard to operations of large scale, especially where fishery resources were concerned. Harvesting of large fractions of stunted fish populations when large areas of shallow lakes are harvested is usually considered to be desirable (Shapiro et al., 1975), but removal of substantial numbers of young game fish could be detrimental. The impacts of harvesting are likely to be greatest if machines inadvertently contribute to propagation of nuisance plants and encourage rapid spread.

Benefits of aquatic-plant harvesting primarily relate to improvements in lake quality for recreational uses such as swimming, boating, and waterskiing, though nutrient removal and protection of the pelagic zone from nutrients released during macrophyte decay may result in some cases. More than seasonal reductions in aquatic-plant biomass (and perceptible levels of nuisance) have been reported. These observations require additional documentation. It was suggested that local circumstances or the depth of mechanical cutting might influence nuisance plant regrowth rates. This area would benefit from more extensive research.

Public perceptions of needs for aquatic plant management have considerable impact on the priority for control, which in turn influences the size and scope of the operation. Little factual information has been published to document public reaction to harvesting or other technologies. The relatively high capital costs of mechanical harvesting equipment exclude most individual groups, so most harvesting programs have been developed by Provincial or State environmental organizations. Often these programs have developed with the support of local ratepayers' groups, and cost-sharing procedures have been developed to spread the costs and aid in the process of setting priorities for control activities. The impact of inflation on harvesting capital equipment was discussed by Smith (1979).

Canellos (1981) reviewed the future of mechanical control equipment and described the relatively limited interest of private industry in developing innovative new equipment for an uncertain market. Demand for equipment on the part of consumers is also limited by lack of familiarity with available methods, concerns about benefits to be derived from their application, and limited budgets to finance purchase and ongoing operation of the equipment.

REFERENCES

Aiken, S., Newroth, P,, and Wile, I. 1979. The biology of Canadian weeds. 34. *Myriophyllum spicatum L. Can J Plant Sci* 59:201–215.

Anderson, P.W. 1984. The Environmental Impacts of Macrophyte Harvesting on a Small Embayment of Wendel R. LaDue Reservoir, Geauga Co., Ohio. M.S. Thesis, Kent State University, Kent, Ohio.

Anonymous, 1969. *Eutrophication: Causes, Consequences, Correctives,* Washington, DC: Natl. Acad. Sci.

Anonymous, 1976. *Making Aquatic Weeds Useful: Some Perspectives for Developing Countries,* Washington, DC: Natl. Acad. Sci.

Anonymous. 1979. *Information Bulletin. Aquatic Plant Management Program. Volume VIII: An Investigation of the Potential for Utilization of Eurasian Water Milfoil in British Columbia,* Victoria, B.C.: Ministry of Environment.

Anonymous. 1980. *Studies on Aquatic Macrophytes. Part XIX. Eurasian Water Milfoil Treatments with 2,4-D in the Okanagan Valley, 1977-78. Volume 3. Herbicide Application Effects on Eurasian Water Milfoil,* Victoria, B.C.: Ministry of Environment, Inventory and Engineering Branch Report No. 2915.

Armour, G., Brown, D., and Marsden, K. 1979. *Studies on Aquatic Macrophytes. Part XV. Bottom Barriers to Aquatic Weed Control*, Victoria, B.C.: Ministry of Environment, Water Investigations Branch Report No. 2801.

Armour, G.D., Hanna, R.S., Walters, I.P., and Maxnuk, M.D. 1980. *Studies on Aquatic Macrophytes. Part XIV. Summary of Mechanical Aquatic Plant Management, Okanagan Valley, 1978*, Victoria, B.C.: Ministry of Environment, Inventory and Engineering Branch Report No. 1917.

Aurand, D. 1982. *Nuisance Aquatic Plants and Aquatic Plant Management Programs in the United States. Volume 2. Southeastern Region*, McLean VA: Mitre Corporation.

Aurand, D. 1983. *Nuisance Aquatic Plants and Aquatic Plant Management Programs in the United States. Volume 4. Northwest Region*, McLean, VA: Mitre Corporation.

Bagnall, L.O. 1980. Bulk mechanical properties of *Hydrilla*. *J Aquat Plant Manage* 18:23-26.

Bagnall, L.O., Shirley, R.L., and Hentges, J.F. 1973. *Processing Chemical Composition and Nutritive Value of Aquatic Weeds*, Publication No. 25, Water Resources Research Center, U. of Florida.

Barko, T.W., and Smart, R.M. 1980. Mobilization of sediment phosphorus by submersed macrophytes. *Freshwater Biol* 10:229-238.

Breck, J.E., Prentki, R.T., and Loucks, O.L. 1979. *Aquatic Plants, Lake Management and Ecosystem Consequences of Lake Harvesting*, Institute for Environmental Studies and Center for Biotic Systems, University of Wisconsin, Madison, Wisconsin.

Bryan, A.D., ed. 1978. *Studies on Aquatic Macrophytes. Part VIII. Experimental Hydraulic Dredging for Aquatic Weed Control in Vernon Arm, Okanagan Lake, 1975*, Victoria, B.C.: Ministry of Environment, Water Investigations Branch Report No. 2727.

Bryan, A., and Armour, G. 1982. *Studies on Aquatic Macrophytes Part XI. Rotovating and Jetting for Aquatic Weed Control in Okanagan and Kalamalka Lakes*, Victoria, B.C.: Ministry of Environment.

Burton, T.M., King, D.L., and Ervin, J.L. 1979. Aquatic plant harvesting as a lake restoration technique. In *Lake Restoration: Proceedings of a National Conference*, U.S. Environmental Protection Agency, EPA-440/5-79-001, 177-185.

Canellos, G. 1981. *Aquatic Plants and Mechanical Methods for their Control*, McLean, VA: Mitre Corporation.

Carpenter, S. 1979. The invasion and decline of *Myriophyllum spicatum* in a eutrophic Wisconsin Lake. In J.E. Breck, R.T. Prentki, and O.L. Loucks, eds., *Aquatic Plants, Lake Management, and Ecosystem Consequences of Lake Harvesting*, Madison, WI: Inst. Environ. Stud., 11-31.

Carpenter, S.R. 1980. Enrichment of Lake Wingra, Wisconsin, by submersed macrophyte decay. *Ecology* 61:1145-1155.

Carpenter, S.R. 1981. Submersed vegetation: An internal factor in lake ecosystem succession. *Am Nat* 118:372-383.

Carpenter, S.R., and Adams, M.S. 1977. *Environmental Impacts of Mechanical Harvesting on Submersed Vascular Plants*, Report 77, Madison, WI: Institute for Environmental Studies, University of Wisconsin.

Carpenter, S.R., and Adams, M.S. 1979. Effects of nutrients and temperature on decomposition of *Myriophyllum spicatum* L. in a hard-water eutrophic lake. *Limnol Oceanogr* 24:520-528.

Carpenter, S.R., and Gasith, A. 1978. Mechanical cutting of submersed macrophytes: Immediate effects on littoral water chemistry and metabolism. *Water Res.* 12:55-57.

Carpenter, S.R., Gurevitch, A., and Adams, M.S. 1979. Factors causing elevated biological oxygen demand in the littoral zone of Lake Wingra. *Hydrobiologia* 67:3-9.

Conyers, D.L. 1983. A comparative study of harvesting and herbicides for the control of aquatic macrophytes. M.S. Thesis, Kent State University, Kent, OH.

Conyers, D.L., and Cooke, G.D. 1983. A comparison of the costs of harvesting and herbicides and their effectiveness in nutrient removal and control of macrophyte biomass. In *Lake Restoration, Protection and Management*, EPA-440/5-83-001, 317-321.

Cottam, G., and Nichols, S.A. 1970. *Changes in Water Environment Resulting from Aquatic Plant Control*, Water Resources Center, University of Wisconsin, Madison.

Crowder, L.B., and Cooper, W.E. 1979. The effects of macrophyte removal on the feeding efficiency and growth of sunfishes: Evidence from pond studies. In J.E. Breck, R.T. Prentki, and O.L. Loucks, eds., *Aquatic Plants, Lake Management, and Ecosystem Consequences of Lake Harvesting*, Madison, WI: Inst. Environ. Stud., 251-268.

Culpepper, M.M., and Decell, J.L. 1978. *Mechanical Harvesting of Aquatic Plants. Report 1. Field Evaluation of the Aqua-Trio System*, Volume I and Volume II, USACE Technical Report A-78-3.

Dunst, R., Born, S.M., Uttormark, P.D., Smith, S.A., Nichols, S.A., Peterson, J.O., Knauer, D.R., Serns, S.L., Winter, D.R., and Wirth, T.L. 1974. *Survey of Lake Rehabilitation Techniques and Experiences*, Technical Bulletin No. 75, Madison, WI: Department of Natural Resources.

Dunst, R., and Nichols, S.A. 1979. Macrophyte control in a lake management program. In J.E. Breck, R.T. Prentki, and O.L. Loucks, eds., *Aquatic Plants, Lake Management, and Ecosystem Consequences of Lake Harvesting*, Madison WI: Inst. Environ. Stud. 411-418.

Erickson, R.T. 1980. Horizontal distribution and flux of phosphorus in Fish Lake, Wisconsin. Ph.D. Dissertation, University of Wisconsin, Madison.

Funk, W.H., and Gibbons, H.L. 1977. Effectiveness of an alum treatment at Liberty Lake, Washington. In *Ann. Pacific Northwest Poll. Contr. Conf.*

Funk, W.H., Gibbons, H.L., and Bailey, G.C. 1978. *Effect of Restoration Procedures upon Liberty Lake*, Third status report, Washington State Univ., Pullman, WA.

Funk, W.H., Gibbons, H.L., and Bailey, G.C. 1980. Lake assessment in preparation for a multiphase restoration treatment. In *Proc. USEPA-OECD Int. Symp. Inland Waters and Lake Restoration,* USEPA 440/5-81-010, 226-237.

Grinwald, M.E. 1968. Harvesting aquatic vegetation. *Hyacinth Control J* 7:31-32.

Haller, W.T., Shireman, J.V., and DuRant, F.F. 1980. Fish harvest resulting from mechanical control of hydrilla. *Trans Am Fish Soc* 109:517-520.

Hill, B.H. 1979. Uptake and release of nutrients by aquatic macrophytes. *Aquatic Bot* 7:87-93.

Hughes, H.R. 1976. *Research into Aquatic Weeds in New Zealand Waterways: A Review*, Inf. Ser. Dep. Sci. Ind. Res. N.Z.

Jacoby, T.M., Lynch, D.D., Welch, E.B., and Perkins, M.A. 1982. Internal phosphorus loading in a shallow eutrophic lake. *Water Res* 16:911-919.

Johnson, R.E., and Bagwell, M.R. 1979. Effects of mechanical cutting on submersed vegetation in a Louisiana lake. *J Aquatic Plant Manage* 17:54-57.

Kern-Hansen, U. 1980. The effect of macrophyte cutting on the invertebrate fauna in a Danish lowland stream. Poster lecture, *XXI S.I.L. Congress,* Kyoto.

Kimball, J.C., and Carpenter, S.R. 1981. Effect of mechanical harvesting on *Myriophyllum spicatum* L. Regrowth and carbohydrate allocation to roots and shoots. *Aquatic Bot* 11:121-127.

King, D.L., and Burton, T.W. 1980. The efficacy of weed harvesting for lake restoration. In *Restoration of Lakes and Inland Waters,* EPA/5-81-010, 158-161.

Koegel, R.G., Bruhn, H.D., and Livermore, D.F. 1972. *Improving Surface Water Conditions Through Control and Disposal of Aquatic Vegetation. Phase I: Processing Aquatic Vegetation for Improved Handling and Disposal or Utilization,* Technical Completion Report OWRR B-018-WIS, Water Resources Center, University of Wisconsin, Madison.

Koegel, R.G., and Livermore, D.F. 1979. Reducing capital investment in aquatic plant harvesting systems. In J.E. Breck, R.T. Prentki, and O.L. Loucks, eds., *Aquatic Plants, Lake Management, and Ecosystem Consequences of Lake Harvesting,* Madison WI: Inst. Environ. Stud., 329-338.

Koegel, R.G., Livermore, D.F., Bruhn, H.D.G. 1977. Costs and productivity in harvesting of aquatic plants. *J Aquatic Plant Manage* 15:12-17.

Koegel, R.G., Livermore, D.F., and Bruhn, H.D. 1978. *Improvement and Evaluation of Techniques for the Mechanical Removal and Utilization of Excess Aquatic Vegetation,* Technical Report 78-02, Water Resources Center, University of Wisconsin, Madison.

Koegel, R.G., Sy, S.H., Bruhn, H.D., and Livermore, D.F. 1973. Increasing the efficiency of aquatic plant management through processing. *Hyacinth Control J* 11:24-30.

Langeland, K.A., and Sutton, D.L., 1980. Regrowth of Hydrilla from axillary buds. *J Aquatic Plant Manage* 18:27-29.

Lie, G.B. 1979. The influence of aquatic macrophytes on the chemical cycles of the littoral. In J.E. Breck, R.T. Prentki, and O.L. Loucks, eds., *Aquatic Plants, Lake Management, and Ecosystem Consequences of Lake Harvesting,* Madison, WI: Inst. Environ. Stud., 101-127.

Little, E.C.S. 1979. *Handbook of Utilization of Aquatic Plants. A Review of World Literature,* FAO Technical Paper 16187, Rome: Food and Agriculture Organization of the United Nations.

Livermore, D.F., and Koegel, R.G. 1979. Mechanical harvesting of aquatic plants: An assessment of the state of the art. In J.E. Breck, R.T. Prentki, and O.L. Loucks, eds., *Aquatic Plants, Lake Management, and Ecosystem Consequences of Lake Harvesting,* Madison, WI: Inst. Environ. Stud., 307-328.

Livermore, D.F., and Wunderlich, W.E. 1969. Mechanical removal of organic production from waterways. In *Eutrophication: Causes, Consequences, Correctives,* Washington, DC: Natl. Acad. Sci., 494-519.

Maxnuk, M. 1979. *Studies on Aquatic Macrophytes. Part XXII. Evaluation of Rotovating and Diver Dredging for Aquatic Weed Control in the Okanagan Valley,* Victoria, B.C.: Ministry of Environment, Water Investigations Branch Report No. 2823.

Mikol, G.F. 1984. Effects of mechanical control of aquatic vegetation on biomass, regrowth rates and juvenile fish populations at Saratoga Lake, New York. In *Lake and Reservoir Management,* EPA 440/5-84-001, 456-462.

Mitchell, D.S., ed. 1974. *Aquatic Vegetation and its Use and Control,* Paris: UNESCO.

Neel, J.K., Peterson, S.A., and Smith, W.L. 1973. *Weed Harvesting and Lake Dynamics,* Ecological Research Series, EPA-660/3-73-001, Washington, DC: U.S. Environmental Protection Agency.

Neil, J.H. 1978. *Report on the Design, Fabrication and Evaluation of an Aquatic Plant Harvester,* Report prepared for Environment Canada, DSS Contract #1SS 77-0071.

Neil, J.H. 1979. New opportunities for lake management through improved harvesting and use. In J.E. Breck, R.T. Prentki, and O.L. Loucks, eds., *Aquatic Plants, Lake Management, and Ecosystem Consequences of Lake Harvesting,* Madison, WI: Inst. Environ. Stud., 339-344.

Neil, J.H. 1980. A computer model and systems cost analysis of the Limnos Aquatic Plant Harvesting System. In *1980 Proceedings, 14th Annual Meeting, Aquatic Plant Control Research Planning and Operations Review,* 26-29 November, 1979, Vicksburg, MS: Environmental Laboratory, U.S. Army Engineer Waterways Experiment Station, Miscellaneous Paper A-80-3, 157-169.

Newroth, P.R. 1979. British Columbia aquatic plant management program. *J Aquatic Plant Manage* 17:12-19.

Newroth, P.R. 1980. Case studies of aquatic plant management for lake preservation and restoration in British Columbia, Canada. In *Restoration of Lakes and Inland Waters,* EPA 440/5-81-010, 146-152.

Nichols, S.A. 1973. The effects of harvesting aquatic macrophytes on algae. *Trans Wisc Acad Sci* 61:165-172.

Nichols, S.A. 1974. *Mechanical and Habitat Manipulation for Aquatic Plant Management,* Madison, WI: Wisc. Dept. Nat. Res. Tech. Bull. 77.

Nichols, S.A., and Cottam, G. 1972. Harvesting as a control for aquatic plants. *Water Res Bull* 8:1205-1210.

Nicholson, S.A. 1981a. Changes in submersed macrophytes in Chautauqua Lake, 1973-1975: *Freshwater Biol* 11:523-530.

Nicholson, S.A. 1981b. Effects of uprooting on Eurasian water milfoil. *J Aquatic Plant Manage* 19:57-59.

Otsuki, A., and Wetzel, R.G. 1974. Release of dissolved organic matter by autolysis of a submersed macrophyte, *Scirpus subterminalis. Limnol Oceanogr* 19:842-845.

Perkins, M.A., and Sytsma, M.D. 1981. Efficacy of mechanical harvesting and its influence upon carbohydrate accumulation in Eurasian water milfoil. In *Proceedings, 15th Annual Meeting, Aquatic Plant Control Research Planning and Operations Review,* 17-20 November, 1980. Environmental Laboratory, U.S. Army Engineer Waterways Experiment Station, Miscellaneous Paper A-81-3, 464-479.

Peterson, S.A., Smith, W.L., and Malueg, K.W. 1979. Full scale harvest of aquatic plants: Nutrient removal from a eutrophic lake. *J Water Pollut Control Fed* 46:697-707.

Prentki, R.T. 1979. Depletion of phosphorus from sediment colonized by *Myriophyllum spicatum* L. In J.E. Breck, R.T. Prentki, and O.L. Loucks, eds., *Aquatic Plants, Lake Management, and Ecosystem Consequences of Lake Harvesting,* Madison, WI: Inst. Environ. Stud. p. 161-176.

Rawls, C.K. 1975. Mechanical control of Eurasian water milfoil in Maryland with and without 2,4-D application. *Chesapeake Sci* 16:266-281.

Rich, P.H., and Wetzel, R.G. 1978. Detritus in the lake ecosystem. *Am Nat* 112:57-71.

Riemer, D.N. 1984. *Introduction to Freshwater Vegetation,* Westport, CT: Avi Publishing.

Robson, T.O. 1974. The control of aquatic weeds. 6.1 Mechanical control. In Mitchell, ed., *Aquatic Vegetation and its Use and Control*, Paris: UNESCO, 72-84.

Sabol, B. 1981. Aquatic disposal of processed *Hydrilla*. In *Proceedings, 15th Annual Meeting, Aquatic Plant Control Research Planning and Operations Review*, 17-20 November, 1980, Environmental Laboratory, U.S. Army Engineer Waterways Experiment Station, Miscellaneous Paper A-81-3, October, 1981, p. 214-223.

Sale, P.J.M., and Wetzel, R.G. 1983. Growth and Metabolism of *Typha* species in relation to cutting treatments. *Aquatic Bot* 15:321-334.

Sassic, N.M. 1982. Harvesting: the future of aquatic plant control? *Aquatics* 4:14-26.

Sculthorpe, C.D. 1967. *The Biology of Aquatic Vascular Plants*, London: Edward Arnold.

Shapiro, J., Lamarra, V., and Lynch, M. 1975. Biomanipulation — An ecosystem approach to lake restoration. In L. Brezonik and J.L. Fox, eds., *Proceedings of a Symposium on Water Quality Management Through Biological Control*, Univ. Florida, Gainesville.

Shireman, J.V., and Maceina, M.J. 1981. The utilization of grass carp, *Ctenopharyngodon idella* Val., for hydrilla control in Lake Baldwin, Florida. *J Fish Biol* 19:629-636.

Smith, G.N. 1979. Recent case studies of macrophyte harvesting costs; options by which to lower costs. In J.E. Breck, R.T. Prentki, and O.L. Loucks, eds., *Aquatic Plants, Lake Management, and Ecosystem Consequences of Lake Harvesting*. Madison, WI: Inst. Environ. Stud., 345-356.

Smith, J.L. 1981. Field tests of the Limnos Mechanical Harvesting System. In *Proceedings, 15th Annual Meeting, Aquatic Plant Control Research Planning and Operations Review*, 17-20 November, 1980, Environmental Laboratory, U.S. Army Engineer Waterways Experiment Station, Miscellaneous Paper A-81-3, 181-196.

Stephenson, W., Alcock, C.M., Armour, G.D., and Baillie, D.D. 1980. *Studies on Aquatic Macrophytes. Part XIII. Aquatic Plant Fragment Barriers in the Okanagan Basin, 1976-1979*, Victoria, B.C.: Ministry of Environment, Inventory and Engineering Branch Report No. 2906.

Storch, T.A., and Winter, J.D. 1983. *Investigation of the Interrelationships Between Aquatic Weed Growth, Fish Communities and Weed Management Practices in Chautauqua Lake*, Environ. Resources Ctr., State Univ. New York, Fredonia.

Trudeau, P.N. 1982. *Nuisance Aquatic Plants and Aquatic Plant Management Programs in the United States. Volume 3. Northeastern and North Central Region*, McLean, VA: Mitre Corporation.

Tyndall, R.W. 1982. *Nuisance Aquatic Plants and Aquatic Plant Management Programs in the United States. Volume 1. Southeastern Region*, McLean, VA: Mitre Corporation.

Welch, E.B., Perkins, M.A., Lynch, D., and Hufschmidt, P. 1979. Internal phosphorus related to rooted macrophytes in a shallow lake. In J.E. Breck, R.T. Prentki, and O.L. Loucks, eds., *Aquatic Plants, Lake Management, and Ecosystem Consequences of Lake Harvesting*, Madison, WI: Inst. Environ. Stud., 81-99.

West, H.W. 1980. Mechanical control technology development. An overview. In *Proceedings, 14th Annual Meeting, Aquatic Plant Control Research Planning and Operations Review*, 26-29 November, 1979, Vicksburg, MS: Environmental Laboratory, U.S. Army Engineer Waterways Experiment Station, Miscellaneous Paper A-80-3, 152-169.

West, H.W. 1981. Mechanical control technology development. An overview. In *Proceedings, 15th Annual Meeting, Aquatic Plant Control Research Planning and Operations Review*, 17-20 November, 1980, Environmental Laboratory, U.S. Army Engineer Waterways Experiment Station, Miscellaneous Paper A-81-3, 179-180.

Wetzel, R.G. 1983. *Limnology*, 2nd Edition, Philadelphia: Saunders.

Wetzel, R.G., and Manny, B.A. 1972. Secretion of dissolved organic carbon and nitrogen by aquatic macrophytes. *Verh Int Ver Limnol* 18:162-170.

Wile, I. 1978. Environmental effects of mechanical harvesting. *J Aquatic Plant Manage* 16:14-20.

Wile, I., Hitchin, G., and Beggs, G. 1979. Impact of mechanical harvesting on Chemung Lake. In J.E. Breck, R.T. Prentki, and O.L. Loucks, eds., *Aquatic Plants, Lake Management, and Ecosystem Consequences of Lake Harvesting*, Madison, WI: Inst. Environ. Stud., 145-159.

Wile, I., Hitchin, G., Lewies, R., Dyke, R., and Painter, S. 1977. *An Assessment of the Practical and Environmental Implications of Mechanical Harvesting of Aquatic Vegetation in Southern Chemung Lake*, Rexdale, Ontario: Ontario Ministry of Natural Resources.

Wile, I., Neil, J., Lumis, G., and Pos, J. 1978. Production and utilization of aquatic plant compost. *J Aquatic Plant Manage* 16:24-27.

13

Biological Controls

Biological controls of aquatic plant nuisances are a developing and promising group of lake management techniques. Their objective is to obtain a more acceptable amount of plant biomass through addition of certain species or by manipulation of the endemic flora and fauna. The appeal of biological control of aquatic vegetation problems is that it not only avoids the introduction of toxic chemicals or expensive machinery, but it also may bring about long-term control of the problem at a moderate cost. This approach may be particularly applicable in cases where it is not possible to treat the causes of the eutrophication problem. There may also be drawbacks to the approach, including introduction of exotic species which themselves could become nuisances.

Knowledge concerning the biological control of nuisance plants in aquatic ecosystems is meager, reflecting our meager knowledge of ecosystems, the paucity of basic research in aquatic ecosystem ecology, and particularly our continued and almost exclusive reliance on engineering, machines, and chemicals for our arsenal of "solutions" to lake problems.

The following paragraphs review the introduction and use of the grass carp (*Ctenopharyngodon idella* Val.) for macrophyte control, and describe the state of progress in biomanipulation and in developing insects and plant pathogens for control of vascular plants. Other reviews of biological control of aquatic vegetation are found in Brezonik and Fox (1975) and Schuytema (1977).

PHYTOPHAGOUS FISH

Phytoplankton Grazers

Several species of fish, including *Tilapia mossambica* (Mozambique mouthbrooder), *Hypothalmicthys molitrix* (silver carp) and *Menidia audens* (Mississippi silversides) have been suggested as potential biocontrols of algae. Preliminary results of field tests are often conflicting (Schuytema, 1977), but suggest that these species might be particularly valuable in managed habitats where escape is impossible and where other controls such as nutrient inactivation and diversion of nutrients are not possible. Since they are restricted to warm-water habitats (always above 10°C), they would have to be restocked annually in most instances. Cooke (1981) has described the circumstantial evidence supporting the belief that control of gnats (*Chaoborus astictopus*) and blue-green algae blooms in Clear Lake, California, is due to the successful introduction of *M. audens* in 1967. It is not clear why there was a decline in algal blooms, but it is speculated that there has been a significant tie-up of nutrients in fish biomass through predation by *M. audens* on the gnat. In the case of *T. mossambica*, stocking at high densities is apparently required, and there could be negative effects on endemic biota, as well as problems in controlling overpopulations (Schuytema, 1977).

The Grass Carp (= white amur, Ctenopharyngodon idella Val.)

The grass carp is widely recognized for its great potential as a biocontroller of nuisance macrophytes, particularly in ponds and other hydrologically closed systems. Intense investigations in Europe and the United States are now underway regarding the efficacy of this animal. In the United States, most studies are being conducted in Alabama, Arkansas, and Florida.

Grass carp have several features that can make them highly desirable for the control or eradication of aquatic vegetation. Their growth rate may be higher than any other fish of comparable size—rates of mature wild fish (6 years) are as high as 2.7 kg^{-1} yr^{-1} according to Russian studies (cited in Smith and Shireman, 1983). A wide range of plant species, including filamentous algae, is consumed by the grass carp at daily rates that can equal the fish's body weight, under warm water conditions (Fisher and Lyakhnovich, 1973). The animal can withstand a wide range of water temperature (1-39°C) and very low dissolved oxygen (lethal level for fry is 0.32-0.60 mg O$_2$ l^{-1}) (Opuszynski, 1972).

Grass carp also have several potentially detrimental features, including destruction of desirable vegetation, introduction of parasites, interference with native fauna, and stimulation of algal blooms. Grass carp do not meet any of the basic criteria of an "ideal" candidate for introduction to an aquatic system, as described by Li and Moyle (1981). These authors consider an "ideal" exotic to be a species that is coadapted with some other species of the new system, has a narrow niche, is easily controlled after escape, and is free of exotic diseases and parasites.

Although grass carp are not operational as macrophyte control agents in most states, due in part to well-founded fears of their escape and reproduction in nontarget waters, their attributes and drawbacks will be summarized here because there is significant pressure to legalize their use and because they may ultimately prove to be both effective and environmentally acceptable.

The controversy in the United States over use of this fish stems in part from the history of its introduction. It was shipped to the Fish Farming Experiment Station in Arkansas and to Auburn University (Alabama) from Malaysia in 1963. Between 1970 and 1976, 115 lakes and ponds in Arkansas were stocked, including Lake Conway, a hydrologically open system stocked in 1971. Free-ranging fish were discovered outside of Arkansas in 1971, all of them of the 1966 age class. The fish has now spread throughout the Mississippi Valley and the Southeast; many of these are of the 1971 age class, corresponding to the release in Lake Conway (Guillory and Gasaway, 1978).

Presumably all major streams in the United States have had introductions, since grass carp are for sale through private hatcheries despite being banned in 26 states (Guillory and Gasaway, 1978; Pierce, 1983). A major scientific effort to understand the beneficial and harmful effects of widespread distribution of grass carp was not launched until after its broadcast to the waters of North America. Our current understanding of these effects is summarized here.

Red Haw Lake, Iowa

Grass carp were introduced to Red Haw Lake, Iowa, a 29-ha, dimictic impoundment with a maximum depth of 12 m and dense growths of *Potamogeton, Najas, Ceratophyllum,* and *Elodea* (Mitzner, 1978). In 1973, 530 animals of mean length 310 mm and weight 380 g were stocked. In 1974, 250 fish of this size range were added. Plant biomass was reduced by 91 percent over three years, with the greatest selection for *Najas* and *Potamogeton*. Mean annual concentrations of nitrates, BOD, and turbidity decreased, while alkalinity and phosphorus increased. Phytoplankton productivity declined. The popularity of the lake for shoreline fishing has increased, and the treatment is considered to be successful.

Lake Baldwin, Florida

Shireman and Maceina (1981) stocked Lake Baldwin, Florida (80 ha, Z_{max} = 7.7 m, $\bar{Z}$ = 4.4 m), with 24 grass carp ha^{-1} (>300 mm TL) in 1978, and then monitored fish density and growth rates and the standing crop of the principal nuisance plant, *Hydrilla verticillata*. Carp grew from a mean body weight of 0.79 kg at stocking to 3.0, 5.25, 7.25, and 9.15 kg in 1/2, 1, 1-1/2, and 2 years after stocking, respectively. Fish greater than 6 kg consumed 26 percent to 28 percent of their body weight in *Hydrilla* per day, and plant biomass was reduced from 781,320 kg in September, 1979, to 32,370 kg in March, 1980. Control of *Hydrilla* was complete at a density of 185 carp ha^{-1}. It should be noted that Shireman et al. (1978) concluded that a successful stocking of grass carp is most likely if the fish are at least 450 mm TL, a size which is not heavily preyed upon by largemouth bass.

Lakes Bell, Clear, and Holden, Florida

These lakes were stocked in 1974 with 50 grass carp ha^{-1} (11 kg ha^{-1}) with an average size of 227 grams (Leslie et al., 1983; Van Dyke et al., 1984). Lake Bell had about half of its surface covered with hydrilla (*Hydrilla verticillata*) in November, 1974. Between 1974 and 1977 vegetation on transects in this lake decreased by 94 percent, and hydrilla was eliminated. Sawgrass (*Cladium jamaicense*) and fragrant water lily (*Nymphaea odorata*) were apparently unaffected. Clear Lake was dominated by hydrilla and coontail (*Ceratophyllum demersum*), which covered one-third of its surface in January 1974. By year three the vegetation was reduced 94 percent, and both plants were eliminated by 1976. *Hydrilla* covered more than one-third of Holden Lake in November, 1974. Grass carp, along with an herbicide treatment of 13 percent of the lake, eliminated the hydrilla. Miller and King (1984) also report the elimination of hydrilla as a nuisance macrophyte in Lake Conway, Florida, after an introduction of grass carp.

These studies add further weight to the conclusion that grass carp introduction provides a relatively slow but highly effective means of eliminating certain nuisance vegetation species, especially hydrilla.

Because other imported animals, such as the common carp, have become nuisances, there have been some concerns about the grass carp, including these questions: What are the effects of this fish on other members of the aquatic community? Will nutrients released from fecal material stimulate further nuisance plant growth? Will the fish reproduce and infest nontarget waters into which it has strayed? Will other, less palatable or unpalatable nuisance plants replace those that have been grazed? Will the grass carp bring disease or parasites to native North American fish? The meager state of our current knowledge of these and related questions is briefly summarized in the following paragraphs. The reader should also consult the proceedings of a recent grass carp symposium (Shireman, 1979) and the detailed review by Smith and Shireman (1983).

Impact of Grass Carp on other Organisms

The effects of grass carp stocking on other fish have been widely investigated, with equivocal results. Some investigators have reported that tests in ponds and lakes indicate no interference with game fish (Bailey, 1978; Baur et al., 1979; Buck et al., 1976; Kilgen, 1978; Lewis, 1978; Rowe, 1984). Others, including Vinogradov and Zolotova (1974), Ware et al. (1975), Merkowsky and Avault (1976), Ball (1977), Bailey (1978), Forester and Lawrence (1978) and Forester (pers. comm.) have reported interference with spawning and loss of fish biomass, particularly in ponds where vegetation is eradicated and small fish become vulnerable to piscivores, or where the macroinvertebrate community is eliminated or changed. Conflicting observations would be expected in studies such as these, where stocking densities vary widely, experimental conditions range from wading pools (e.g., Buck et al., 1976) to lakes (e.g., Bailey, 1978), where data are often obtained at widely spaced intervals (months or years), and where measurements of actual productivity and fish species diversity were omitted.

Bailey's (1978) study, for example, dealt with the response of vegetation and certain fish species in 31 Arkansas lakes stocked with grass carp in densities ranging from 3-140 fish ha^{-1} with the objective of eradicating vegetation. Fish were sampled annually by the addition of rotenone to a bay. No trends in fish biomass attributable to grass carp were observed, although no replicate samples were taken and sampling variances are unknown. Biomass of shad, sunfish, crappies, and young-of-the-year largemouth bass and sunfish varied greatly from year to year, apparently with no relationship to the number of grass carp. In 8 of 12 lakes designated as having moderate vegetation cover (10-40%), the number of catchable largemouth bass declined from prestocking years. This study is encouraging on concerns about the negative impacts on game fish, since there appeared to be no relationship between grass carp density and biomass of most game fish. The sampling frequency, however, was annual, stocking densities were highly variable, and the extreme variances in the data are not attributed to any measured variable. Few firm conclusions can be drawn from this work, even though it is one of the most detailed studies on this question to date.

Additional very intensive studies of the effect of grass carp on other fish have been made in Florida, where it was concluded that gamefish in particular may have been negatively affected (Ware and Gasaway, 1976).

Conflicting results among investigators strongly suggest that determination of impact on sport fish populations needs to be carried out in a more rigorous way. When vegetation is severely reduced or removed, as will occur in lakes and ponds that have rapid growth in biomass of grass carp or which are intentionally overstocked, energy flow may also be sharply curtailed or diverted into epipelic and planktonic algae. The absence of rooted plants may also alter habitat conditions for fish sufficiently to produce undesired population changes. Rowe (1984) reports an incident in New Zealand in which elimination of weeds by grass carp increased predation by birds on rainbow trout (*Salmo gairdneri*). Predation, combined with low dissolved oxygen, and not direct competitive effects of the carp, brought about a sharp decline in trout abundance.

There apparently has been no study on the relationships between grass carp and abundance and diversity of other fish species in which the effect of long-term and short-term changes in energy flow, cover, and competitive relationships, as well as effects on spawning behavior, reproductive success, and embryonic development, were considered.

What are the effects of grass carp on other lake species besides fish? Even less is known about the answer to this question, except to state that the animal apparently is an omnivore (Fischer and Lyakhnovich, 1973; Vinogradov and Zolotova, 1974). Lewis (1978) found that after a pond was grazed clear of vegetation the grass carp then switched to fingernail clams. Forester and Avault (1978) report that yields of crayfish were reduced in ponds with grass carp and express concern that crayfish production areas in the Mississippi River may be affected if the fish reproduces there. Other workers (Forester and Lawrence, 1978; Kilgen and Smitherman, 1971; Miller and King, 1984; Mitzner, 1978; Shireman et al., 1983; Van Zon, 1979) have found little or no consumption of benthic invertebrates or significant changes in their abundance.

Impact of Grass Carp on Nutrient Recycling

Benthivorous fish, notably the common carp (*Cyprinus carpio*) and the brown bullhead (*Ictalurus nebulosus*) have been shown to be significant sources of nutrients, through their digestive activities, to the water columns of experimental enclosures (Keen and Gagliardi, 1981; LaMarra, 1975). These animals may have significant eutrophying effects on some lakes. Many comparable experiments on phytophagous fish such as the grass carp are needed, particularly in view of the observation that the short gut of this fish allows 50 percent or more of food ingested to leave as fecal material in the form of microscopic organic particles (Michewicz et al., 1972; Opuszynski, 1972). Since this animal may ingest plant material at rates of 20-100 percent of body weight day^{-1}, depending on animal size and water temperature, tremendous quantities of finely divided particulate organic matter may be released, along with urinary nitrogen. The impact of such grazing on nutrient concentration and oxygen demand in a lake will depend on lake morphometry, sediment chemistry, wash-out characteristics, the level of external loading, and fish density. Rich and Wetzel (1978) provide a model of the impact of detritus on aquatic ecosystems. Their work, plus our knowledge of the effect of detritus and nutrients from decaying macrophytes (e.g., Carpenter, 1981; Pomogyi et al., 1984) strongly suggests that significant inputs of fine particulate organic matter may help to close a positive feedback loop in some lakes and thereby increase the lake's degree of eutrophication through establishment of reducing conditions and through stimulation of algal blooms.

Despite these predicted implications of the impact of extensive grass-carp grazing on lake nutrient cycling and algal growth, there have been only a few studies about this problem, and their results are equivocal. Mitzner (1978), Lembi et al. (1978), and Mitchell et al. (1984), for example, noted little change in nutrient concentrations. Their sampling frequencies were widely spaced and their measurements did not examine nutrient dynamics nor identify whether the nutrients studied were those that limited productivity. Similar findings for soft-water acidic ponds were made by Terrell (1982). Van Zon (1977) suggested that the fish may actually benefit the lake by becoming a sink for nutrients. Canfield et al. (1983) provide one of the very few studies of grass carp in which nutrient dynamics were studied. Their data, from Lake Baldwin, Florida, support Van Zon's hypothesis. They suggest that grass carp have reduced the phosphorus concentration below the lake's geological potential by storage in their tissues, and have thus contributed to reduced potential chlorophyll in the water column. Their conclusion is similar to that made by Cooke (1981) to explain the reduction in algal blooms in Clear Lake, California, after the introduction of the planktivorous Mississippi silversides (*Menidia audens*) to control gnats. Canfield et al. (1983) believe that the increased planktonic chlorophyll and reduced water transparency in Lake Baldwin, after *Hydrilla* elimination by grass carp, was due in some unknown way to the absence of macrophytes and not to their removal by grass carp.

Richard et al. (1984), however, found sharply increased density of blue-green algae in three of four Florida lakes following reduction or elimination of *Hydrilla* biomass with grass carp, with herbicides, or a combination of treatments. They suggest that elimination of significant competitor (macrophytes) for nutrients permits greater accessibility to nutrients by phytoplankton. Thus while carp activities may not increase nutrient concentration *per se*, Richard et al. (1984) contend that there still may be increased phytoplankton productivity.

Leslie et al. (1983) and Van Dyke et al. (1984) have examined the impact of grass carp introduction in four Florida lakes over a four-year period. The fish were very effective in removing hydrilla, and the plant may have been eradicated from the three lakes in which it was a nuisance. In the fourth lake, *Potamogeton illinoiensis* was eliminated two years after stocking and, later, Eurasian water milfoil was greatly reduced, following a resurgence in its biomass. Turbidity increased from two- to threefold in all lakes, perhaps due to elimination by the carp of marginal plants so that waves mixed littoral sediments. Despite modest nutrient enrichment of the water column, chlorophyll *a* decreased significantly. This may have been due to light limitation. Leslie et al. (1983) point out that, unlike herbicide treatments, where a sharp pulse of nutrients and a phytoplankton bloom may occur after application (e.g., Hestand and Carter, 1978; James, 1984), the release of nutrients by grass carp may be much slower. This slow release would allow adsorption by sediments, precipitation with calcium, or other mechanisms to buffer the effect on the water column.

The Lake Conway, Florida study (Crisman and Kooijman, 1980; Miller and Potts, 1982; Miller and Boyd, 1983; Miller and King, 1984) is one of the most detailed and intensive examinations of the impact of grass carp on lake water quality. They have observed, compared to prestocking baseline data, a decrease in mean filterable and total phosphorus concentrations, a slight decrease in mean BOD, and an increase in ammonia and chlorophyll a in the water column. Algal populations in any given month are at least double those of comparable months before stocking.

As with most of the other significant questions regarding the impact of grass-carp stocking on lake water quality, we are not certain at this time whether the activities of these animals will have a eutrophying effect on the lake through the enhancement of nutrient cycling and the stimulation of algal blooms. Further research on this question must emphasize rates of nutrient release, tracing of the fates of the nutrient limiting to algal productivity, and the establishment of parallel studies of biomass and productivity of algae and macrophytes. Widely spaced sampling intervals are unlikely to provide sufficient data, and such studies must also include data on external loading and wash-out dynamics of the lake or reservoir.

Reproduction of Grass Carp

The major issue in introduction and widespread release of grass carp is whether these fish will reproduce and infest downstream reservoirs, sloughs, marshes, and lakes where vegetation may be desirable for many reasons, including the maintenance of habitat for migratory birds. The criteria for successful reproduction are stringent (reviewed by Stanley et al., 1978), and it was assumed by importers that reproduction could not occur outside the native range. Spawning is elicited by a sharp rise in water level and by temperatures above $17°C$. The eggs must remain in suspension, and currents of about 0.6 m sec^{-1} are needed. Stream length required to carry eggs to hatching increases with decreasing temperature. Larvae hatch in quiescent nursing areas (oxbows, sloughs) where they feed on zooplankton.

Despite the assumption that reproduction would not occur outside the native range, there have been many reported instances—in areas of diverse topography and latitude, ranging from the USSR to Japan, Taiwan, the Philippines, and Mexico—where introduced grass carp have spawned successfully. Rivers in the United States that appear to be likely sites include the Mississippi and Arkansas, and the canals of the Imperial Valley in California (Stanley et al., 1978). Stanley (1976) provides additional details about reproductive requirements and the geographic distribution of spawning populations.

There is now direct evidence that grass carp have reproduced in the Mississippi and Atachfalaya Rivers in Louisiana and Arkansas (Connor et al., 1980). It is not now known whether grass carp populations will increase and spread, nor whether they will have destructive effects similar to those of another imported species, the common carp.

Impact of Preferential Plant Removal by Grass Carp

Since grass carp apparently do not find all aquatic vegetation to be equally palatable, there has been concern that some plants will become nuisances due to selective feeding by the fish. Fowler and Robson (1978) describe an incident of this type, in which the unpalatable plants *Potamogeton natans* and *Myriophyllum spicatum* became dominant in a pond in the year after grass carp had removed the palatable species (*Chara* sp., *P. pectinatus*, *P. berchtoldii*, and *Elodea canadensis*). *M. spicatum* was reluctantly consumed later in the summer of the second year of the experiment. Kobylinski et al. (1980) and Van Dyke et al. (1984) also suggest that *M. spicatum* could become a nuisance in grass-carp-stocked lakes. Deer Point Lake, Florida (a reservoir), was infested with a mixed stand of Illinois pondweed (*Potamogeton illinoiensis*) and milfoil over more than one-third of its surface in 1974. From 1975 to 1978 the reservoir was stocked with a total of 61 grass carp ha^{-1} (7 kg ha^{-1}). Vegetation on the transects was reduced 97 percent by year six (1981) but began to recover, following water-level drawdown, in year seven. Milfoil had the greatest increase; Illinois pondweed was eliminated. Figure 13-1 illustrates the preference of the carp for the pondweed and the reexpansion of the milfoil in 1982. Note that milfoil actually increased after year three, when the pondweed abundance fell. Only after the pondweed was eliminated did the biomass of milfoil fall. The increase of milfoil in 1982 corresponds to a decrease in fish abundance and a decrease in their feeding rate. A similar trend is seen with hydrilla and coontail at Clear Lake (Figure 13-1). The authors believe that the milfoil will become a major nuisance again, due to reduced consumption of food per fish as the fish age (Osborne, 1981) and as fish are lost over the spillway.

Eurasian water milfoil is a serious problem in North American lakes and reservoirs. These results on the impact of grass-carp food preferences are of great significance because they suggest that this plant could become a serious nuisance in grass-carp-stocked lakes unless the fish are stocked at high densities.

Introduction of Parasites by Grass Carp

Fisheries biologists have been greatly concerned about the parasites carried by grass carp. *Bothriocephalus acheilognathi* (= *gowkongensis*), a cestode, was introduced to Russia from China with the grass carp and has become a serious problem in cyprinids in Europe, particularly in Germany (Bardach et al., 1972). Riley (1978) examined 163 grass carp from Florida ponds and lakes and found no exotic parasites. Riley also reports (personal communication) that *B. acheilognathi* has been found in grass carp from fish farms in Arkansas. Riley recommends that further imports to Florida should be quarantined until checked for parasites and disease. Pool et al. (1984) report that a dose of 105 mg kg^{-1} body weight of praziquantel, administered in fish pellets over a three-day period to 30,000 infected grass carp (15 cm total length, 25-30 gm each) was successful in eliminating the parasite. Thus it should be possible to screen all imported grass carp and to treat them for this parasite.

The preceding paragraphs provide abundant evidence that our knowledge of the grass carp is inadequate, given that it is an exotic and that the purpose of its introduction is either weed control or weed eradication, and not fish culture for food. The fish has been released in the United States and is being commercially promoted without sufficient assurances, based upon detailed and well-controlled experiments, that the

Biological Controls 323

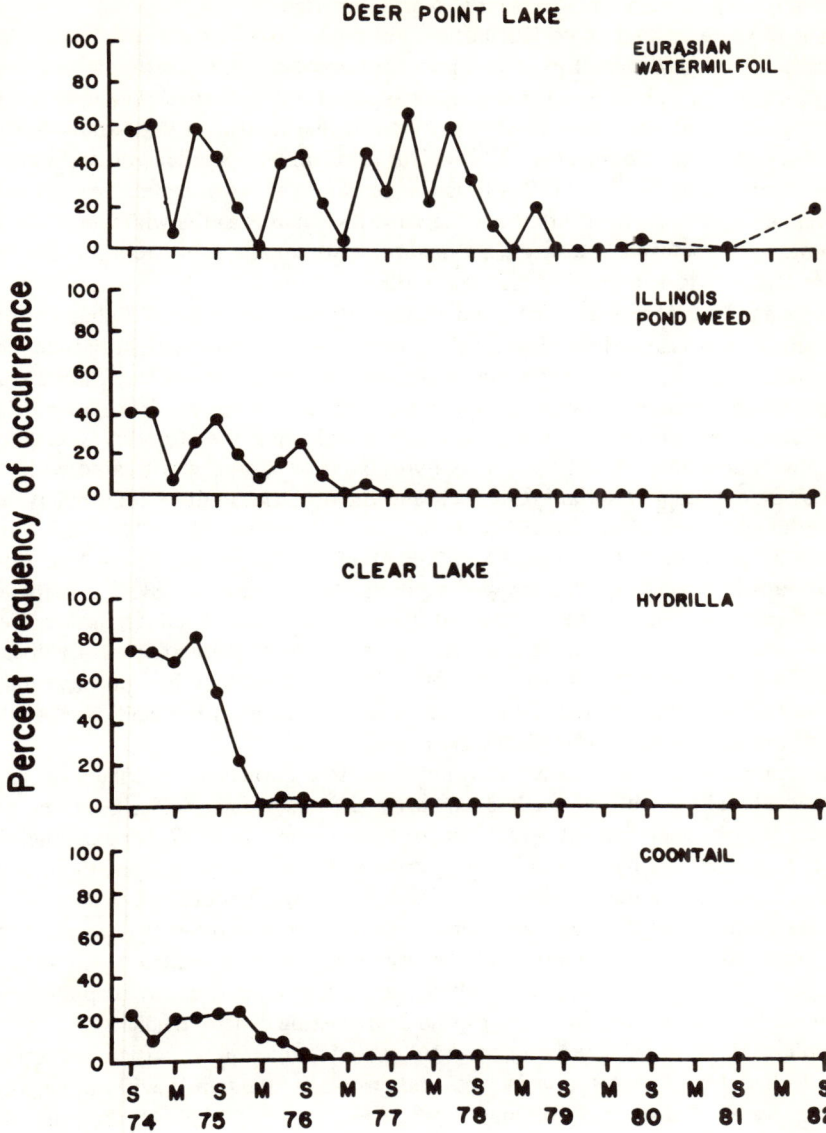

Figure 13-1 Percent frequency of occurrence of Eurasian water milfoil and Illinois pondweed on transects in Deer Point Lake (Fla.) and hydrilla and common coontail on transects in Clear Lake (Fla.) (from Van Dyke et al., 1984).

fish will not create more problems than it is supposed to solve. We will be fortunate if the animal is both useful and innocuous in North America.

It is too late to control its distribution, although the use of sterile animals might eliminate some occurrences of its spread into nontarget waters. Unfortunately, at least one sterile, triploid hybrid (a cross of a male bighead carp, *Aristochthys nobilis*, and a female grass carp) has been found to be fairly ineffective in vegetation control (Osborne, 1982, 1985; Young et al., 1983). This problem with sterile hybrids should be solvable. Theriot and Sanders (1975) describe a hybrid carp (white amur male crossed with a female *Cyprinus carpio*) that has the same food habits as the white amur. If the white amur is to be used in a way least disruptive to aquatic ecosystems, sterile fish should be the only type permitted for future introductions.

Shireman and Maceina (1981) and Pierce (1983) have suggested that an integrated control approach utilizing low stocking densities, combined with initial chemical or mechanical control, could be effective and could circumvent the ecologically disruptive use of high densities followed by plant eradication and the need to remove fish. Unfortunately, this idea could not be supported following the study at Lake Pearl, Florida, by Shireman et al. (1983). This hydrilla-infested lake was treated with the herbicides diquat, copper sulfate (Cutrine), and the potassium salt of endothal at two-month intervals. Carp were stocked at 6 ha^{-1} in July 1980, 12 ha^{-1} in July 1981, 12.5 ha^{-1} in December 1981, and 12.5 ha^{-1} in March 1982. During the stocking interval, continuous herbicide application was required. Only in August 1982 was sufficient density of carp reached. At this point, hydrilla was eradicated. The investigators were unable to attain a specific level of control, such as 15 percent plant cover. Until carp were stocked at a density sufficient to produce eradication, they had no measurable impact on the macrophytes and there could be no reduction in herbicide use. At this level of stocking grass carp require intensive management.

Shireman (1982) has pointed out, however, that control of hydrilla with grass carp in Florida lakes would be far less expensive than use of herbicides has been, and far more effective, since control would last for several years. In 1977, for example, the cost of chemically treating 15,000 ha of hydrilla was about $9.1 million; the cost of grass carp, stocked at a density of 35 fish ha^{-1} (>300 mm), would have been $1.71 million. As pointed out by Miller and King (1984), there is another significant difference between chemical treatments and the use of a biological control agent such as grass carp. Chemicals offer temporary relief, and therefore many costly reapplications are required. Grass carp provide a slow but far more lasting control of plants.

While there have been several successful cases in which grass carp introductions have controlled aquatic vegetation in lakes and reservoirs, apparently without negative changes in water quality (e.g., Mitzner, 1978; Van Dyke et al., 1984), their use as a general tool for vegetation control in North American aquatic ecosystems is not now recommended. We need a sounder scientific basis for their use in most lakes. Their use in open systems, especially reservoirs, is highly questionable because of the possibility of undesired vegetation removal in downstream ecosystems. In-depth, long-term studies are now under way at Lake Conway, Florida (Buglewicz, 1980), and Lake Baldwin, Florida (Shireman and Maceina, 1981), and these will provide some of the guidelines needed for the use of grass carp. Further studies on northern waters are needed.

Tilapia zillii

Tilapia zillii has been successfully introduced for macrophyte control in California irrigation canals, without apparent effect on native fish (Hauser, 1975; Schuytema, 1977). Childers and Bennett (1967) found that a *Tilapia* density of 2500 ha^{-1} was sufficient to control *Chara, Potamogeton foliosus,* and filamentous algae in an Illinois pond. Schiller (1984) reports that *T. zillii* was chosen to control *Utricularia vulgaris* (bladderwort) in a cooling-water impoundment in North Carolina. About 50,000 fish were stocked annually in this 445-ha lake (124 ha^{-1}) beginning in 1980, and no further bladderwort problems were experienced, though elimination of bladderwort may have allowed invasion of other plants. In 1982 and 1983, blue-green algae and macrophytes invaded the lake. In another reservoir, Hyco, a *Tilapia* density of 741 ha^{-1} was used to control *Elodea densa*. Restocking was required each year since this species does not tolerate extended periods of temperature below 10°C.

CYANOPHAGES AND BACTERIA

Safferman and Morris (1964) were apparently the first to suggest the possibility of using viruses to control blue-green algae. These cyanophages have some nearly ideal properties (Desjardins, 1981). They are selective and specific for the nuisance, nontoxic to other microorganisms in the food chain, harmless to animals, and harmless to equipment. They have no direct effect on water quality, and increase during use, rather than decrease. Considerable research has resulted in the identification of phages and their modes of action specific for blue-green algae such as *Plectonema, Anabaena, Lyngbya,* and *Anacystis*. Problems with the development of host resistance and with the effect of environmental factors on cyanophages have been studied, but as yet there has been no field-scale test of the effectiveness of phages in controlling blue-green algae. Shapiro et al. (1982) believe that phage activity may account for the success of some artificial circulation experiments.

The lytic ability of myxobacteria toward blue-green algae has been known since the reports of Shilo (1967; 1970). Several preliminary investigations of their use as biocontrol agents have been made (see Burnham, 1981; Burnham and Fraleigh, 1983; Stewart and Daft, 1977), but to date there have been no field applications.

At present, the use of viruses and bacteria is in the laboratory stage of development. Considerable research will be required before any conclusions can be drawn about their efficacy as biological control agents for aquatic plants.

PLANT PATHOGENS AND INSECTS

The successful development of biological controls for water hyacinth (*Eichhornia crassipes*) in the southern United States was accomplished by a systematic evaluation of several organisms, under strict quarantine rules, by the U.S. Department of Agriculture, the U.S. Army Corps of Engineers, and the University of Florida. This work, in striking contrast to the unregulated broadcast of grass carp, has resulted in introduction to Louisiana waters of a fungus (*Cercospora rodmanii*) and two beetles (*Neochetina bruchi* and *N. eichhorniae*) that are specific in their life cycles to the host plant (see review by Perfetti, 1983). Development of this biological control unit is especially instructive because it is the model after which other organisms will be investigated.

According to Zetter and Freeman (1972), Freeman et al. (1975), and Freeman (1977), plant pathogens have five attributes that make them ideal for the control of rooted aquatic plants: Pathogens of plants are numerous and diverse; about 100,000 plant diseases are caused by viruses, bacteria, fungi, algae, mycoplasmodia, and nematodes. These diseases are often host specific. They are easily disseminated and are self-maintaining once they are established. Plant diseases exert a limiting influence, but seldom eradicate a plant. Finally, only a few plant diseases have been shown to be dangerous to man and domestic animals.

Freeman et al. (1981) and Perfetti (1983) summarize the history of the development of *Cercospora rodmanii* as a pathogen of water hyacinth and describe its pathological characteristics and the field application methodology. Briefly, a sharp decline in the abundance of hyacinth, an incredible nuisance of lakes, reservoirs, and waterways in the southern states, was noted in Rodman Reservoir, Florida, in 1971. This was attributed to a fungus. *C. rodmanii* Conway was isolated, and laboratory and field tests ensued. A pilot field test in Louisiana was held in 1977-80. A commercial formulation was developed by Abbott Laboratories and has since been applied on a large scale to two Louisiana sites in 1980 and 1981.

Earlier, three insects were imported from Argentina, a native habitat of water hyacinth. These animals, *Neochetina bruchi, N. eichhorniae* (Coleoptera: Curculionidae), and *Sameodes albiguttalis* (Lepidoptera: Pyralidae), were studied under quarantine to assess their specificity to the target plant and to ensure that they were free of parasites and disease. *N bruchi*, for example, is found on plants of only two genera, its distribution does not exceed that of its host plant *Eichhornia*, and its life cycle can be completed only on water hyacinth (Perkins and Maddox, 1976). *N. eichhorniae* is obligatorily monophageous on hyacinth. The beetles were released in 1972 and 1974, and the moth in 1977 (Center, 1981).

The moth, *Sameodes albiguttalis*, has proven to be an effective control of water hyacinth when the plant is small, luxuriantly growing, has large leaf petioles, and is not part of a mat. The moth is most effective in an integrated management program where the water hyacinth is continually perturbed through chemicals or harvesting (Center et al., 1984).

The effectiveness of the fungus *Cercospora rodmanii* in the two large-scale treatment sites in Louisiana cannot be isolated from that of the beetles, *Neochetina bruchi* and *N. eichhorniae*, and from some climatic changes. The beetles have spread throughout the state and have exerted the greatest impact on water hyacinth. This plant has undergone a major decline in abundance in Louisiana, from a state-wide peak of 1.7 million acres (690,000 ha) in 1975 to 0.3 million acres (122,000 ha) in 1980. The dense infestation of 1975 was caused by floods in 1973 and 1974, which washed plants out of backwaters, and by mild winters, which prevented dieback. The mild winters also aided in development of beetle populations. By 1980 the beetle populations reached swarming levels. Unusually dry weather concentrated the plants in 1980, and beetle predation caused plant elimination in some areas and has sharply reduced the need for applications of herbicides (Perfetti, 1983). Goyer and Stark (1984) found that treatments as low as one *N. eichhorniae* per plant had an adverse effect, and that there was a direct relation between insect density and plant mortality at southern Louisiana sites. Further observation of this fungus-beetle-plant community will be of great interest since long-term control is the objective. Already the use of expensive, temporary controls in the form of the herbicide 2,4-D have been reduced, demonstrating that biological controls may become cost-effective for water hyacinth control.

Haag (1985 and personal communication) has examined the effects of herbicide applications on *Neochetina bruchi* and *N. eichhorniae*. This is a matter of some concern since chemical use, especially diquat and 2,4-D, is widespread and could either be directly toxic to the weevils or could secondarily reduce their population density by producing a loss of habitat and a subsequent interference with their life cycle. Her preliminary studies have found that neither herbicide is toxic at low doses and that adult weevils will migrate to untreated plants. This suggests that a controlled chemical treatment program which leaves areas of unsprayed plants might be successful in producing both immediate relief from water hyacinth while allowing enough plants to remain to permit development of the insect density to a level which will exert long-term biocontrol.

Del Fosse et al. (1976) have demonstrated, in laboratory tests, that the combination of grass carp, *Ctenopharyngodon idella*, and the mottled hyacinth weevil, *Neochetina eichhorniae*, can be used more effectively together for hyacinth control than separately. Pools that had either grass carp or weevils exhibited more plant biomass than combination pools. While hyacinth is not a preferred species of the grass carp's diet, many lakes in Florida, Louisiana, and other areas of this latitude have near monocultures of hyacinth. This suggests that the use of the insect and carp together could be very effective. Or, since the moth *Sameodes albiguttalis* has been found to be most effective against water hyacinth when the plants are perturbed (Center et al., 1984), it might be possible to combine them with grass carp.

At present, no other plant pathogens are operational as biological controls of nuisance aquatic plants. *Fusarium roseum* holds great promise for the control of hydrilla (Freeman et al., 1980). Andrews and Hecht (1981) have discovered a fungal pathogen of *Myriophyllum spicatum*, *F. sporotrichioides*. Unfortunately, the fungus is apparently not highly pathogenic, and seems to penetrate the plant only through wounds. This suggests, however, that integrated control methods could enhance effectiveness. For example, the fungus could be used in combination with harvesting or insects, which could create conditions favorable to heavy fungal infestation. Carpenter (1980) has described the unexplained decline of *M. spicatum* in Madison, Wisconsin, lakes and speculates that a parasite or pathogen, in combination with other unknown factors, could be involved. It is now clear that there are insect predators on *M. spicatum*, especially the midge genus *Cricotopus* (Diptera: Chironomidae), according to Center and Balciunas (1980) and Kangasniemi (1983). These insects could be responsible for plant damage that would allow the establishment of an *F. sporotrichioides* population of sufficient size to bring about the decline of the plants in Carpenter's lakes. Since *M. spicatum* is a serious nuisance of economic consequences, this possibility should be explored. It may be possible to develop a control situation similar to that used for water hyacinth. Presently, the moth *Paraponyx stratiotata* (Lepidoptera: Pyralidae) is being investigated in Italy prior to its importation to the United States as a biological control for *M. spicatum* (Habeck, 1983).

Other insects have been imported, evaluated under quarantine, and pilot-tested. The first insect released in the United States for aquatic plant control was the alligator weed (*Alternanthera philoxeroides*) flea beetle *Agasicles hygrophila* (Coleoptera: Chrysomelidae). This animal is effective but has no diapause in its life cycle and is thus sensitive to freezing (Spencer and Coulson, 1976). Cold-tolerant individuals, collected from the coldest part of their range in Argentina, have also been introduced in pilot studies, in the hope that overwintering may occur. Overwintering success appears to be limited (Buckingham and Boucias, 1982). The most promising insect predator of alligator weed appears to be *Vogtia malloi* (Lepidoptera: Pyralidae), because of its ability to overwinter. Spencer and Coulson (1976) predict that the insect stress placed on alligator weed will be sufficient over the years to eliminate it as a nuisance.

Monophagous insects and host-specific plant pathogens are an extremely promising approach to the control of nuisance aquatic plants and deserve careful further evaluations. The costs for locating and testing these organisms are high, but the long-range economics of their use could make them far better than repeated applications of expensive toxic chemicals. As well, the use of insects and pathogens with other techniques in the form of an integrated control approach may prove to be cost-effective. Much additional research is needed in this area.

BIOMANIPULATION

Lake restoration has long been considered to be primarily a matter of chemistry and engineering. Clearly, in-lake techniques to lower nutrient concentration or to remove sediments or rooted plants can be effective in improving lake trophic state following diversion of silt and nutrients. These approaches, however, usually ignore the biological interactions of the lake, interactions which themselves may be responsible for low water transparency, high internal nutrient release, and the frequently observed slow response to nutrient diversion.

Shapiro et al. (1975) and Shapiro (1978) suggested the term "biomanipulation" to include lake improvement procedures that alter the food web to favor grazing on algae by zooplankton, or that eliminate fish species that recycle nutrients. Biomanipulation is new to the lake management community, particularly to those without training in limnology, and is based upon only a few demonstrations and experiments to date, primarily by Shapiro at the University of Minnesota. As he has pointed out, this paucity of experience in manipulating lake biota to effect an improvement in lake trophic state reflects the paucity of our knowledge of lake communities and emphasizes the narrowness of our current popular view that lake problems are mainly problems with algae and nutrients to be solved by chemistry and engineering. There seems to be little question that the biomanipulation approach will become increasingly popular due to its lower cost, the absence of machinery and toxic chemicals (in some cases), and to its effectiveness.

Biomanipulation involves eliminating certain fish species or restructuring the fish community to favor the dominance of piscivorous fish instead of planktivorous fish. It should be clear that these manipulations will be most successful in closed lakes, such as seepage lakes, where invasion of undesired animal species is difficult. The use of biomanipulation in reservoirs may not be practical, although Chapter 11 of this text describes the use of water-level drawdown to restructure the fish community. What follows here describes the evidence for the applicability of biomanipulation to improve lake-water clarity and to reduce internal nutrient loading, and describes some procedures for changing the structure of fish communities.

Figure 13-2 summarizes the model used by Shapiro and colleagues (Shapiro et al., 1982) for biomanipulation. Planktivorous fish feed, in some cases, by sight (e.g., bluegills, pumpkinseed sunfish), and in others by the "sieve" action of gillrakers (e.g., alewives). It is clear, based upon many observations and experiments (e.g., Brooks and Dodson, 1965; Hrbacek et al., 1961; Hutchinson, 1971), that size-selective predation occurs, in which planktivorous fish preferentially remove the largest zooplankton species, based either upon body size or amount of pigmented area (ephippia, eye spots). Ultimately such predation may change the zooplankton species composition of the lake, eliminating such large-bodied species as *Daphnia pulex, D. galeata, D. retrocurva, Limnocalanus macrurus*, or *Diaptomus sicilis*, and favoring small species such as *Bosmina longirostris* (see review by Nilsson, 1978). Removal of the large-bodied species is thought to allow phytoplankton blooms, through reduced grazing pressure. In the absence of fish predation, it is believed that large zooplankton predominate over small, due to their greater grazing efficiency and their ability to select particles over a wider range. Algal blooms in lakes with these zooplankton species should be less. Size-selective predation and size-efficiency hypotheses were apparently first formulated by Brooks and Dodson (1965) and are the basis of the fish planktivory–phytoplankton abundance portion of Shapiro's model. The biomanipulation model is also based upon several experiments demonstrating that benthivorous fish may be significant as internal sources of nutrients.

Experimental Evidence

How are these hypotheses related to improvements in lake water transparency? Shapiro (1979) and Lynch and Shapiro (1981; 1982) describe the several experiments and experiences to support the idea that elimination of intense planktivory will favor the return of large-bodied zooplankton, and with this return more intense grazing on phytoplankton and clearer water.

In winter-kill lakes transparency in subsequent summer months may be very high, due to zooplankton grazing in the absence of fish predation. Similar results occur when fish have been removed with rotenone (Henrickson et al., 1980; Reinersten and Olsen 1984; Shapiro et al., 1982; Stenson et al., 1978).

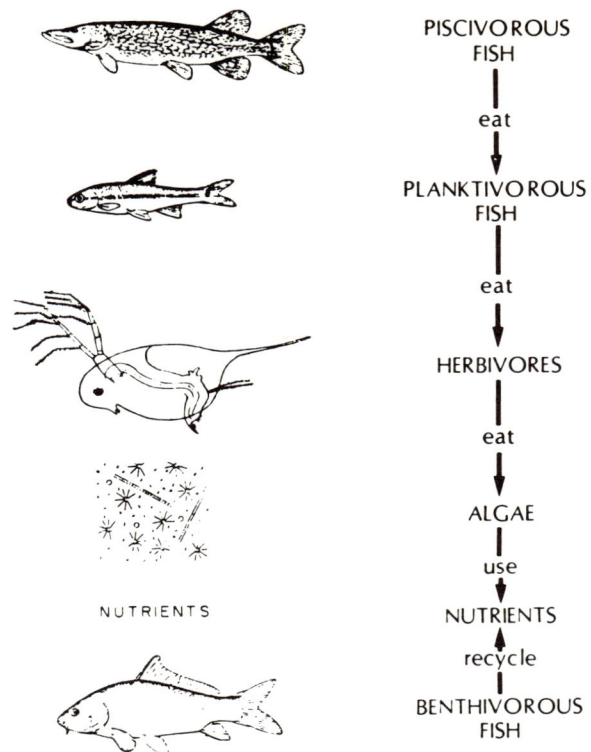

Figure 13-2 The aquatic food chain (from Shapiro et al., 1982).

Lynch and Shapiro (1981; 1982) conducted a series of experiments in Pleasant Pond, Minnesota, to test the effect of planktivorous fish on phytoplankton abundance and species composition. Twelve enclosures, 1.0 by 0.8 m, were suspended in deep water. Ten were stocked with five different densities of bluegill (*Lepomis macrochirus*). Figure 13-3 illustrates the effects of planktivory on zooplankton. *Daphnia pulex, D. galeata mendotae*, and *Chaoborus*, all large-bodied species, were eliminated from the fish enclosures. Algal volumes in fish enclosures were as much as 16 times those of fishless controls and were dominated by green algae. At the highest levels of fish predation a dense population of single filaments of *Aphanizomenon flos-aquae* developed, and *Anabena circinalis* bloomed. In fishless enclosures the phytoplankton biomass was small and dominated by flagellates and heavily armored species. Three levels of nitrogen and phosphorus enrichment in fish and fishless enclosures were studied to test whether cycling of nutrients by fish could account for the algae bloom in fish enclosures. Algal volume remained higher in the enclosures that contained fish, supporting the conclusion that zooplankton removal and not nutrients was involved.

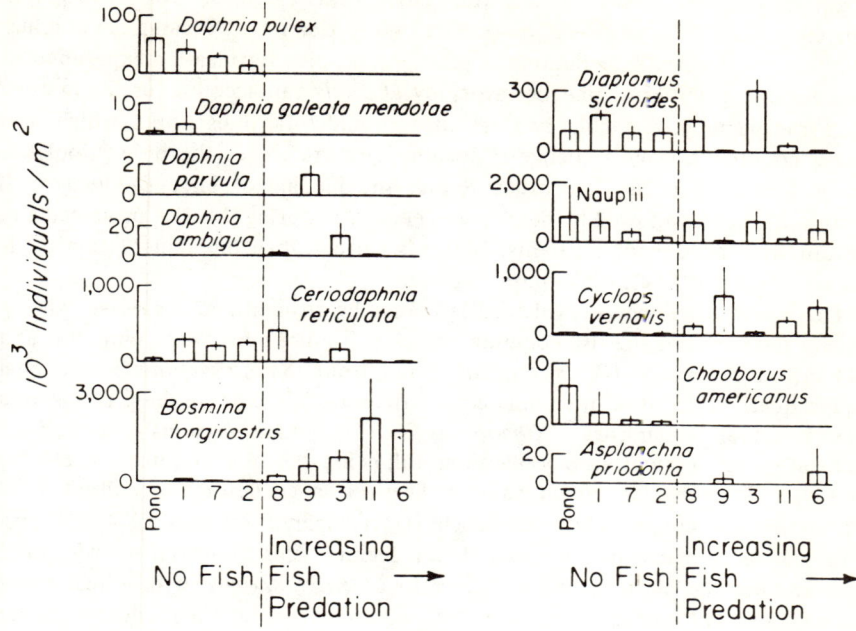

Figure 13-3 Mean and range for abundance of crustaceans, *Chaoborus* and *Asplanchna*, for last three sampling dates of enclosure experiment (from Shapiro et al., 1982).

Pleasant Pond was divided in half; one half was essentially fishless and the other contained fathead minnows (*Pimephales promelas*). In midsummer, algal volume in the fish half was five times that of the fishless half, even though carbon fixation on the fishless side was as great or greater than in the fish-containing half.

Further evidence of the effectiveness of restructuring the fish community has been provided by the experiments of Shapiro and Wright (1984), Benndorf et al., (1984), and Spencer and King (1984).

Round Lake, Minnesota (12.6 ha, Z_{max} = 10.5 m, $\bar{Z}$ = 2.9 m), a dimictic, eutrophic lake, was treated with rotenone to eliminate a fish community dominated by planktivorous bluegill (*Lepomis macrochirus*) and black crappie (*Pomoxis nigromaculatus*), and the benthivorous black bullhead (*Ictalurus melas*). The lake was restocked with piscivorous largemouth bass (*Micropterus salmoides*) and walleye (*Stizostedion vitreum*) in a ratio of 2.2 bluegills to 1 of the piscivorous fish. Before the application of rotenone, the ratio had been about 165 planktivores to 1 piscivore. Channel catfish (*Ictalurus punctatus*) were added to prevent reestablishment of the black bullhead, an animal capable of producing a significant rate of internal nutrient loading. The mean Secchi disc transparency in the summer prior to rotenone application (1980) was 2.1 m. In summer 1981, following biomanipulation in September and restocking in October, 1980, the mean transparency was 4.8 m. In 1982 mean summer transparency was 4.7 m. *Daphnia pulex*, rare before treatment, became the dominant microcrustacean species. Calculations of grazing rate, plus fertilization experiments in enclosures, strongly suggest that herbivory by *Daphnia* can account for the improvement in transparency. A decline in total nitrogen and total phosphorus, which could not be accounted for by a change in loading, occurred after the biomanipulation. Wright and Shapiro (1984) hypothesize that the feeding of large *Daphnia* in the epilimnion at night, followed by descent to deep water during the day, produced a net downward translocation of nutrients. If this is correct, then biomanipulation of this type may bring about lasting lake improvement.

The use of rotenone, as explained later, can be difficult and expensive. Another option for restructuring the fish community is to introduce a piscivore. Benndorf et al. (1984) introduced perch (*Perca fluviatilis*) and trout (*Salmo gairdneri*) to a small, eutrophic quarry to control or eliminate planktivorous fish. Even in the presence of an efficient invertebrate planktivore (*Chaoborus flavicans*), there was a major reduction in the mortality of herbivorous zooplankton following the biomanipulation. *Daphnia hyalina* and *D. culcullata*, which had not been observed before, accounted for 30 percent of the microcrustacea biomass. Secchi disc transparency (Figure 13-4) improved, even though phytoplankton biomass did not decline. This improvement was due to grazing, because the phytoplankton shifted to *Chlamydomonas* (with a high growth rate to compensate for grazing) and *Oocystis* (with a morphology that inhibits grazing). The disappearance of *Bosmina* and *Thermocyclops* may have been due to predation by *Chaoborus*. Benndorf et al. (1984) provide a pictoral model (Figure 13-5) to describe their view of biomanipulation.

Spence and King (1984) have examined the effect of zooplanktivory to fish on the flora of shallow, nutrient-rich ponds. Ponds with fathead minnows (*Pimephales promelas*) and brook stickleback (*Culaea inconstans*) were dominated by intense algal blooms and had very low zooplankton biomass. A pond without fish and a pond with a dense population of large (99% > 200 mm) largemouth bass (*Micropterus salmoides*) had very low phytoplankton biomass, but supported dense populations of submerged macrophytes due to the high water transparency.

Lynch and Shapiro (1981) concluded that the response of lakes to nutrient enrichment is as dependent on the fish community as on the nutrients. Phytoplankton in planktivore-free lakes will be less sensitive to enrichment if herbivores keep algal biomass below levels at which nutrients are critical. A lake's response to fish removal will depend upon the level of nutrient enrichment. In eutrophic, highly enriched lakes, in which phytoplankton are less sensitive or limited by nutrients, there will be higher sensitivity to changes in planktivory than to further nutrient enrichment. This suggests that lakes with large nutrient incomes, in which diversion is not possible or too expensive, would be good candidates for a fish removal or restructuring manipulation.

Further evidence of the significance of fish to phytoplankton biomass was obtained by Leah et al. (1980). A man-made lake was separated into two basins, both of which received highly fertile river water. In one basin fish density (dominated by the roach *Rutilus rutilus*) was high, and so was phytoplankton biomass, while in the

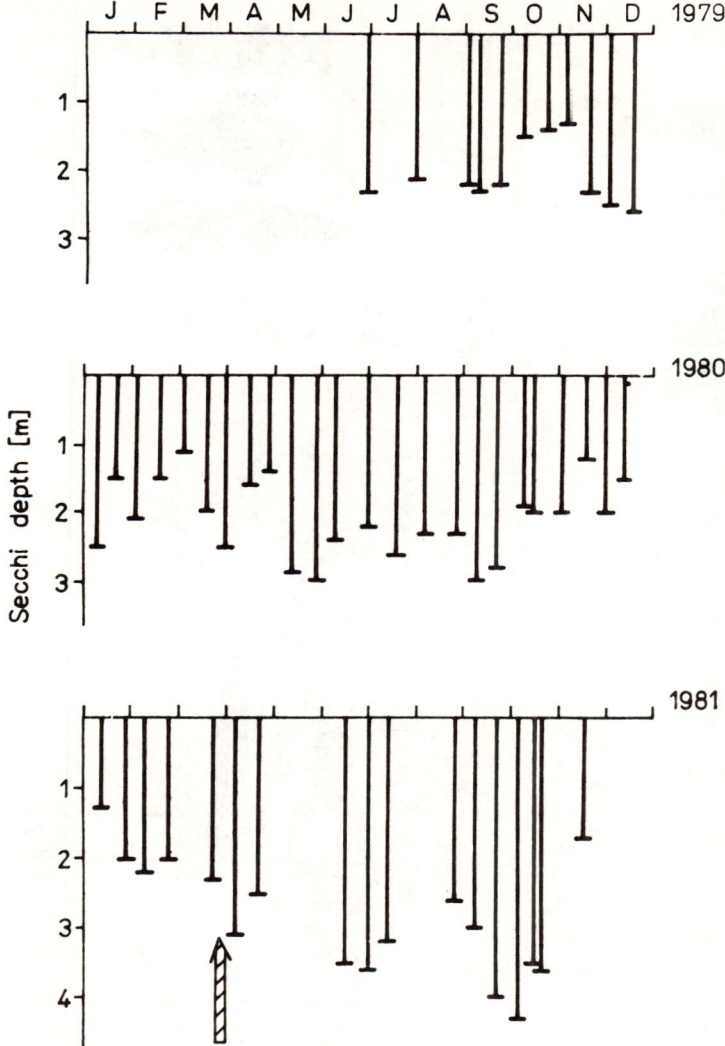

Figure 13-4 Variation of Secchi depth (thick arrow ← stocking with predacious fishes) from Benndorf, et al., 1984.).

other basin fish were essentially absent, and grazing of Cladocera (*Daphnia longispina, Bosmina longirostris*) brought about greatly reduced phytoplankton abundance. Like the ponds described by Spencer and King (1984), water clarity was so high in the fishless basin that macrophytes developed.

The impact of planktivory on phytoplankton biomass may also be related to the distribution of dissolved oxygen with depth and thus the ability of zooplankton to

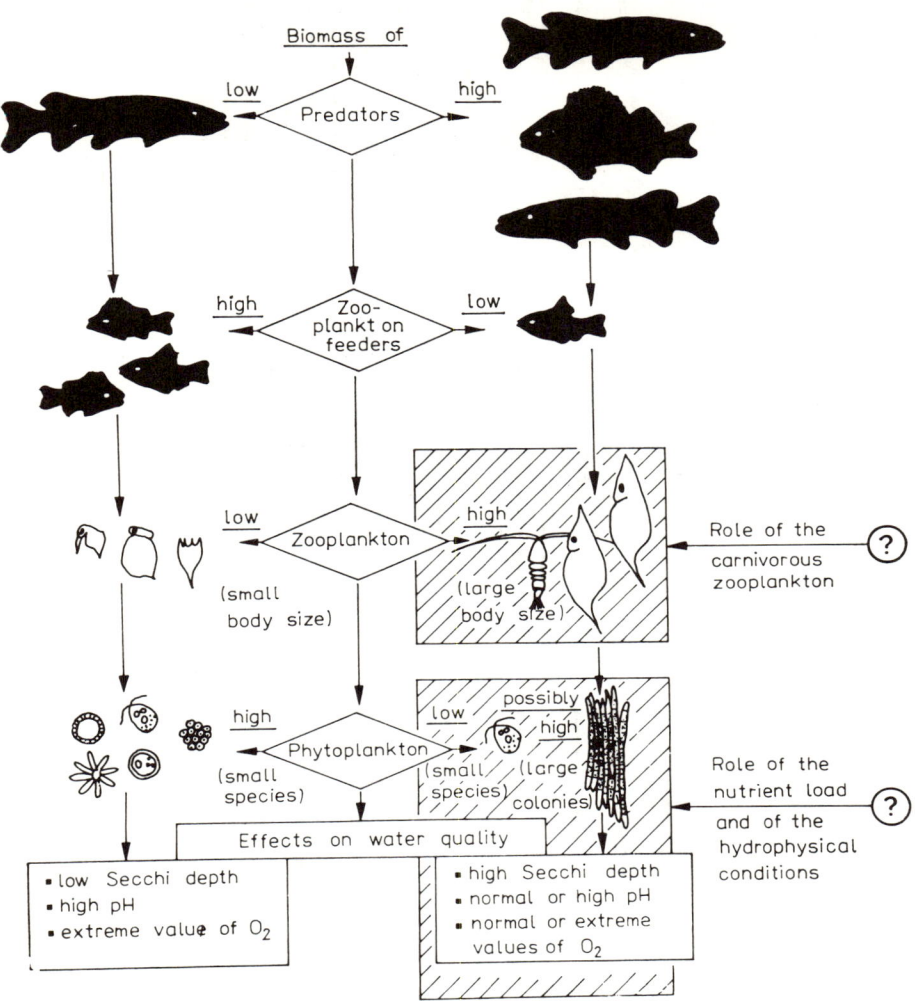

Figure 13-5 Hypothetical scheme showing the connections involved in biomanipulation; shaded area = connections as yet unsure (from Benndorf et al., 1984).

migrate to deep water. It has long been known that many species of zooplankton ascend to surface waters at night to graze and descend to dark, deep waters during the day. There is evidence (e.g., Vuorinen et al., 1983) that such vertical migration is adaptive, in that it may permit zooplankton to escape sight predation by using the dark metalimnion as a refuge during the day. Dissolved oxygen depletion of this layer will eliminate this refuge and allow size-selective predation to occur. Shapiro et al.

(1982) believe that when enrichment of the metalimnion with allochthonous and autochthonous organic matter is sufficient to promote metalimnetic oxygen depletion, then the elimination of large-bodied zooplankton will proceed rapidly and there will be even more intense algal blooms, whose decomposition will continually reinforce the oxygen loss. This phenomenon may then be another aspect of the closing of a positive feedback loop in eutrophication, similar to that of the development of the littoral zone (Carpenter, 1981). An obvious solution to the problem, in addition to biomanipulation, would be to install an aeration or circulation system (Shapiro et al., 1982), as described in earlier chapters.

Timms and Moss (1984) describe another refuge for large-bodied zooplankton, which allows them to escape planktivory by fish and to exert control of phytoplankton biomass. They describe two shallow basins, both of which receive their water from a highly enriched river. One basin, Hudsons Bay, supported a large stand of water lilies and had clear water in the summer, despite high nutrient levels. Evidence from bioassays showed that the water could support phytoplankton. Hoveton Great Broad, an adjacent basin, had very high chlorophyll concentrations. The zooplankton of the basins differed, despite their interconnectedness. Rotifers, Cladocera (particularly *Bosmina longirostris*), and the copepod *Cyclops vernalis americanus* were very abundant in both basins, but Hudsons Bay had a high relative abundance of large-bodied, weed-associated Cladocera, which were essentially absent from Hoveton Great Broad. Experiments with enclosures that had no fish, and observations of demographic data, suggested that the absence of large-bodied Cladocera in Hoveton Great Broad was due to fish predation. Diurnal studies of zooplankton and phytoplankton strongly suggested that the cladoceran grazers of Hudsons Bay are confined to the water lilies during the day, a refuge that apparently lowers the impact of fish predation. Extensive nighttime grazing controls the phytoplankton standing crop.

There may be another significant source of zooplankton mortality of direct interest to lake managers. Shapiro (1979) points out that agricultural runoff is often contaminated with significant amounts of pesticides, which are extremely toxic to zooplankton, even in trace amounts. He lists ten commonly used insecticides for which the 48-hour LC_{50} to zooplankton is less than $1.0\ \mu g\ l^{-1}$ and notes several cases in which elimination of zooplankton by pesticides was associated with algal blooms.

Copper is the most common algicide. It is also very toxic to zooplankton (see review by Demayo et al., 1982). Winner and Farrell (1976) and Winner et al. (1977) report significant reductions in survival for *Daphnia magna, D. pulex, D. parvula*, and *D. ambigua* at copper concentrations in excess of $40\ \mu g\ l^{-1}$, a concentration well below that recommended for algal control (1-2 mg Cu l^{-1}). Chelates of copper (nitrilotriacetic acid) are far less toxic to *D. magna*, according to Biesinger et al. (1974). The well-known "rebound" effect after adding copper sulfate to lakes to control planktonic algae may therefore be caused in part by mortality of herbivorous zooplankton and the subsequent bloom of algae.

There is no complete agreement among limnologists regarding the causes of the generally observed phenomenon that eutrophic lakes have small-bodied zooplankton in the epilimnion during summer algal blooms, while oligotrophic lakes often have the large-bodied species. Z.M. Gliwicz (personal communication) has asked, "Do algal blooms fail to occur because of large *Daphnia*, or do we have an absence of *Daphnia* because of large algal species?" Webster and Peters (1978), Porter (1981), and Gliwicz and Siedlar (1980), among others, have evidence that large-bodied zooplankton may be absent from eutrophic lake waters because of the presence of large, inedible, sometimes toxic blue-green algae. Smaller zooplankton, which cannot filter these large particles and are thus confined to detritus and nannoplankton, have higher fecundity during summer algal blooms. Mortality of large-bodied zooplankton is not as much due to fish grazing as to reduced fecundity, in the view of these researchers. Algae such as *Anabaena flos-aquae* (Porter and Orcutt, 1980), *A. affinis* (Porter, 1973), *Anacystis nidulans, Gloeocapsa* (Arnold, 1971), *Oscillatoria, Lyngbya* (Burns, 1968), and flakes of *Aphanizomenon flos-aquae* (Shapiro et al., 1982) have been found to be either toxic, unmanageable, or of poor nutritional value to *Daphnia*.

Other investigators, using laboratory and field experiments, have found that *Daphnia* can ingest blue-green algae and that their grazing can have a measurable effect on water transparency. Schoenberg and Carlson (1984) have provided some direct evidence of the differential grazing effects of the smaller *Bosmina longirostris* versus the larger-bodied *Daphnia galeata mendotae*. Their work supports the view that *Daphnia* can have a significant impact on the density of the colonial blue-green *Microcystis*, and that this grazing, in the absence of fish predation, could bring about increased transparency. Holm et al. (1983) found that while *D. pulex* cannot graze large *Aphanizomenon* flakes, it will ingest small (less than 1.5 mm) flakes and single filaments, and that flake formation may be an adaptive response to *D. pulex* grazing. Webster and Peters (1978) found that with *Lyngbya* as a food, larger zooplankton (*D. pulex, D. ambigua,* and *Simocephalus vetulus*) in high concentrations of filaments still produce more eggs per unit time than small-bodied animals (*Bosmina*). Thus, even with the reduced filtering rates and increased rejection of food associated with large filamentous algae, the reduced populations of large-bodied zooplankton in summer apparently are not explained by the poor-food-quality argument.

Planktonic microcrustacea are not the only group of organisms in the plankton that may alter algal density through grazing. Brabrand et al. (1983) have examined the relationship between the density of the ciliated protozoan *Nassula* and blooms of *Oscillatoria*. Laboratory observation revealed that *Nassula* could remove large numbers of filaments and has the potential of preventing *Oscillatoria* blooms. However, the scarcity of *Nassula* in lakes appears to be due to copepod predation. Brabrand et al. suggest that *Oscillatoria* blooms could be controlled, at least in small lakes, by inoculation of *Nassula*, particularly when copepod density is low. The practicality and effectiveness of this suggestion awaits further research.

The effect of maintaining low planktivory and high herbivory seems clear from the experiments now available. Herbivores may increase transparency, primarily through greatly increased removal of detritus and nannoplankton. With the exception of Lynch and Shapiro's (1981) work, however, there are few studies to document the species-by-species response of phytoplankton to high herbivory (Z.M. Gliwicz, personal communication). Even though some algae appear to be detrimental to zooplankton, intense grazing appears to increase transparency. While more research is needed on the size-efficiency hypothesis and the causes for small numbers of large-bodied zooplankton in the epilimnion of lakes with algal blooms, it seems clear that control or elimination of planktivorous fish can bring about increased transparency in nutrient-enriched lakes and may offer an alternative to nutrient diversion where such diversion is cost prohibitive. Obviously, control or elimination of planktivorous fish in drainage lakes and in reservoirs may be difficult or impossible, and may even be undesirable where the lake is used for fishing. Timms and Moss (1984) suggest an alternative to fish elimination in shallow lakes. Their work provides evidence that clear water in the nutrient-enriched Hudsons Bay was due to grazing by large-bodied zooplankton, which had escaped fish predation by taking refuge during the day in water lily beds. They suggest that it might be possible to mimic this refuge in macrophyte-free lakes by adding some type of artificial "structure" that could be used as a littoral refuge by zooplankton. This idea is similar to the use of aeration to open the dark hypolimnion as a refuge from fish visual predation.

Nutrient Regeneration by Fish

It has been pointed out repeatedly in this text that lakes often respond very slowly to nutrient diversion. Models that predict lake response time based on wash-out phenomena often fail to describe actual lake response because of internal nutrient loading. One significant source of that loading can be benthivorous fish. LaMarra (1975) was among the first to report that common carp (*Cyprinus carpio*) excretion could add phosphorus to the water column at significant rates. In one lake, phosphorus loading from carp was calculated to be 2.2 mg P M^{-2} day^{-1} at 22°C, and 200 kg fish ha^{-1}. Keen and Gagliardi (1981) found similar results with the brown bullhead *Ictalurus nebulosus*). Smeltzer and Shapiro (1982) calculated the phosphorus loading rate of carp and bullheads in Union Lake, Minnesota to be 88 mg P M^{-2} $year^{-1}$, or 100 percent of the external loading rate. Andersson et al. (1978), using benthivorous fish enclosure-exclosure studies, also found a "eutrophying" effect in fish enclosures and "oligotrophication" in exclosures. Significant lake improvement may therefore be achieved by control or removal of benthivorous fish.

Biological Evaluation of Lakes and Ponds

It should be clear that a biological evaluation of a problem lake is as essential as the traditional physical and chemical one. An expensive diversion or in-lake manipulation may not produce the desired results simply because the biota of the lake are dominated by certain benthivorous or planktivorous fish that have stimulated algal growth through nutrient recycling or by suppression of zooplankton grazing. An examination of the phytoplankton-zooplankton-fish community is therefore essential. Lakes with summertime dominance of small-bodied zooplankton (e.g., *Bosmina*), low transparency, dense algal blooms, and large populations of carp and bullheads or stunted panfish may show only slight recovery after expensive diversions. If a lake is used primarily for recreational fishing, a choice may have to be made by lake users between algal blooms and high density of centrarchid fish (bass, bluegill, pumpkinseed) on the one hand, and clear water on the other. The ideal solution may be to restructure the fishery for a balance of piscivorous and planktivorous fish, following a fish elimination.

Methods of Restructuring Fish Communities

How can the fish community be removed or restructured? At least five methods, used singly or in combination, have been suggested. They are stocking of piscivores, fish winter kill, rough fish removal, use of fish poisons, and water-level drawdown. These are discussed in the following paragraphs:

Bennett (1970), based upon his own work and that of many others (e.g., Lagler and Deroth, 1953), concluded that the addition of piscivorous fish to a lake or pond may cause almost no change in the fish community. It may not be possible to add enough fish to make an impact on rough fish populations or on stunted panfish, both of which may have significant effects on nutrient cycling and size-selective predation. The addition of piscivorous fish to improve lake trophic state, rather than to improve sport fishing *per se*, has received almost no attention and deserves additional consideration before it is rejected as a means of reducing the negative impacts of rough fish or stunted panfish. Benndorf et al. (1984) provide one of the first reports on stocking of predaceous fish to improve lake trophic state. One way to increase the probability of success of such a restocking program would be a partial fish removal, through use of rotenone, a partial fish winter kill, or a water-level drawdown. Success is also related to maintenance or increase in the density of the fish that are stocked. Keup (1979) has pointed out that spawning habitats for walleyes (*Stizostedion vitreum*) or northern pike (*Esox lucius*) may be minimal or absent in many eutrophic lakes. Installation of rock rubble or gravel for walleyes and access to marshy areas for northern pike will help ensure their propagation.

Fish winter kill provides a natural means of thinning or eliminating fish. For very small lakes, this process could be enhanced by covering the ice with sand or other opaque material for the entire winter. It should be remembered that the least desirable fish (e.g., common carp, *Cyprinus carpio*; brown bullhead, *Ictalurus nebulosus*) may survive prolonged periods at very low oxygen concentrations, whereas other fish are filled. A winter-killed lake may then be no better than before. Thus, following a winter kill, the actual survival of fish species should be checked with gill nets, seining, spot poisoning, or electroshocking. Finally, the decomposition of fish following winter kill could provide a significant pulse of nutrients to the lake in early spring, and thus produce an algal bloom in later months. Removal of dead fish should be attempted.

Programs of rough fish removal through intense seining have been attempted in several midwestern states, with limited success. Seining requires that the bottom be smooth so that the net does not snag and damage the nets, thus allowing fish to escape. Many lakes and reservoirs have obstructions such as macrophytes, stumps, and boulders. Seining is also confined to shallow water. The process has low efficiency and is labor-intensive and therefore expensive.

Fish poisons are often recommended as a means of restructuring or eliminating a fish community. Rotenone (powdered derris root in a 5% formulation) is a frequently used piscicide. According to Bennett (1970), a dose of at least 1.0 mg l^{-1} will give a complete fish kill, including bullheads. The material is applied during warm weather (water temperatures of at least 20°C), because effectiveness is limited in cold water and because cold-water-treated lakes may remain toxic for long periods. While no toxicity may remain after seven days, a fish bioassay, in which live fish are kept in cages in the treated lake at several depths, is recommended before restocking.

Partial poisoning with rotenone can be used to remove fish such as gizzard shad (*Dorosoma cepedianum*) and small sunfish (Centrarchidae), both of which are very sensitive to low doses (Bennett, 1970). This is accomplished by adding rotenone along the shoreline at a dose of 0.25 mg l^{-1}. If this is done prior to the spawning of largemouth bass (*Micropterus salmoides*) in the spring or after bluegills (*Lepomis macrochirus*) have completed spawning in early October, then many bluegills and other centrachids will be eliminated. This may sharply reduce bluegill predation on the eggs and fry of largemouth bass, and thus enhance reproduction of this predatory fish.

Rotenone is also very toxic to other lake fauna, particularly zooplankton such as *Daphnia*. Several months may be required before zooplankton return to pretreatment densities (Anderson, 1970; Kiser et al., 1963; Serns, 1979). Grazing on phytoplankton, one object of such treatment, will be delayed.

There are presently only a few studies that have carefully examined phytoplankton and zooplankton responses to rotenone treatment for the purpose of improving lake trophic state. This drastic treatment of a lake is therefore highly experimental at this time.

The fish community of artificial ponds and reservoirs can be altered effectively through water-level drawdown. For example, Nimrod Lake, Arkansas was drawn down between 1 October and 1 December 1955 from 1460 ha to 285 ha, then 39,000 kg of commercial fish were seined. A large increase in game fish occurred in the following season, associated with an increase in Secchi disc transparency from 0.3 m to 1.2 m (Hulsey, 1957). Since there were no plankton data, it is not clear whether this change in transparency is associated with piscivory.

There are numerous examples of enhancement of the predatory fish population through water-level drawdown (e.g., Lantz, 1974; Pierce et al., 1963). During the drawdown process, small fish may be caught in vegetation or in temporary pools and killed. Forage-size bluegills, which prey on zooplankton, bass eggs, and bass fry decline conspicuously, apparently due to predation in the remaining pool (Bennett, 1970; Heman et al., 1969). Lantz et al. (1964) found that winter drawdown was effective in removing gizzard shad and sunfish, while summer drawdown prevented spawning of these species. Shields (1958) used drawdown to severely limit common carp reproduction. Carp spawning is induced at temperatures of 10-15°C. Water drawdown at this time was used to expose eggs.

The ideal reservoir for fish control through drawdown should have a deep channel or a harvest basin capable of supporting fish during low water level. Hulsey (1958) recommends that in new reservoirs a harvest basin free of stumps and obstructions be built, so that drawdown can be used in conjunction with a rough fish removal program, such as seining or low doses of rotenone. There should also be an adequate watershed for refilling of the reservoir.

A complete fish kill can occur following drawdown in reservoirs with dense macrophyte beds or high sediment oxygen demand in the remaining pool. Geagan (1960) reported an incidence of this, in which the planktivore gizzard shad failed to reproduce after a dramatic population decline due to low dissolved oxygen in the pool remaining after a summer drawdown. Game fish, particularly largemouth bass, did reproduce successfully after the reservoir was refilled.

The restocking of a lake following fish elimination should be accomplished with the objectives of the lake's users in mind. Some fisheries experts recommend the addition of forage fish, such as shad, to increase the growth of game fish. If fishing is the only use of the lake or reservoir, then this might be done. It should be kept in mind that these fish are planktivorous and may be involved in nutrient recycling, and the lake may revert to a trophic state similar to premanipulation years. A better tactic, based upon Shapiro's hypothesis and the current evidence, will be to keep rough fish, including gizzard shad, out of the lake as much as possible. Minnows used for bait, and emigration via streams may make this strategy difficult. State and local fish and game personnel are one source of information about the best ratios of game fish to stock in a lake following partial or complete fish removal. The types of fish that will be successful will vary geographically.

SUMMARY

Biological control of nuisance aquatic vegetation has not yet reached the operational level of mechanical and chemical techniques. Several insect species and a plant pathogen have been successfully introduced to control alligator weed and water hyacinth. The white amur or grass carp (*Ctenopharyngodon idella* Val.) has been widely disseminated, but the possibility of significant negative environmental impact has not been ruled out and considerable study of the ecology of this fish is needed before it can be further introduced to the nation's lakes and reservoirs. Spawning of escaped individuals in major waterways could pose a threat to fisheries and other resources and thus research and stocking programs should emphasize the sterile hybrids. The manipulation of food webs to enhance grazing of algae and the successful elimination of fish such as the common carp may be long-lasting, inexpensive lake management techniques. At present our knowledge about these techniques is insufficient, but rapid progress has been made in recent years. A description of techniques used by fishery biologists to restructure fish communities of lakes and ponds is presented here.

The state of our knowledge of the biological control of nuisance aquatic vegetation is undeveloped, due in part to the appeal of "quick-fix" techniques such as herbicides, but largely to our concentrated efforts to understand the causes and controls of excessive plant growth by studying only the nutrient-autotroph portion of aquatic ecosystems while largely ignoring the role(s) of other components.

REFERENCES

Anderson, R.S. 1970. Effects of rotenone on zooplankton communities and a study of their recovery patterns in two mountain lakes in Alberta. *J Fish Res Can* 27:1335-1356.

Anderson, G., Berggen, H., Cronberg, G., and Gelin, C. 1978. Effects of planktivorous and benthivorous fish on organisms and water chemistry in eutrophic lakes. Hydrobiol. 59:9-16.

Andress, J.H., and Hecht, E.P. 1981. Evidence for pathogenicity of *Fusarium sporotrichoides* to Eurasian water milfoil *Myriophyllum spicatum*. *Can J Bot* 59:1069-1077.

Arnold, D.E. 1971. Ingestion, assimilation, survival and reproduction by *Daphnia pulex* fed seven species of blue-green algae. *Limnol Oceanogr* 16:906-920.

Bailey, W.M. 1978. A comparison of fish populations before and after extensive grass carp stocking. *Trans Am Fish Soc* 107:181-206.

Ball, R.L. 1977. Effects of white amurs (*Ctenopharyngodon idella*) on production and food habits of sport fishes in Indiana ponds. Paper presented at 39th Midwest Fish and Wildlife Conf., Madison, Wisconsin, 1977.

Bardach, J.E., Ryther, J.H., and McLarney, W.O. 1972. *Aquaculture. The Farming and Husbandry of Freshwater and Marine Organisms*, New York: Wiley-Interscience.

Baur, R.J., Buck, H.D., and Rose, C.R. 1979. Production of age-0 largemouth bass, small mouth bass and bluegills in ponds stocked with grass carp. *Trans Am Fish Soc* 108:496-498.

Benndorf, J., Kneschke, H., Kossatz, K., and Penz, E. 1984. Manipulation of the pelagic food web by stocking with predacious fishes. *Int Rev Ges Hydrobiol* 69:407-428.

Bennett, G.W. 1970. *Management of Lakes and Ponds*, 2nd ed., New York: Van Nostrand Reinhold.

Biesinger, K.E., Andrew, R.W., and Arthur, J.W. 1974. Chronic toxicity of NTA (nitrilotriacetate) and metal-NTA complexes to *Daphnia magna*. *J Fish Res Can* 31:486-490.

Brabrand, A., Faafeng, B.A., Kallqvist, T., and Nilssen, J.P. 1983. Biological control of undesirable cyanobacteria in culturally eutrophic lakes. *Oecol* 60:1-5.

Brezonik, P.L., and Fox, J.L., eds. 1975. *Water Quality Management through Biological Control*. Proceedings of a Symposium, Gainesville, FL: Department of Environmental Engineering Sciences, Univ. Florida.

Brooks, J.L., and Dodson, S.I. 1965. Predation, body size, and composition of plankton. *Science* 150:28-35.

Buck, D.H., Baur, R.J., and Rose, C.R. 1976. Comparison of the effects of grass carp and the herbicide Diuron in densely vegetated pools containing golden shiners and bluegills. *Progr Fish-Cult* 37:185-190.

Buckingham, G.R., and Boucias, D. 1982. *Release of Potentially Cold-Tolerant Alligatorweed Flea Beetles (Agasicles hygrophila Selmon and Vogt) into the United States from Argentina*, Misc. Paper A-82-2, Vicksburg, MS: U.S. Army Corps of Engineers.

Buglewicz, E.G. 1980. Large-scale operations management test using the white amur at Lake Conway, Florida. An overview. In *Proc. 14th Ann. Mtg. Aquatic Plant*

Control, Research Planning and Operations Review, Vicksburg, MS: U.S. Army Corps of Engineers. Misc. Pap. A-80-3, 246-248.

Burnham, J.C. 1981. The utilization of bacteria in managing cyanobacterial populations: A review and update. In *Proceedings of a Workshop on Algal Management and Control,* Tech. Rept. E-81-7, Vicksburg, MS: U.S. Army Corps of Engineers, 230-258.

Burnham, J.C., and Fraleigh, P.D. 1983. Predatory myxobacteria: Lytic mechanisms and prospects as biological control agents for Cyanobacteria (blue-green algae). In *Lake Restoration, Protection and Management,* EPA-440/5-83-001, 249-256.

Burns, C.W. 1968. Direct observations of mechanisms regulating feeding behavior of *Daphnia* in lakewater. *Int Rev Ges Hydrobiol* 53:83-100.

Canfield, D.E. Jr., Maceina, M.J., and Shireman, J.V. 1983. Effects of *Hydrilla* and grass carp on water quality in a Florida lake. *Water Res Bull* 19:773-778.

Carpenter, S.R. 1980. The decline of *Myriophyllum spicatum* in a eutrophic Wisconsin lake. *Can J Bot* 58:527-535.

Carpenter, S.R. 1981. Submersed vegetation: An internal factor in lake ecosystem succession. *Am Nat* 118:372-383.

Center, T.D. 1981. *Release and Establishment of Sameodes albiguttalis for Biological Control of Water Hyacinth,* Tech. Rept. A-81-3, Vicksburg, MS: U.S. Army Corps of Engineers.

Center, T.D., and Balciunas, J.K. 1980. Biological control technology development. Biological control of water hyacinth, hydrilla, and Eurasian water milfoil. In *Proc. 14th Ann. Mtg., Aquatic Plant Control. Research Planning and Operations Review*, Vicksburg, MS: U.S. Army Corps of Engineers, Misc. Pap. A-80-3, 134-138.

Center, T.D., Durden, W.C., and Corman, D.A. 1984. *Efficacy of Sameodes albiguttalis as a Biocontrol of Water Hyacinth,* Tech. Rept. A-84-2, Vicksburg, MS: U.S. Army Corps of Engineers.

Childers, W.F., and Bennett, G.W. 1967. Experimental vegetation control by largemouth bass—*Tilapia* combinations. *J Wildlife Manage* 31:401-407.

Connor, J.V., Gallagher, R.P., and Chatry, M.F. 1980. Larval evidence for natural reproductions of the grass carp (*Ctenopharyngoden idella*) in the Lower Mississippi River. In *Proc. 14th Ann. Larval Fish Conf.,* Ann Arbor, MI: U.S. Fish Wildlife Serv., Biol. Sci. Program, Nat'l Power Plant Team, FWS/UBS-80/43.

Cooke, S.F. 1981. The Clear Lake example: An ecological approach to pest management. *Environment* 23:25-30.

Crisman, T.L., and Kooijman, F.M. 1980. Large-scale operations management test using the white amur at Lake Conway, Florida. Benthos. In *Proc. 14th Ann. Mt., Aquatic Plant Control Research and Operations Review,* Vicksburg, MS: U.S. Army Corps Engineers Misc. Pap. A-80-3, 298-304.

Del Fosse, E.S., Sutton, D.L., and Perkins, B.D. 1976. Combination of the mottled water hyacinth weevil and the white amur for biological control of water hyacinth. *Jour Aquatic Plant Manage* 14:64-67.

Demayo, A., Taylor, M.C., and Taylor, K.W. 1982. Effects of copper on humans, laboratory and farm animals, terrestrial plants, and aquatic life. *CRC Crit Rev Environ Control* 12:183-255.

Desjardins, P.R. 1981. Cyanophages—Are they potential biological control agents of nuisance blue-green algae? In *Proc. Workshop on Algal Management and Control,* Tech. Rept. E-81-7, Vicksburg, MS: U.S. Army Corps of Engineers, 198-229.

Fischer, Z., and Lyakhnovich, V.P. 1973. Biology and bioenergetics of grass carp *Ctenopharyngodon idella*. *Pol Arch Hydrobiol* 20:521-557.
Forester, J.S., and Avault, J.W. 1978. Effects of grass carp on freshwater red swamp crawfish in ponds. *Trans Am Fish Soc* 107:156-160.
Forester, T.S., and Lawrence, J.M. 1978. Effects of grass carp on populations of bluegill and largemouth bass in ponds. *Trans Am Fish Soc* 107:172-175.
Fowler, M.C., and Robson, T.O. 1978. The effects of food preferences and stocking rates of grass carp (*Ctenopharyngodon idella* Val.) on mixed plant communities. *Aquatic Bot* 5:261-276.
Freeman, T.E. 1977. Biological control of aquatic weeds with plant pathogens. *Aquatic Bot* 3:175-184.
Freeman, T.E., Charudattan, R., and Conway, K.E. 1975. Use of plant pathogens for bioregulation of aquatic macrophytes. In P.L. Brezonik and J.L. Fox, eds., *Water Quality Management through Biological Control*. Gainesville, FL: Department of Environmental Engineering Sciences, Univ Florida, 20-23.
Freeman, T.E., Charudattan, R., Conway, K.E., Cullen, R.E., Martyn, R.D., McKinney, D.E., Olexa, M.T., and Reese, D.F. 1981. *Biological control of aquatic plants with pathogenic fungi*, Tech. Rept. A-81-1, Vicksburg, MS: U.S. Army Corps Engineers.
Freeman, T.E., Charudattan, R., and Cullen, R.E. 1980. Biological control of water hyacinth and hydrilla using plant pathogens. In *Proc. 14th Ann Mtg. Aquatic Plant Control Research Planning and Operations Review*, Misc. Pap. A-80-3, Vicksburg, MS: U.S. Army Corps of Engineers, 141-147.
Geagan, D. 1960. A report of a fish-kill in Chicot Lake, Louisiana during a water-level drawdown. *Proc La Acad Sci* 23:39-44.
Gliwicz, Z.M., and Siedlar, E. 1980. Food size limitation and algae interfering with food collection in *Daphnia*. *Arch Hydrobiol* 88:155-177.
Goyer, R.A., and J.D. Stark, 1984. The impact of *Neochetina eichhorniae* on water hyacinth in southern Louisiana. *J Aquatic Plant Manage* 22:57-61.
Guillory, V., and Gasaway, R.D. 1978. Zoogeography of the grass carp in the United States. *Trans Am Fish Soc* 107:105-112.
Haag, K.H. 1985. Does herbicide application affect water hyacinth weevils? *Aquatics* 7:13-15.
Habeck, D.H. 1983. The potential of *Paraponyx stratiotata* L. as a biological control agent for Eurasian water milfoil. *J Aquatic Plant Manage* 21:26-30.
Hauser, W.J. 1975. *Tilapia* as biological control agents for aquatic weeds and noxious aquatic insects in California. *Proc Papers Annu Conf Calif Mosquito Control Assoc* 43:51-53.
Heman, M.L., Campbell, R.S., and Redmond, L.C. 1969. Manipulation of fish populations through reservoir drawdown. *Trans Am Fish Soc* 98:293-304.
Henrikson, L., Nyman, H.G., Oscarson, H.G., and Stenson, J.A.E. 1980. Trophic changes without changes in the external nutrient loading. *Hydrobiol* 68:257-263.
Hestand, R.S., and Carter, C.C. 1978. Comparative effects of grass carp and selected herbicides on macrophyte and phytoplankton communities. *J Aquatic Plant Manage* 16:43-50.
Holm, N.P., Ganf, G.G., and Shapiro, J. 1983. Feeding and assimilation rates of *Daphnia pulex* fed *Aphanizomenon flos-aquae*. *Limnol Oceanogr* 28:677-687.
Hrbacek, J., Dvorakova, M. Korinek, V., and Prochazkova, L. 1961. Demonstration of the effect of the fish stock on the species composition of zooplankton and the intensity of metabolism of the whole plankton association. *Verh Int Ver Limnol* 14:192-195.

Hulsey, A.H. 1957. Effects of a fall and winter drawdown on a flood control lake. *Proc Southeast Assoc Game Fish Comm* 10:285-289.

Hulsey, A.H. 1958. A proposal for the management of reservoirs for fisheries. *Proc Southeast Assoc Game Fish Comm* 12:132-143.

Hutchinson, B.P. 1971. The effect of fish predation on the zooplankton of ten Adirondack lakes, with particular reference to the alewife, *Alosa pseudoharengus*. *Trans Am Fish Soc* 100:325-335.

James, W.F. 1984. *Effects of Endothall Treatment on Phosphorus Concentration and Community Metabolism of Aquatic Communities*, Misc. Paper A-84-1, Vicksburg, MS: U.S. Army Corps of Engineers.

Kangasniemi, B.J. 1983. Observations on herbivorous insects that feed on *Myriophyllum spicatum* in British Columbia. In *Lake Restoration, Protection and Management*, EPA-440/5-83-001, 214-218.

Keen, W.H., and Gagliardi, J. 1981. Effect of brown bullheads on release of phosphorus in sediment and water systems. *Prog Fish-Cult* 43:183-185.

Keup, L.E. 1979. Fisheries in lake restoration. *Fisheries* 4:7-10.

Kilgen, R.H. 1978. Growth of channel catfish and striped bass in small ponds stocked with grass carp and water hyacinth. *Trans Am Fish Soc* 107:176-180.

Kilgen, R.H., and Smitherman, R.D. 1971. Food habits of the white amur stocked in ponds alone and in combination with other species. *Progr Fish-Cult* 33:123-127.

Kiser, R.W., Donaldson, J.R., and Olson, P.R. 1963. The effect of rotenone on zooplankton populations in freshwater lakes. *Trans Am Fish Soc* 92:17-24.

Kobylinski, G.J., Miley, W.W. II, Van Dyke, J.M., and Leslie, A.J. Jr. 1980. *The Effects of Grass Carp (Ctenopharyngodon idella Val.) on Vegetation, Water Quality, Zooplankton, and Macroinvertebrates of Deer Point Lake, Bay County, Florida*, Tallahassee, FL: Florida Dept. Nat. Res.

Lagler, K.F., and Deroth, G.C. 1953. Populations and yields to anglers in a fishery for largemouth bass. *Papers Mich Acad Sci* 38:241-250.

LaMarra, V.J. Jr. 1975. Digestive activities of carp as a major contributor to the nutrient loading of lakes. *Verh Int Ver Limnol* 19:2461-2468.

Lantz, K.E. 1974. *Natural and Controlled Water Level Fluctuation in a Backwater Lake and Three Louisiana Impoundments*, Baton Rouge, LA: Louisiana Wildlife and Fisheries Comm.

Lantz, K.E., Davis, J.T., Hughes, J.S., and Schafer, H.E. 1964. Water level fluctuation—Its effects on vegetation control and fish population management. *Proc Southeast Assoc Game Fish Comm* 18:483-494.

Leah, R.T., Moss, B., and Forrest, D.E. 1980. The role of predation in causing major changes in the limnology of a hyper-eutrophic lake. *Int Rev Ges Hydrobiol* 65:223-247.

Lembi, C.A., Ritenour, B.G., Iverson, E.M., and Forss, E.C. 1978. The effects of vegetation removal by grass carp on water chemistry and phytoplankton in Indiana ponds. *Trans Am Fish Soc* 107:161-171.

Leslie, A.J. Jr., Nall, L.E., and VanDyke, J.M. 1983. Effects of vegetation control by grass carp on selected water-quality variables in four Florida lakes. *Trans Am Fish Soc* 112:777-787.

Lewis, W.M. 1978. Observations on the grass carp in ponds containing fingerling channel catfish and hybrid sunfish. *Trans Am Fish Soc* 107:153-155.

Li, H.W., and Moyle, P.B. 1981. Ecological analysis of species introductions into aquatic systems. *Trans Am Fish Soc* 110:772-782.

Lynch, M., and Shapiro, J. 1981. Predation, enrichment, and phytoplankton community structure. *Limnol Oceanogr* 26:86-102.

Lynch, M., and Shapiro, J. 1982. Manipulations of planktivorous fish—Effects on zooplankton and phytoplankton. In J. Shapiro et al., *Experiments and Experiences in Biomanipulation*, Interim Rept. No. 19, Limnological Research Center, University of Minnesota, Minneapolis, Minn., 158-189.

Merkowsky, A., and Avault, J.W. Jr. 1976. Polyculture of channel catfish and hybrid grass carp. *Prog Fish-Cult* 38:76-77.

Michewicz, J.E., Sutton, D.L., and Blackburn, R.D. 1972. Water quality of small enclosures stocked with white amur. *Hyacinth Control J* 10:22-25.

Miller, H.D., and Boyd, J. 1983. *Large-Scale Management Test of the Use of the White Amur for Control of Problem Aquatic Plants; Rept. 4, Third Year Poststocking Results. Vol. VI: The Water and Sediment Quality of Lake Conway, Florida*, Tech. Rept. A-78-3, Jacksonville, FL: U.S. Army Corps of Engineers.

Miller, A.C., and King, R.H. 1984. *Large-Scale Operations Management Test of Use of the White Amur for Control of Problem Plants. Report 5. Synthesis Report*, Tech. Rept. A-78-2, Vicksburg, MS: Aquatic Plant Control Program, U.S. Army Corps of Engineers.

Miller, H.D., and Potts, R. 1982. *Large-Scale Operations Management Test of the Use of the White Amur for Control of Problem Aquatic Plants; Report 3, Second Year Poststocking Results, Volume VI: The Water and Sediment Quality of Lake Conway, Florida*, Tech. Rept. A-78-2, Vicksburg, MS: U.S. Army Corps of Engineers.

Mitchell, C.P., Fish, G.R., and Burnet, A.M.R. 1984. Limnological changes in a small lake stocked with grass carp. *NZ J Mar Freshwater Res* 18:103-114.

Mitzner, L. 1978. Evaluation of biological control of nuisance aquatic vegetation by grass carp. *Trans Am Fish Soc* 107:135-145.

Nilsson, N.A. 1978. The role of size-biased predation in competition and interactive segregation in fish. In S.D. Gerking, ed., *Ecology of Freshwater Fish Production*, New York: Wiley, 303-325.

Opuszynski, K. 1972. Use of phytophagous fish to control aquatic plants. *Aquaculture NL* 1:61-74.

Osborne, J.A. 1981. The size of grass carp as a factor in the control of *Hydrilla*. *Aquatic Bot* 11:129-136.

Osborne, J.A. 1982. The potential of the hybrid grass carp as a weed control agent. *J Freshwater Ecol* 1:353-360.

Osborne, J.A. 1985. A preliminary study of the efficacy of hybrid grass carp for hydrilla control. *J Aquatic Plant Manage* 23:16-20.

Perfetti, P.B. 1983. *Use of Insects and Pathogens for Management of Water Hyacinth in Louisiana*, Info. Exchange Bull. A-83-1, Vicksburg, MS: U.S. Army Corps of Engineers.

Perkins, B.D., and Maddox, D.M. 1976. Host specificity of *Neochetina bruchi* Hustache (Coleoptera, Curculionidae), a biological control agent for water hyacinth. *J Aquatic Plant Manage* 14:59-64.

Pierce, B.A. 1983. Grass carp status in the United States: A review. *Environ Manage* 7:151-160.

Pierce, P.C., Frey, J.E., and Yawn, H.M. 1963. An evaluation of fishery management techniques utilizing winter drawdowns. *Proc Southeast Assoc Game Fish Comm* 17:347-363.

Pool, D., Ryder, K., and Andrews, C. 1984. The control of *Bothriocephalus acheilognathi* in grass carp, *Ctenopharyngodon idella*, using Praziquantel. *Fish Manage* 15:31-33.
Pomogyi, P., Best, E.P.H., Dassen, J.H.A., and Boon, J.J. 1984. On the relation between age, plant composition and nutrient release from living and killed *Ceratophyllum* plants. *Aquatic Bot* 19:243-250.
Porter, K.G. 1973. Selective grazing and differential digestion of algae by zooplankton. *Nature* 244:179-180.
Porter, K.G. 1981. Limits to the control of algal populations by grazing zooplankton: The environmental theater and the ecological play. In *Proceedings of Workshop on Algal Management and Control*, Tech. Rept. A-81-7, Vicksburg, MS: U.S. Army Corps of Engineers, 121-130.
Porter, K.G., and Orcutt, J.D. 1980. Nutritional adequacy, manageability, and toxicity as factors that determine the food quality of green and blue-green algae for *Daphnia*. In W.C. Kerfoot, ed., *Evolution and Ecology of Zooplankton Communities*, Hanover, NH: Univ. Press of New England, 268-281.
Reinersten, H., and Olsen, Y. 1984. Effects of fish elimination on the phytoplankton community of a eutrophic lake. *Verh Int Ver Limnol* 22:649-657.
Rich, P.H., and Wetzel, R.G. 1978. Detritus in the lake ecosystem. *Am Nat* 112:51-57.
Richard, D.I., Small, J.W. Jr., and Osborne, J.A. 1984. Phytoplankton responses to reduction and elimination of submerged vegetation by herbicides and grass carp in four Florida lakes. *Aquatic Bot* 20:307-319.
Riley, D.M. 1978. Parasites of grass carp and native fishes in Florida. *Trans Am Fish Soc* 107:207-212.
Rowe, D.D. 1984. Some effects of eutrophication and the removal of aquatic plants by grass carp (*Ctenopharyngodon idella*) on rainbow trout (*Salmo gairdneri*) in Lake Parkinson, New Zealand. *NZ J Mar Freshwater Res* 18:115-127.
Safferman, R.S., and Morris, M.-E. 1964. Control of algae with viruses. *J Am Water Works Assoc* 56:1217-1224.
Schiller, D.H. 1984. A major utilities program to manage aquatic weeds. In *Proceedings, 18th Annual Meeting, Aquatic Plant Control Research Program*. Misc. Pap. A-84-4, Vicksburg, MS: U.S. Army Corps Engineers, 179-182.
Schoenberg, S.A., and Carlson, R.E. 1984. Direct and indirect effects of zooplankton grazing on phytoplankton in a hypereutrophic lake. *Oikos* 42:291-302.
Schuytema, G.S. 1977. *Biological Control of Aquatic Nuisances — A Review*, EPA-600/3-77-084.
Serns, S.L. 1979. Effects of Pro-Noxfish® on the benthos and zooplankton of Bug Lake, Forest County, Wisconsin. *Water Res Bull* 15:1385-1393.
Shapiro, J. 1978. The need for more biology in lake restoration. In *Lake Restoration*, EPA-440/5-79-001, 161-167.
Shapiro, J. 1979. The importance of trophic level interactions to the abundance and species composition of algae in lakes. SIL Workshop on Hypertrophic Ecosystems. *Dev Hydrobiol* 2:105-116.
Shapiro, J., Forsberg, B., La Marra, V., Lindmark, G., Lynch, M., Smeltzer, E., and Zoto, G. 1982. *Experiments and Experiences in Biomanipulation*, Interim Rept. No. 19, Limnological Research Center, Univ. of Minnesota, Minneapolis, Minn.
Shapiro, J., La Marra, V., and Lynch, M. 1975. Biomanipulation: An ecosystem approach to lake restoration. In P.L. Brezonik and J.L. Fox, eds., *Water Quality Management through Biological Control*, Gainesville, FL: Department of Environmental Engineering Sciences, Univ. Florida, 85-96.

Shapiro, J., and Wright, D.I. 1984. Lake restoration by biomanipulation: Round Lake, Minnesota, the first two years. *Freshwater Biol* 14:371-383.
Shields, J.T. 1958. Experimental control of carp reproduction through water drawdowns in Fort Randall Reservoir, South Dakota. *Trans Am Fish Soc* 87:23-33.
Shilo, M. 1967. Formation and mode of action of algal toxins. *Bacteriol Rev* 31:180-193.
Shilo, M. 1970. Lysis of blue-green algae by Myxobacter. *J Bacteriol* 104:453-461.
Shireman, J.V., ed. 1979. *Proceedings of the Grass Carp Conference*, Aquatic Weeds Research Center, Univ. of Florida, Gainesville, Florida.
Shireman, J.V. 1982. Cost analysis of aquatic weed control: Fish versus chemicals in a Florida lake. *Prog Fish-Cult* 44:199-200.
Shireman, J.V., Colle, D.E., Rottman, R.W. 1978. Size limits to predation on grass carp by largemouth bass. *Trans Am Fish Soc* 107:213-215.
Shireman, J.V., Haller, W.T., Colle, D.E., Watkins, C.E. II, Durant, D.F., and Canfield, D.E. 1983. *Ecological Impact of Integrated Chemical and Biological Aquatic Weed Control*, EPA-660/3-83-098.
Shireman, J.V., and Maceina, M.J. 1981. The utilization of grass carp, *Ctenopharyngodon idella* Val., for hydrilla control in Lake Baldwin, Florida. *J Fish Biol* 19:629-636.
Smeltzer, E., and Shapiro, J. 1982. Biological effects on the size of the nutrient pool. The role of benthivores in Lake Marion, Minnesota. In J. Shapiro et al., *Experiments and Experiences in Biomanipulation*, Interim Rept. No. 19, Minneapolis, MN: Limnological Research Center, University of Minnesota, 12-29.
Smith, C.R., and Shireman, J.V. 1983. *White Amur Bibliography*, Jacksonville, FL: U.S. Army Corps of Engineers. Misc. Pap. A-83-7.
Spencer, N.R., and Coulson, J.R. 1976. The biological control of alligatorweed, *Alternanthera philoxeroides*, in the United States of America. *Aquatic Bot* 2:177-190.
Spencer, C.N., and King, D.L. 1984. Role of fish in regulation of plant and animal communities in eutrophic ponds. *Can J Fish Aquatic Sci* 41:1851-1855.
Stanley, J.G. 1976. Reproduction of the grass carp (*Ctenopharyngodon idella*) outside its native range. *Fisheries* 1:7-10.
Stanley, J.G., Miley, W.W. II, and Sutton, D.L. 1978. Reproductive requirements and likelihood for naturalization of escaped grass carp in the United States. *Trans Am Fish Soc* 107:119-128.
Stenson, J.A.E., Bohlin, T., Henrikson, L., Nilsson, B.I., Nyman, H.G., Oscarson, A.G., and Larsson, P. 1978. Effects of fish removal from a small lake. *Verh Int Ver Limnol* 20:794-801.
Stewart, W.D.P., and Daft, M.J. 1977. Microbial pathogens of cyanophycean blooms. In M.R. Droop and H.W. Kannasch, eds., *Advances in Aquatic Microbiology. Volume I*. London: Academic Press, 177-218.
Terrell, T.T. 1982. Responses of plankton communities to the introduction of grass carp into some Georgia ponds. *J Freshwater Ecol* 1:395-406.
Theriot, R.F., and Sanders, D.R. Jr. 1975. Food preferences of yearling hybrid carp. *Hyacinth Control* 13:51-53.
Timms, R.M., and Moss, B. 1984. Prevention of growth of potentially dense phytoplankton populations by zooplankton grazing, in the presence of zooplanktivorous fish, in a shallow wetland ecosystem. *Limnol Oceanogr* 29:472-486.
Van Dyke, J.M., Leslie, A.J. Jr., and Nall, L.E. 1984. The effects of the grass carp on the aquatic macrophytes of four Florida lakes. *J Aquatic Plant Manage* 22:87-95.

Van Zon, J.C.J. 1977. Grass carp *Ctenopharyngodon idella* in Europe. *Aquatic Bot* 3:143-156.
Van Zon, J.C.J. 1979. The grass carp for weed control. *Proc. First British Freshwater Fisheries Conference*, Liverpool, England.
Vinogradov, V.K., and Zolotova, Z.K. 1974. The influence of the grass carp on aquatic ecosystems. *Hydrobiol J* 10:72-78.
Vuorinen, I., Rajasilta, M., and Salo, J. 1983. Selective predation and habitat shift in a copepod species — Support for the predation hypothesis. *Oecol* 59:62-64.
Ware, F.J., and Gasaway, R.D. 1976. Effects of grass carp on native fish populations in two Florida lakes. *Proc Southeast Assoc Game and Fish Comm* 30:324-334.
Ware, F.D., Gasaway, R.D., Martz, R.A., and Drda, T.F. 1975. Investigations of herbivorous fishes in Florida. In P.L. Brezonik and J.L. Fox, eds., *Water Quality Management through Biological Control*, Gainesville, FL: Department of Environmental Engineering Sciences, Univ Florida, 79-84.
Webster, K.E., and R.H. Peters, 1978. Some size-dependent inhibitions of larger cladoceran filterers in filamentous suspensions. *Limnol Oceanogr* 23:1238-1245.
Winner, R.W., and Farrell, M.P. 1976. Acute and chronic toxicity of copper to four species of *Daphnia. J Fish Res Bd Can* 33:1685-1691.
Winner, R.W., Keeling, T., Yeager, R., and Farrell, M.P. 1977. Effect of food type on the acute and chronic toxicity of copper to *Daphnia magna. Freshwater Biol* 7:343.
Wright, D.I., and Shapiro, J. 1984. Nutrient reduction by biomanipulation: An unexpected phenomenon and its possible cause. *Verh Int Ver Limnol* 22:518-524.
Young, L.M., Managhan, J.P. Jr., and Heidinger, R.C. 1983. Food preferences, food intake, and growth of the F_1 hybrid of grass carp ♀ × bighead carp ♂. *Trans Am Fish Soc* 112:661-664.
Zetter, F.W., and Freeman, T.E. 1972. Plant pathogens as biocontrols of aquatic weeds. *Annu Rev Phytopathol* 10:455-470.

14

Surface and Sediment Covers for Macrophyte Control

Numerous attempts have been made to control rooted aquatic vegetation by covering it with materials such as sand, gravel, or clay. Nearly all of these applications have been unsuccessful because root systems remain to produce shoots that can eventually penetrate earthen covers, and because many aquatic plants can reestablish infestation through the growth of fragments carried to the treated site from other lake areas. More recently, sheeting and screening materials such as polyethylene have been evaluated for their usefulness in controlling plant growth. These compounds in some cases have proven to be extremely effective, at least in the short term, but all are expensive, and their effectiveness seems to be highly correlated with application techniques as well as with the type of material. Like the application of herbicides the use of these materials addresses only the symptoms of a lake problem. But unlike the introduction of chemicals, screens have little negative impact on the lake.

Engel (1982) has reviewed the use of some screening materials for macrophyte control. He lists these advantages of sediment covers:

1. Their use is confined to specific lake areas.
2. Screens are usually out of sight and thus create no disturbance on shore.
3. They can be installed in places where harvesters or sprayer boats cannot gain access.
4. No toxic substances are released.
5. They usually require no permit or license.
6. They are easy to install over small areas.

There are these disadvantages to the use of sediment covers, according to Engel (1982):

1. They fail to correct the cause of the problem.
2. They are expensive.
3. They are difficult to apply over large areas or on sites with obstructions.
4. They may slip on steep grades or float to the surface after trapping gases beneath them.

5. They can be difficult to remove or relocate.
6. They may rip during application.
7. Some materials are degraded by sunlight.

The objectives of this chapter are to describe several cases of the use of screening materials, and to provide estimates of their costs and directions for their successful application. These topics have been reviewed by Armour et al. (1979), Cooke (1980; 1983), and Nichols and Shaw (1983).

CASE STUDIES

The number of applications of screening material to the bottoms of ponds and lakes must be very large, but only a few of these applications have been described in the literature. Prior to their use for rooted plant control, synthetic materials were applied to ponds and reservoirs used for waste disposal, brine evaporation, etc. Kumar and Jedlicka (1973) provide a useful description of the properties of these materials, including those used for weed control, such as nylon, polyethylene, polypropylene, polyvinyl (PVC), and hypalon. Table 14-1 is a summary of Kumar and Jedlicka's data. Briefly, hypalon is resistant to weather and chemicals, resists puncturing, tearing, and ultraviolet radiation, has comparatively low tensile strength, and is expensive. Polyethylene is tough, flexible, has poor weatherability and puncture resistance, is relatively inexpensive, and floats. Polypropylene is strong, flexible, and also has a specific gravity less than 1.0. Polyvinyl chloride is strong, has good weatherability, and is relatively inexpensive. Nylon is flexible, has high tensile strength, but fair to poor weatherability.

Polyethylene

One of the earliest applications of sediment covers to control nuisance macrophytes was the treatment of 10 ha (2.5 acres) of Marion Millpond, Wisconsin, with 0.1-mm (4-mil) black polyethylene. The project is described in the reports of Born et al. (1973), Peterson et al. (1974), Nichols (1974), Engel (1982), and Engel and Nichols (1984).

The pond had experienced a severe macrophyte infestation for many years and had been treated with sodium arsenite and copper sulfate. In 1969-70 the pond was drained and the basin cleared of stumps and debris. Polyethylene was laid over the sediments and covered with 7-15 cm of sand and gravel. Application in the soft mud was tedious, so enough water was added to just cover the treatment area, and the water was allowed to freeze. Sheeting and the sand and gravel cover were placed on the ice and the water withdrawn. The ice cracked and the materials fell to the pond's bottom. The pond was refilled in 1971, and good control of macrophytes was achieved in 1971-72. Nichols (1974) reported that *Chara* and filamentous algae covered the treated area in 1973, and herbicides were applied annually until 1978, when a harvester was purchased. Engel (1982) and Engel and Nichols (1984) examined the treated areas in 1978 and found that they were covered with macrophytes, although biomass was

Table 14-1 Properties of Pond Lining Materials (from Kumar and Jedlicka, 1973)

	High-Density Polyethylene	Polyvinyl Chloride	Polypropylene	Nylon	Hypalon
Specific gravity	0.94–0.96	1.20–1.50	0.9–0.91	1.08–1.40	
Weatherability	poor	good	poor	fair	excellent
Time to crack, hrs.	300	2,500	100	1,200	

about half that of untreated areas. The polyethylene sediment cover exerted little control of macrophyte biomass after 1973 because plants like *Najas, Myriophyllum,* and some species of *Potamogeton* were able to colonize the sand and gravel cover over the sheeting. Born et al. (1973) estimated the cost (1970) to be $81 ha^{-1} ($206 ha^{-1} in 1983 prices) for materials. Labor was donated by the lake users.

Armour et al. (1979) also reported on the use of black polyethylene for macrophyte control. Their experiences were largely negative. Since the material floats, it is difficult to apply and it must be covered. They applied the sheeting by unrolling it over the bottom, anchoring with sand bags, and then covered it with 1 cm of sand or gravel. Adjacent sheets were overlapped by 30 cm. At Skaha Lake, British Columbia, 0.43 ha (1.06 acres) were treated, but most of the sheeting washed up during storms. Treatments at other sites were effective in controlling plants, but weeds grew through the holes that had been punched to allow escape of gases, or grew on accumulated sediment. Armour et al. concluded that while polyethylene will control plants, it is easily dislodged by wave action. Other negative features include:

1. It is difficult to apply over an irregular bottom or over a high density of weeds.
2. Gas forms under the sheets, even with 1.2-cm holes.
3. It is not feasible to move and relocate the sheets.
4. It slides down steep inclines.
5. It is deteriorated by sunlight (about 1 year in direct sun).
6. Its buoyancy makes it difficult to handle.

The application at Skaha Lake in 1976 cost $2491 ha^{-1} ($4609 ha^{-1} in 1984 prices) for materials. Total cost, including labor and materials, was $12,558 ha^{-1} ($23,232 ha^{-1} 1984 prices).

Hypalon

Armour et al. (1979) applied small (91 m^2) sheets of Hypalon, a synthetic rubber, which had been drilled with 1.2-cm holes to allow gas to escape. The material was 100 percent effective, was not damaged by waves due to its strength and weight, was more easily applied than polyethylene due to its high specific gravity, and did not rip. As with other materials, plants regrew on sediments that accumulated over the sheeting. Gas bubbles formed, even with escape holes, and weeds grew through the holes. A major drawback to Hypalon is its expense. Material costs were $40,876 ha^{-1} ($59,270 ha^{-1} in 1984 prices), and total cost was $59,452 ha^{-1} ($86,205 ha^{-1} in 1984 prices).

Polyvinyl (PVC)

An area of 36 m^2 of lake bottom was covered with this material by Armour et al. (1979). Holes were punched to allow gas to escape. The cost of the material was high ($15,275 ha^{-1}; $22,149 ha^{-1} in 1984 prices), and it was easily crumpled or folded and dislodged by waves. Since polyvinyl is heavier than polyethylene, it was easier to position. It was effective for two years in keeping the treated area free of plants.

Polypropylene

Armour et al. (1979), Cooke and Gorman (1980), Lewis et al. (1983) and Engel (1984) have evaluated polypropylene, a material sold under various names, such as Permealiner, Typar (DuPont), and Terratrack (Terratrack Ltd., Rexdale, Ontario). Polypropylene is a black, woven, semipermeable sheeting often used as a soil stabilizer. It has a specific gravity less than one, and thus requires a covering to prevent bulges and floating, but it is permeable to gases and does not need slits or holes.

Armour et al. (1979) found that this material was effective in controlling *Myriophyllum spicatum*, but roots of plant fragments grew on the sediments that accumulated over the screen, and could penetrate through the screen. Lewis et al. (1983), in contrast, found that Terratrack, anchored with concrete blocks, did not allow root penetration by *M. spicatum* over three summers. A weed-free water column was evident, even though plant fragments did appear on the sediments that accumulated on the screen. Cooke and Gorman (1980) found Typar, also anchored with cement blocks, to be completely effective in preventing the growth of *Najas flexilis, Potamogeton gramineus, P. crispus, P. foliosus,* and *P. pusillus* for one year in an area that had been drawn down and exposed to freezing. No evaluation was made in subsequent years. Small growths of *N. flexilis* and filamentous algae were evident on the screens. Engel (1984) found that Typar controlled vegetation during the first year, but that it was difficult to remove and clean so that plants regrew on the accumulated sediments. Only Engel (1984) has reported any problems with gas accumulation and "ballooning" of the polypropylene screens. Costs of this material were $27,000 ha^{-1} ($39,150 ha^{-1} in 1984 prices) for Permealiner (Armour et al., 1979) and $5900 ha^{-1} ($6254 ha^{-1}, 1984 prices) for Terratrack (Lewis et al., 1983). Cooke and Gorman (1980) reported the cost of Typar to be $2600 ha^{-1} ($3328 ha^{-1}, 1984 prices), while Engel (1984) gives a price of $8,000 ha^{-1} based on 1982 prices.

Aquascreen

Aquascreen is a fiberglass screen coated with polyvinyl chloride (Menardi-Southern Division of U.S. Filter Corp., Augusta, Ga.), that has been intensively investigated for its effectiveness in controlling aquatic weeds. The material is flexible and heavier than water (specific gravity = 2.54), and is sold with a mesh size of 62 apertures cm^{-2} (400 $inch^{-2}$). The standard roll size is 7 ×100 feet and costs $140 per roll ($21,500 ha^{-1}) plus installation charges (1984 prices).

Every investigation of Aquascreen has shown it to be completely effective in eliminating nuisance aquatic plants, at least for the season of application. Most investigators have also noted that the accumulation of sediments on the screen will eventually allow new plants to root, forcing the user to remove and clean the screens annually. This, however, is easily accomplished (Engel, 1984). Mayer (1978) was the first to describe the use of this product. He examined screens of three mesh sizes as a means of controlling *Myriophyllum spicatum* and *Potamogeton crispus* in Lake Chautauqua, New York. Screens with 39 to 62 apertures cm^{-2} brought about decomposition of plants in two to three weeks, with the best control achieved with the finest mesh. Sedimentation on the screens was significant after two to three years of placement, but Mayer found that autumnal removal of the screens and repositioning in the spring maintained 95 percent control of nuisance weeds.

A more detailed study of Aquascreen was carried out by Perkins et al. (1980) and Boston and Perkins (1982). The first study's objective was to determine the relation between coverage time and effective control of *M. spicatum*. Test areas 9 × 24 m were set out in shallow (0.5-2.0 m) and deep (2-3 m) plots in Union Bay, Lake Washington, and panels were then removed at 1-month, 2-month, and 3-month intervals. The results are illustrated in Figure 14-1. One month of coverage, compared to controls, produced decreases of 25 percent and 35 percent for shallow and deep plots, respectively. One month after the panels had been removed, plant regrowth was small. Relative to controls, panels in place for two months produced decreases of 78 percent in shallow plots and 56 percent in deep. Regrowth was minimal. Plant biomass and regrowth after three months of coverage was very small. Perkins and his colleagues concluded that the screen was most effective where there was good contact with the lake bottom. Boston and Perkins (1982) studied the effect of Aquascreen installation on dissolved oxygen and phosphorus in aquaria. They found that plant death would be sufficiently slow to prevent dissolved oxygen loss and phosphorus accumulation under the typical field conditions.

Engel (1982; 1984) examined Aquascreen's effectiveness in control of rooted plants, as well as its impact on benthic macroinvertebrates in Cox Hollow Lake, Wisconsin. As did other investigators, Engel found that the screen was easily applied, removed, cleaned, and stored. Macrophyte growth was prevented regardless of when the screen was installed during the summer, and few plants could be found under firmly anchored screen. There was considerable growth under loosely applied screen. There was little control in the second season after application unless the screens were removed, cleaned, and repositioned. Lewis et al. (1983) made a similar finding in

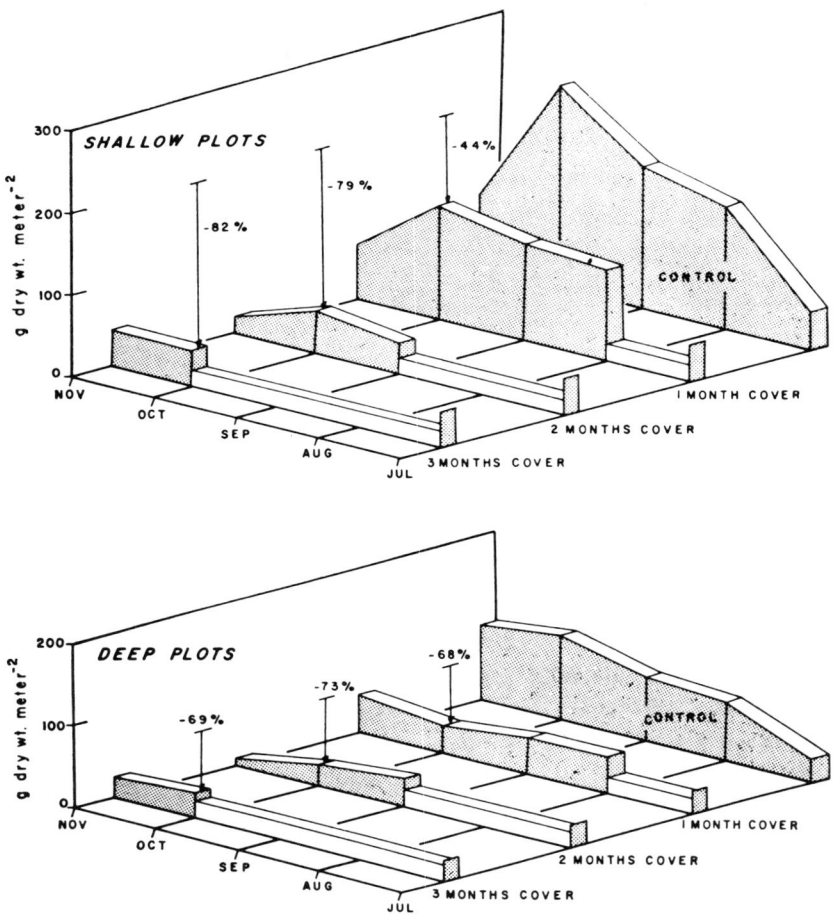

Figure 14-1 Results of aquascreen applications in shallow and deep-water test plots. The percent reductions in dry-weight biomass relative to untreated controls are shown for the October sampling date (from Perkins et al., 1980).

Buckhorn Lake, Ontario. Macroinvertebrates were eliminated by Aquascreen panels in Cox Hollow Lake, apparently because of poor circulation and low dissolved oxygen over the one-year application period (Figure 14-2).

Dartek

Perkins (1984) has evaluated Dartek (DuPont, Canada), a black-pigmented (2 % carbon black), nylon material, which is commercially available in 2-mil-thick sheets on

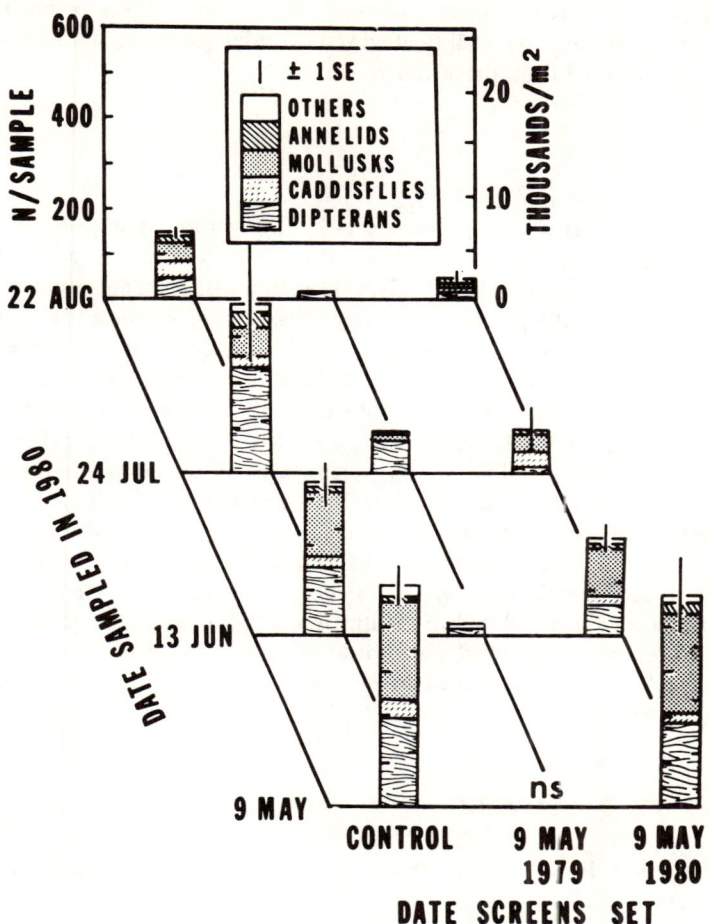

Figure 14-2 Benthic macroinvertebrates sampled in 1980 on control sites and under Aquascreen set in May of 1979 or 1980. No samples (ns) were collected on one date under the May 1979 screen (from Engel, 1984).

2.5 × 30.5-m rolls at about \$0.80 m^{-2} (\$8,000 ha^{-1}) plus installation (1983 prices). One roll was placed over soft, unconsolidated organic muck in Green Lake, Washington, a test area dominated by *Elodea canadensis*. Four panels were placed over consolidated sand and gravel sediments in Lake Washington in plots dominated by species of *Potamogeton*. Panels without vents and panels with diagonal slits ballooned and lifted off the sediments. Panels with 12 cross-hatch slits m^{-2} lifted 20 cm, at maximum, from the sediments. Several slits (about 6%) had plant growth extending as high as 30 cm above the panels. After 35 days in place, plant decomposition beneath the panel was nearly complete, and Perkins concluded that Dartek is "highly effective in removing nuisance aquatic growth."

Dartek has been tested in British Columbia (P.R. Newroth, pers. comm.) and was found to be more difficult to install than woven material; gases that accumulate under it tend to lift the material from the sediment surface.

Burlap

Jones and Cooke (1984) applied 10-ounce yd^{-2} (340-gm m^2) burlap to two sites in Lake Rockwell Reservoir, Ohio. The burlap at one site was lightly pretreated with Netset (Nichols Net and Twine Co., E. St. Louis, Illinois), a sealant used to preserve nets and seines. Despite the porosity of burlap, "ballooning" occurred at the site with unconsolidated organic muck sediments due to the difficulty of securely anchoring the material in the highly fluid mud. At both locations plant growth was effectively controlled over the growing season, but both treated and untreated burlap rotted by the end of three months of placement. This material is thus useful for one season at best. Removal and reapplication in subsequent seasons would be impossible. P.R. Newroth (pers. comm.) has reported that extensive testing of untreated burlap in British Columbia has revealed that this material is cost-effective. Plant growth was controlled for two to three years, after which rotting and sediment accumulation over the burlap curtailed effectiveness. The difference with regard to rotting in these two studies of burlap may be attributed to the highly organic, biologically active sediments of the Ohio Lake, where microbial decomposition of the burlap would be expected to be rapid. Burlap's cost is low ($0.34 m^{-2}; $3400 ha^{-1}), suggesting that it could be useful in selected areas.

SEDIMENT COVERS—SUMMARY

Perkins (1984) has listed the attributes of the "ideal" sediment covering for macrophyte control:

1. sufficiently opaque to block photosynthetically active radiation
2. durable during and subsequent to application
3. negatively buoyant, to aid in installation and to impede lifting or "ballooning"
4. vented, to allow escape of gases
5. smooth on upper surface, to inhibit fragment rooting
6. competitively priced.

There is no material on the market that meets all of these attributes sufficiently to be called the ideal sediment cover. Sediment accumulates on all of them, at rates that vary from lake to lake and from site to site within a lake. Therefore, another valuable attribute is ease of installation, removal, storage, and reinstallation. A new substance for limnological applications is Texel (Newroth, pers. comm.). Although there are no published data on its characteristics and effectiveness, it may prove to be an excellent sediment cover for macrophyte control.

Polyethylene clearly is of little value in control of macrophytes due to its lightness and due to the problems of application and of securing it to the sediment. Hypalon is effective and easier to install, but is far too costly to warrant further consideration. Polyvinyl chloride has the same shortcoming.

Aquascreen, of the remaining materials, has the fewest problems and many positive attributes. Its cost, however, will prevent it from being used over large areas. Typar needs further evaluation, since its cost is lower, but unfortunately it is buoyant. Dartek, while far less expensive than Aquascreen, has problems with venting and lifting and requires very secure anchoring. Any of these three materials will prove to be completely effective in providing at least seasonal control of macrophytes, but this effectiveness will be strongly related to the quality of the application process.

APPLICATION PROCEDURES FOR SEDIMENT COVERS

The most important factor in using sediment covers for the symptomatic control of nuisance plants, other than the selection of materials, is the technique of application.

The cover should be applied as close to the sediments as possible. There should be no "ballooning" or pockets after installation. In the muck-type unconsolidated sediments of many eutrophic lakes, this ideal cannot be met because stakes to secure the material may never be held by the soil and bricks or cement blocks will anchor just the areas where they have been placed. Dense plant infestations may also make it difficult to apply the screen tightly to the sediment. There must be no gaps between strips, since plants will grow there. Finally, the ideal installation will be one in which the covering material is anchored firmly enough that it cannot lift, but can still be removed or repositioned if desired.

The first step in installing a sediment cover is to survey the lake bottom to locate major obstructions and to test the sediment for its ability to hold a stake. In the case of highly fluid, unconsolidated sediment, very long stakes will be required, and these should be tested for their holding ability prior to actual application. If the sediments are too flocculent, bricks or cement blocks will have to be used, or link chain can be sewn into the edges of the fabric. Steel stakes usually cannot be used in gravel, hard clay, or rock bottoms, because they cannot be pushed deeply enough by the diver to hold the screen.

A very satisfactory stake can be made from 1/4-in-diameter steel reinforcing bar. Bend the bar at one end into an "L"-shaped handle. The length of the stake will vary with the softness of the sediments. Sharpen the long end to ease penetration of the screen and the sediments. Drive the stake through a doubled layer of the screen until the "L" end is flush with the sediments.

One of the easiest ways to apply sediment screening is to build a reel in the stern of a rowboat. Two applicators, one on each side of the screen, unroll it and place stakes opposite each other every 1–2 meters. If the screening is applied directly over lush vegetation, the stakes will probably have to be placed at 1-m intervals to prevent lifting. In deeper water, where SCUBA is used, another helper in addition to the rower will be needed to hand stakes to the divers and to assist in any diving emergency. It should be kept in mind that divers will disturb the sediments and that visibility will be low. Dartek is applied by filling the tube upon which it is rolled with sand or gravel. The screen is then soaked for 12-24 hrs. to make it more flexible, tear-resistant, and manageable, then the loose end is securely anchored and the screen unrolled and staked.

In some cases application can be improved by returning to the site in 15-20 days, when plants beneath the screen have started to decompose, to flatten bulges in it. A second trip to the lake with divers will increase considerably the cost of application.

The ideal time for application is prior to weed growth, in early summer. As Perkins et al. (1980) and Engel (1982) have shown, screens in place for about two months can be removed and used elsewhere in the lake. That means that sites covered in May and June can be uncovered and the screen moved for July and August coverage of another location. Application can be greatly facilitated by lake-level drawdown, followed by application of screen to frozen lake sediments. Harvesting of dense infestations will also ease the difficulty of covering the sediments. Herbicides cannot be used for this purpose unless the divers wait until the recommended no-water-contact period is over.

It is conceivable that complete coverage with screen of a dense infestation of a small pond could produce a sharp decline in dissolved oxygen, if the screening process is done over a very short period. This idea has been examined in laboratory experiments (Boston and Perkins, 1982), but all field observations of dissolved oxygen changes have been with large, open systems. Caution is recommended when covering the sediments of small ponds.

SHADING OF MACROPHYTES WITH SURFACE COVERS

This procedure to reduce macrophyte biomass through shading has received little attention because a surface cover obviously denies use of that area of the lake or pond, and because such covers can be dislodged by wind or by vandals.

Mayhew and Runkel (1962) described the use of black polyethylene sheeting as a surface cover to control nuisance plants in a pond. Polyethylene sheets were floated (sp. gr. = 0.92) over 2000-ft^2 (186-m^2) plots, and the corners were anchored to prevent dislodging by the wind. Eight similar plots, populated by different dominant species, were studied. All species of *Potamogeton* were controlled for the entire summer if a cover of 15-21 days was carried out before the plants came to maturity. Normally, this would mean that covers should be applied in May in the north temperate latitudes of the United States. *Ceratophyllum demersum* was controlled by a continuous cover of 18-28 days. In areas where plants were controlled, filamentous algae invaded and revegetated the plots. The covers were unsuccessful in controlling *Chara vulgaris, Sagittaria latifolia*, and various emergent species.

This procedure clearly needs further evaluation. For example, swimming areas could be covered in early May and the covers allowed to remain in place for 25-35 days. This would probably not deny the area to swimming, since water temperatures usually do not reach the comfortable range until mid-June. If the sheeting was removed carefully, it could be reused in subsequent seasons, and the expense and dangers of herbicide use might be reduced or eliminated.

Eicher (1947) was among the first to suggest the use of dyes to suppress plant growth. A commercial product, Aquashade (Aquashade, Inc., Eldrod, NY), is designed specifically to shade plants in hydrologically closed systems such as ponds. The dye is added as a concentrate, and winds disperse it throughout the pond, turning the pond blue. The manufacturer claims that the material is effective against *Elodea, Potamogeton, Najas, Myriophyllum, Hydrilla, Chara*, and various filamentous algae, without toxicity to aquatic life. Swimming is permitted immediately after application, but the material cannot be used in a potable water source. There is insufficient published information at this time to evaluate commercial dyes, but initial experiments (Spencer, 1984; Manker and Martin, 1984) suggest that the mode of action is light limitation and not direct toxicity to the plants. The cost of Aquashade is approximately $50 per gallon, an amount sufficient to treat four acre-feet (4900 m^3) of water at the manufacturer's recommended dose of 1.0 ppm.

REFERENCES

Armour, G.D., Brown, D.W., and Marsden, K.T. 1979. *Studies on Aquatic Macrophytes. Part XV. An Evaluation of Bottom Barriers for Control of Eurasian Water Milfoil in British Columbia*, Water Investigations Branch, Province of British Columbia, Vancouver, B.C., Canada.

Born, S.M., Wirth, T.L., Brick, E.M., and Peterson, J.P. 1973. *Restoring the Recreational Potential of Small Impoundments. The Marion Millpond Experience*, Tech. Bull. No. 71, Department of Natural Resources, Madison, Wisconsin.

Boston, H.L., and Perkins, M.A. 1982. Water column impacts of macrophyte decomposition beneath fiberglass screens. *Aquatic Bot* 14:15-27.

Cooke, G.D. 1980. Covering bottom sediments as a lake restoration technique. *Water Res Bull* 16:921-926.

Cooke, G.D. 1983. Review of lake restoration techniques and an evaluation of harvesting and herbicides. In *Lake Restoration, Protection and Management*, EPA-440/5-83-001, 257-266.

Cooke, G.D., and Gorman, M.E. 1980. Effectiveness of DuPont Typar sheeting in controlling macrophyte regrowth after overwinter drawdown. *Water Res Bull* 16:353-355.

Eicher, G. 1947. Aniline dye in aquatic weed control. *J Wildlife Manage* 11:193-197.

Engel, S. 1982. *Evaluating Sediment Blankets and a Screen for Macrophyte Control in Lakes*, Office of Inland Lake Renewal, Department of Natural Resources, Madison, Wisconsin.

Engel, S. 1984. Evaluating stationary blankets and removable screens for macrophyte control in lakes. *J Aquatic Plant Manage* 22:43-48.

Engel, S., and Nichols, S.A. 1984. Lake sediment alteration for macrophyte control. *J Aquatic Plant Manage* 22:38-41.

Jones, G.B., and Cooke, G.D. 1984. Burlap screening for the control of macrophyte growth. *Ohio J Sci* 84:246-251.

Kumar, J., and Jedlicka, J.A. 1973. Selecting and installing synthetic pond linings. *Chem Eng* 5:67–70.

Lewis, D.H., Wile, I., and Painter, D.S. 1983. Evaluation of Terratrack and Aquascreen for control of macrophytes. *J Aquatic Plant Manage* 21:103–104.

Manker, D.C., and Martin, D.F. 1984. Investigation of two possible modes of action of the inert dye Aquashade® on hydrilla. *J Envron Sci Health* A 19(b): 725--753.

Mayer, J.R. 1978. Aquatic weed management by benthic semi-barrier. *J Aquatic Plant Manage*, 1631–1633.

Mayhew, J.K., and Runkel, S.T. 1962. The control of nuisance aquatic vegetation with black polyethylene plastic. *Proc Ia Acad Sci* 69:302–307.

Nichols, S.A. 1974. *Mechanical and Habitat Manipulation for Aquatic Plant Management. A Review of Techniques*, Tech. Bull. No. 77, Department of Natural Resources, Madison, Wisconsin.

Nichols, S.A., and Shaw, B.H. 1983. Review of management tactics for integrated aquatic weed management of Eurasian water milfoil (*Myriophyllum spicatum*), curly-leaf pondweed (*Potamogeton crispus*) and elodea (*Elodea canadensis*). In *Lake Restoration, Protection and Management*, EPA-440/5-83-001, 181–192.

Perkins, M.A. 1984. An evaluation of pigmented nylon film for use in aquatic plant management. In *Lake and Reservoir Management*, EPA 440/5-84-001, 467–471.

Perkins, M.A., Boston, H.L., and Curren, E.F. 1980. The use of fiberglass screens for control of Eurasian water milfoil. *J Aquatic Plant Manage* 18:13–19.

Petersen, J.O., Born, S., and Dunst, R.C. 1974. Lake rehabilitation techniques and experiences. *Water Res Bull* 10:1228–1245.

Spencer, D.F. 1984. Influence of Aquashade® on growth, photosynthesis, and phosphorus uptake of microalgae. *J Aquatic Plant Manage* 22:80–84.

IV

Acidified Lakes

15

Liming Acidified Lakes

ACIDIFICATION OF LAKES

Thousands of lakes in Europe and North America are being acidified by the deposition of acid in precipitation and dry fall. Acid precipitation gained prominence, beginning in the late 1960s, as a global pollution problem. It has become the focal point of much controversy in the 1980s, with one side calling for more research to prove cause and effect while the other side is content that cause and effect are sufficiently established to institute emission controls. Restoring acidified lakes by liming has been more or less accepted by both sides as an interim, stopgap measure to save the most valuable lakes and their fisheries.

Precipitation has often been considered to be acid when its pH falls below 5.6, the level at which CO_2 in water is in equilibrium with the atmosphere. Charlson and Rodhe (1982) suggest, however, that average pH in remote areas may be naturally acidic, with an average pH about 5.0. Nevertheless, the pH of precipitation over large expanses of Europe and North America has been reduced considerably below "normal" and is now consistently around 4.0 to 4.5 in markedly affected areas (Likens et al., 1979).

The cause of acid precipitation is generally agreed to be fossil fuel combustion. Strong mineral acids result from the oxidation of SO_2 and NO_x, which are products of combustion, to sulfuric and nitric acid, respectively. Likens et al. (1979) have summarized data available for Europe and North America and have shown a substantial increase in precipitation acidity over a 20-year interval from the mid-1950s to the mid-1970s. With demands for energy increasing, along with the declining favor of nuclear power, emphasis is being placed more and more on fossil fuel, especially Western U.S. coal reserves. Unless additional controls are instituted, the acidity of precipitation and the resulting potential for lake acidification may worsen.

When acid in precipitation and dry fall deposits on a watershed, it tends to become neutralized by weathering processes (Kramer and Tessier, 1982). The hydrogen ions are consumed by bases in the soil. In watersheds dominated by sedimentary bedrock rich in $CaCO_3$, precipitation acidity is readily neutralized. In watersheds dominated by igneous bedrock such as granite and basalt and overlain by a thin soil mantle, however, the supply of basic neutralizing substances through chemical weathering is low. In such environments the available base supply or alkalinity entering the lake or stream becomes depleted. Furthermore, as the ratio of lake area to watershed area increases, the effect of alkalinity depletion or acidification becomes even greater, because less of the precipitation acidity is neutralized by the soil (Wright and Henriksen, 1978).

Acidification of surface waters is considered to be "the result of a large scale acid-base titration; bases that have been set free by weathering primary rock are titrated with acids that are deposited from the atmosphere" (Henriksen, 1980). Thus acidification is the removal of HCO_3^- in a linear fashion as acid (H^+) is added. Bicarbonate is replaced by SO_4^{2-} and NO_3^-, which are the anions of strong mineral acids. As in an alkalinity titration, lake pH drops sharply into a range causing ecological damage only after most alkalinity has been removed. Even after substantial acidification the lake pH may remain above 6.0; Henriksen refers to these as bicarbonate lakes. At some point, however, when alkalinity has been nearly depleted, the pH may fluctuate and at times reach critically low values. That can be recognized as a transition phase, in which the lake pH may fall to within 4.7 and 5.3 on the average. Once bicarbonate is removed the pH will remain below 4.7 and the lake can be referred to as acid.

ECOLOGICAL EFFECTS

A large number of lakes over wide areas in the world have been acidified because of the pervasive nature of acid precipitation. Several thousand lakes have been severely acidified in Scandinavia (Almer et al., 1974; Miljöakuellt, 1981; Sevaldrud et al., 1980) as have hundreds of lakes in the eastern U.S. and Canada (Harvey, 1980; Schofield, 1976). As many as 250,000 lakes may be affected in Ontario (Fraser et al., 1982). The ecological effects are numerous and include:

1. Acid-sensitive species give way to acid-tolerant ones, resulting in a decrease in diversity at all trophic levels, from bacteria through fish (Almer et al., 1974; Wright et al., 1976).

2. Decreased diversity may result in slower rates of organic matter decomposition and recycling of nutrients, resulting in accumulations of detritus and ungrazed plants such as sphagnum moss, filamentous algae and even rooted macrophytes (Almer et al., 1974; Gran et al., 1974; Hendrey, 1976).

3. Complexation by aluminum, mobilized from the soil through acidification, can effectively remove phosphorus from the water column in the pH range 5-6, and this can be a cause for the reduced productivity observed in some situations (Dickson, 1978).

4. Fish may be the trophic level most sensitive to acidification, with thousands of lakes having either lost their fish populations or had them depleted as a consequence of acidification (Harvey, 1980; Johansson and Nyberg, 1980; Schofield, 1976; Sevaldrud et al., 1980; Wright and Snekvik, 1978). Fish loss can be due primarily

to ion loss from the blood, accentuated by Al^{3+}, or by direct toxicity of Al^{3+} or $Al(OH)_3$ (Baker and Schofield, 1982). Chronic effects to fish are considered to begin in the pH range 5.5-6.0, with mortality occurring at 5.0 and below. With Al present, however, mortality can occur around pH 5.5. Henriksen (1980) has shown that the fish population status in 684 Norwegian lakes relates well with the classification scheme of bicarbonate, transition and acid lakes.

LIMING PRACTICES

Restoration of acidified lakes requires replenishment of lost alkalinity, so that pH will be maintained above critical levels. Alkalinity can be restored to acidified lakes by either reducing the deposition rate of acid in precipitation and dry fall or by adding alkalinity to overcompensate for acid deposition. Because alkalinity is a conservative property, lakes can be expected to respond to reduced acid input and recover their alkalinity in direct proportion to their flushing rates.

Although reducing inputs of acid is the most logical approach to restoring lakes acidified to critical levels, it is presently fraught with controversy over policy and technical understanding, and immediate action has been considered necessary to preserve important recreational and commercial fisheries. Thus, the addition of alkalinity through liming has been the principal tool with which to combat lake acidification. Liming is nevertheless considered to be a short-term corrective measure only, with long-term, permanent recovery and protection expected largely from emission controls.

Liming of lakes has been practiced most extensively in Sweden, but there have been and continue to be liming activities in most areas where acidification is pronounced, including Norway, Eastern Canada, and New York State. Liming of colored lakes to increase productivity has been tried in Wisconsin and Michigan. Sweden spent about $\$10 \times 10^6$ from 1975 to 1980 to distribute 180×10^3 tons of lime into lakes and streams (Miljöaktuellt, 1981). To cope with the problem of acidification in 20,000 lakes, Sweden planned to spend $\$200 \times 10^6$ over the five-year period from 1981 through 1985.

Materials and Methods

Fraser et al. (1982) have thoroughly reviewed the practice of liming acidified waters; their findings will be summarized here. The three most popular materials used for liming are pulverized limestone ($CaCO_3$), slaked or hydrated lime ($Ca(OH)_2$), and unslaked lime or quicklime (CaO). Dolomites are not considered practical because of their high content of magnesium, which reacts slowly. Limestone is preferred because it is less caustic and less expensive, but limestone is less reactive than hydrated lime or quicklime and therefore more is required to achieve the same results—five times more limestone may be required than hydrated lime. Table 15-1 shows a comparison of the characteristics of lime and limestone.

Table 15-1 Comparison of Lime and Limestone (from Fraser et al., 1982)

	Lime		Limestone ($CaCO_3$)	
Characteristics	Calcium Oxide (CaO) (Quicklime)	Calcium Hydroxide ($Ca(OH)_2$) (Hydrated Lime)	Pulverized Limestone (Aglime)	Limestone Rock
Solubility	instantaneous	instantaneous	intermediate	slow
Dissolution effect on pH	rapid rise in pH	rapid rise in pH	less rapid rise in pH than lime	less rapid rise in pH than aglime
Corrosivity	extremely caustic	less caustic than CaO	not caustic	not caustic
Reactivity in water	high reactivity; less quantity required than limestone	high reactivity; less quantity required than limestone	lower reactivity than lime[a]	lower reactivity than lime[a]
Length of effectiveness	short	short	longer than lime	longer than lime
Resource requirements[b]		1/5 quantity of aglime	0.6 to 2.24 metric tons/ surface hectare	depends on water flow

[a]Uncertainty exists as to whether the $CaCO_3$ that does not go into solution becomes reactive at a later date or becomes nonreactive.
[b]Assuming direct lake liming.

Dose requirements are difficult to obtain from case studies, but Fraser et al. (1982) indicate that a range of 0.6 to 2.24 tons of agricultural limestone may be appropriate. Dose can be calculated from chemical equilibrium equations, as Menz and Driscoll (1983) did for the Adirondack lakes. Target acid-neutralizing capacity (ANC) was calculated from an equation assuming equilibrium between the aqueous system, atmospheric carbon dioxide, and solid-phase aluminum trihydroxide. Sediment base demand was estimated from a Langmuir-type adsorption model. Base addition requirement was the amount necessary to raise ANC from existing to target levels, or base-neutralizing capacity (BNC), and was estimated by multiplying BNC by net watershed area, watershed runoff, and the years of proposed treatment. The authors discuss the precautions and risks involved in such estimates.

Results of applying limestone at a rather low rate (0.5 tons ha^{-1}) to a Swedish lake are shown in Figure 15-1. A rapid and continued response resulted from a single direct application in this lake, which had a long water renewal time (FS/SNV, 1979). Direct application to the lake surface is recommended if lake renewal times are at least two to four years. During 1977-78, 24 percent of the limestone used in Sweden was applied directly to the lake surface, at an average dose of 40-50 g m^{-3} (FS/SNV, 1979).

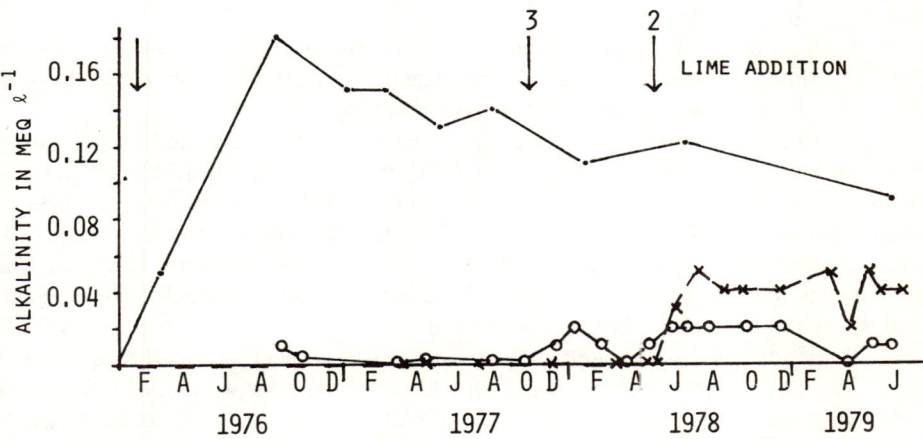

Figure 15-1 Results of three different methods of liming (CaCO$_3$) in terms of lake alkalinity (from FS/SNV, 1979). Doses are as follows:

	tons ha^{-1} watershed	gms m^{-3} lake	tons ha^{-1} limed area
1 — Direct to lake	0.5	33	
2 — 0-20 m from water line	0.5	60	10-20
3 — 0-30 m from water line	0.3	15	5.5

Treatment 1, solid circles, treatment two x, and treatment 3, open circles.

368   *Acidified Lakes*

Limestone is also applied to the shore areas around lakes, either in accessible locations or in a narrow band (30 m) around the entire lake. This application method is considered most appropriate for lakes with relatively short water renewal times. In such situations the effect of raised alkalinity would tend to persist longer than if applied directly to the lake water. A less pronounced but more persistent effect from single applications in lakes with short renewal times is shown in Figure 15-1. Lake shore application rates are on the order of 5-10 tons ha^{-1}. About 12 percent of the limestone applied in Sweden during 1977-1978 was by this technique.

Application of limestone to watersheds rather than directly to the lakes has been practiced, but the outcome may be uncertain. Although the effect may last longer than direct treatment of the lake, the dose must be much larger. Also, increasing the pH of coniferous forest soil may not be desirable (Fraser et al., 1982).

Fraser et al. (1982) have summarized the application methods for lime (Figure 15-2). Truck transport for direct application to lakes accessible by roadway is a popular method. The actual application can be achieved by spreading the lime onto ice or blowing it from roadways. Large amounts of limestone can be distributed in a short time by these methods. Application by boat for accessible lakes is also a common practice. Although rather slow, it is nonetheless inexpensive. For remote situations, the application of dry limestone dispersed over the lake surface from the air is probably the most cost-effective method. Large amounts can be distributed over large areas in a short time.

Silos are used for the application of limestone to running water courses. Dispersal from the silo is automatically activated by streamflow or pH. These units are relatively expensive, however, and require continual maintenance.

Another basic material, soda ash, is under investigation as an additive to restore acidified lakes. By injecting soda ash (Na_2CO_3) into sediments of acidified lakes, the ion exchange capacity of the humates could be utilized as a neutralizing process to exchange incoming H^+ for Na^+ (Lindmark, 1981). This technique would be especially attractive in relatively small lakes with short renewal times and rather high content of humates. On the other hand, most of the high-elevation acidified lakes are clear, and the method would not be appropriate in these.

Soda ash is presently being evaluated to restore a small, acidified Swedish lake (Lilla Galtsjön). Forty tons of soda ash, diluted with lake water to a 10-percent solution, were injected up to 25 cm into the sediments over 25 percent of the lake's area (Ripl, 1981). (See Chapter 7, Sediment Oxidation, for a description of the injection device.) Results of this method appear to be favorable. Average alkalinity and pH in the lake during the critical spring runoff period have, respectively, increased from zero to greater than 120 meq l^{-1} and from less than 5.0 to 6.7 (Lindmark, 1982).

Costs

Fraser et al. (1982) found that costs varied from $22 to $77 per ton of lime applied, depending on the type of lime. For powdered limestone, at 2.2 tons ha^{-1}, the cost would average about $100 ha^{-1}. If application costs were included, the cost would be about 50 percent more. The cost for lime application to Swedish lakes (FS/SNV, 1979) was about $40 per ton, exclusive of the method of application.

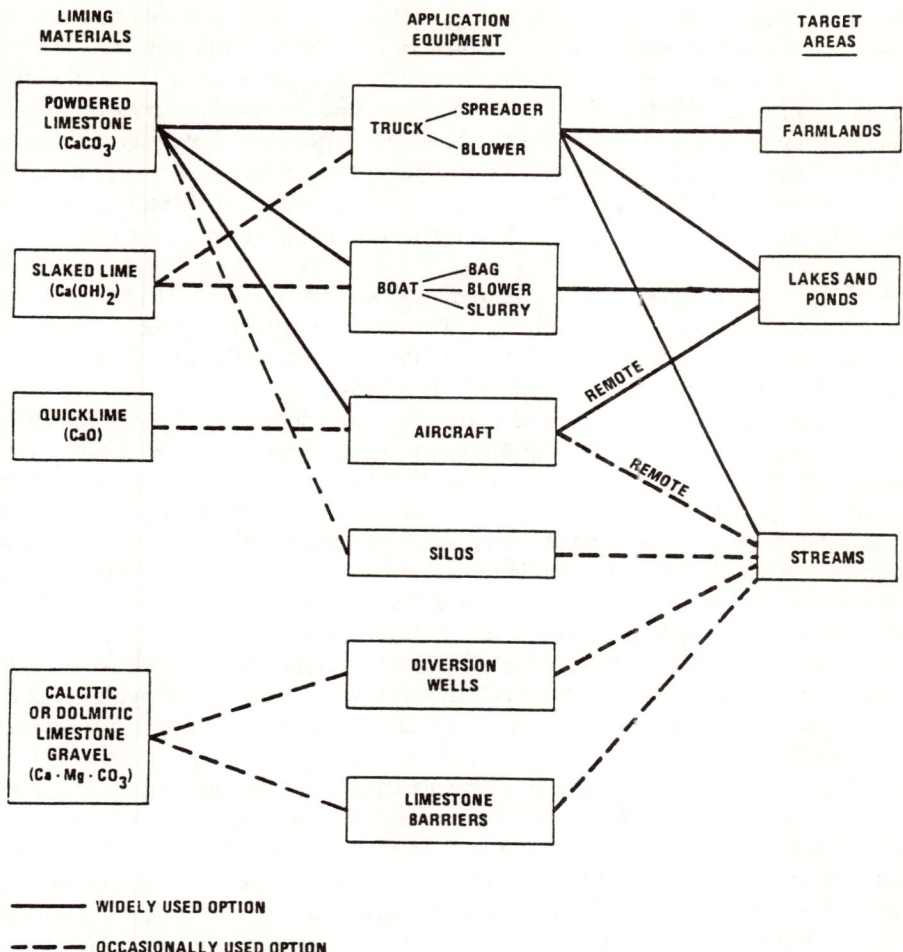

Figure 15-2 Summary of techniques used in the application of lime (from Fraser, et al., 1982).

Estimated costs for a five-year program to lime 777 acidified Adirondack lakes ranged from $2 to $4 × 10^6 (1982 dollars). The cost for accessible lakes would be about $40 to $75 ha^{-1} and for remote lakes $450-675 ha^{-1} (Menz and Driscoll, 1983).

Effects

Adding lime to acidified lakes has resulted in increased alkalinity, pH, and calcium content. For each mole of limestone ($CaCO_3$) dissolved there is an increase in alkalinity of two moles and a decrease in acidity of one mole, with another mole of acidity being lost with the evolution of a mole of CO_2 (Pearson and McDonnel, 1975).

There is disagreement on the longevity of the chemical effect of liming. Swedish scientists contend that dissolution of $CaCO_3$ continues for several years after liming. Canadian researchers believe that $CaCO_3$ becomes nonreactive after a very short time, through either complexing with sediments or coating with metals or hydroxides (Fraser et al., 1982). This uncertainty about the base demand of acidic sediments and the long-term effectiveness of limestone dissolution affects the reliability of dose estimates.

Fraser et al. (1982) have reviewed the recovery of several lakes, primarily Canadian and Swedish, following liming, and noted that recovery was more pronounced in lakes that were highly acidic (pH 4.8-5.5) than in those with intermediate acidity (pH 5.6-5.7). The effects are summarized in Table 15-2.

Eriksson et al. (1982) have also reported on results following the liming of several Swedish lakes. The phytoplankton of their limed lakes were observed to change rather slowly, from dominance by dinophyceans to dominance by diatoms and chrysophyceans. The principal change in macrophytes was a significant decline in *Sphagnum*, which had become very abundant when the lakes were acidified. Zooplankton species composition improved after liming but was retarded where planktivores also increased. The benthos was found to be rather sparse when the lakes were acidified, with species of *Asellus* and *Leptophlebia* the principal representatives. Those species, as well as a few others, increased in abundance following liming. In contrast, the plankton and benthos in Ontario and Swedish lakes cited by Fraser et al. (1982) (different lakes than in Eriksson et al.) declined immediately after liming but gradually recovered.

Fish reproduction is successful following liming. For example, arctic char were reintroduced in 1976, as fingerlings, into a previously acidified lake in which perch reproduction had ceased and char had been eliminated (FS/SNV, 1979). Within two years, after the pH increased to around 6.0, the catch rates of the two species increased markedly and the char were reproducing (Figure 15-3). Changes in plankton species composition in limed lakes have apparently benefitted fish survival (Eriksson et al., 1982).

Although liming has been shown to be an effective means of neutralizing acidified lakes, there are nevertheless potential adverse effects on biota, as noted in Table 15-2. Most important among adverse effects is the toxicity of aluminum, which is considered greatest during its hydrolysis. Fish kills have occurred during liming, probably due to aluminum toxicity (Eriksson et al., 1982). Schofield and Trojnar (1980) found that aluminum was most toxic when water pH was increased from 4.4 to 5.2, and at supersaturated Al concentrations toxicity was greatest in the pH range of 5.2 to 5.4 (Baker and Schofield, 1982). This was thought to be due to insolubility of $Al(OH)_3$ in that pH range and its precipitation and accumulation on gill surfaces. Thus, lime dosage should be adequate for neutralization and that state should be maintained to minimize recurring aluminum toxicity (Driscoll et al., 1982).

Table 15-2 Major Biological Changes in Acidified and Limed Aquatic Ecosystems (from Fraser et al., 1982)

Water Status	Bacteria	Phytoplankton and Primary Producers	Zooplankton	Benthic Macroinvertebrates	Fish
Acidified	Decrease in numbers	Decrease in species and numbers	Limitation of species and simplification of communities	Limitation of species and simplification of communities	Decline in populations
		Shift in taxonomic composition toward acid-tolerant green algae and flagellates	Shift in taxonomic composition toward acid-tolerant species	Shift in taxonomic composition toward acid-tolerant species	Decrease in recruitment through mortality of eggs and larvae and disturbance of adult reproductive physiology or behavior
		Encroachment of sphagnum moss			Fish mortality due to physiological disturbance in osmoionic regulation of fish
		No change in primary productivity			Fish mortality due to elevated metal concentrations
Limed; preneutralization pH 4.0–5.0, very high metal concentrations (e.g., Middle (M), Hannah (H), and Lohi (L) Lakes near Sudbury, Ontario)	Aerobic heterotrophic bacteria increased several orders of magnitude (M, L)	Initial decline in phytoplankton biomass (M,H,L)	Immediate decline in biomass (M,H,L)	Immediate decline in biomass (M,H,L)	Stocked smallmouth bass (M) and stocked brook trout (L) eradicated, due to metal toxicity*
		Phytoplankton biomass recovered in several months almost to preliming levels (M,H,L)	Gradual reestablishment of biomass, almost to preliming levels, in several years (L)		Experiments with stocked rainbow trout also resulted in mortality due to metal toxicity (M, L)*

Table 15-2 *(continued)*

		Biological Parameters			
Water Status	Bacteria	Phytoplankton and Primary Producers	Zooplankton	Benthic Macroinvertebrates	Fish
	Taxonomic composition changed to more acid-tolerant species (M,L)	Species changed from acid-tolerant dinoflagellates to less acid-tolerant chrysophytes (M,H,L)	Species changed from acid-tolerant cladocerans to less acid-tolerant copepods (M,H,L)		
		Addition of phosphorus enhanced phytoplankton recovery and caused an increase in biomass (M,H)*	Addition of phosphorus enhanced recovery of biomass (M,H)*		
Limed; preneutralization pH 4.0–5.5, low metal concentrations (e.g., Lysevaten (Ly), Stensjön (S), and Bredvatten (B) Lakes in Sweden)	Populations increased, resulting in fin rot (Mersey Fish Hatchery, Nova Scotia)	Initial decline in biomass (Ly, S,B)	Immediate decline in biomass (Ly, S,B)	Immediate decline in biomass (Ly, S, B)	Aluminum toxicity lethal to trout and Salmon (Sweden)*
		Change in speciation from acid-tolerant to less acid-tolerant species (Ly,S,B)	Gradual recovery of biomass, up to or occasionally greater than prelimimg levels, in several years (Ly,S,B)	Acid-tolerant chironomids replaced by less acid-tolerant species in several years (Ly,S,B,)	Reproduction of acid-sensitive salmon, arctic char, brook trout, brown trout, and sea trout (Sweden)
		Biomass recovered in several months, almost to prelimimg levels (Ly,S,B)	Species changed from acid-tolerant cladocerans to less acid-tolerant copepods (Ly,S,B)	Reintroduced species generally survived (S, B)	Incubation of rainbow trout, lake trout, and brook trout within crushed limestone substrates greatly enhanced hatching success and sac fry survival (George Lake, Ontario)

Table 15-2 (continued)

Water Status	Biological Parameters				
	Bacteria	Phytoplankton and Primary Producers	Zooplankton	Benthic Macroinvertebrates	Fish
		Sphagnum mosses and benthic algal mats severely damaged or eradicated (Ly,S,B)			Raising pH of hatchery water by limestone filters increased survival of Atlantic salmon parr during third and fourth weeks after first feeding (Mersey, Nova Scotia)
Limed; preneutralization pH 5.6–5.7, average or slightly high metal concentrations (e.g., Nelson Lake (N) near Sudbury, Ontario)		No change (N)	Biomass relatively constant (N)		No effects on established inshore fish communities (N) Reintroduced smallmouth bass spawned successfully (N)

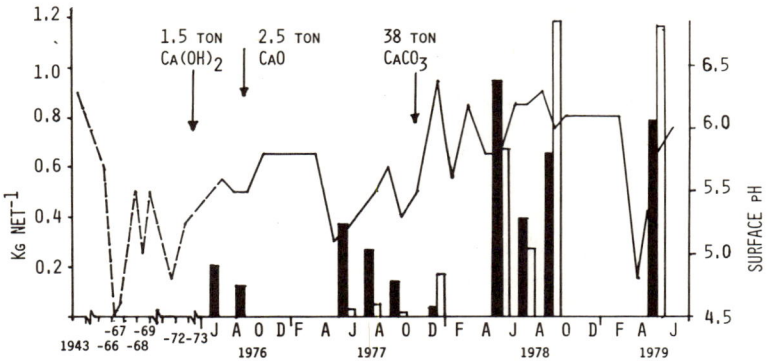

Figure 15-3 Introduction of arctic char in a limed water body that previously supported char and yellow perch (FS/SNV, 1979). Open bar is char; closed is yellow perch.

REFERENCES

Almer, B., Dickson, W., Ekström, C., Hörnström, E., and Miller, U. 1974. Effects of acidification on Swedish lakes. *Ambio* 3:30–36.
Baker, J.P., and Schofield, C.L. 1982. Aluminum toxicity to fish in acid waters. *Water Air Soil Pollut* 18:209–309.
Charlson, R.J., and Rodhe, H. 1982. Factors controlling the acidity of natural rainwater. *Nature* 295:683–685.
Chen, C.W. 1981. Acid rain and liming lakes—What is in it for lime? Paper presented at 79th Annual Convention, National Lime Assoc., April.
Dickson, W. 1978. Some effects of the acidification of Swedish lakes. *Verh Int Ver Limnol* 20:851–856.
Driscoll, C.T., White, J.R., and Schafran, G.C. 1982. $CaCO_3$ neutralization of acidified surface waters. *J Environ Eng Div ASCE*.
Eriksson, F., Hörnström, E., Mossberg, P., and Nyberg, P. 1982. *Ekologiska effekter av Kalkning i Försurade Sjöar och Vattendrag*, Sötvattens–Laboratoriet, Drottningholm, Sweden, Nr. 6.
Fraser, J., Hinckley, D., Burt, R., Severn, R.R., and Wisniewski, J. 1982. *A Feasibility Study to Utilize Liming as a Technique to Mitigate Surface Water Acidification*, Rept. No. RP 1109-14, General Res. Corp. for Elect. Power Res. Inst.
FS/SNV (Fiskeristyrelsen/Statens Naturvårdverk). 1979. *Kalkning av Sjöar och Vattendrag*, Sövattena–Laboratoriet, Drottningholm, Sweden, Nr 8.
Gran, O., Hullberg, H., and Landner, L. 1974. Oligotrophication—A self-accelerating process in lakes subjected to excessive supply of acid substances. *Ambio* 3:93–94.
Harvey, H.H. 1980. Some effects of the acidification of Swedish lakes. *Verh Int Ver Limnol* 29:851–856.
Hendrey, G.R. 1976. *Effects of Low pH on the Growth of Periphytic Algae in Artificial Stream Channels*, SNSF Proj. 1432 Aas-NLH, Norway.

Henriksen, A. 1980. Acidification of fresh waters — A large scale titration. In Dr. Drablos and A. Tollan, eds., *Proc. Int. Conf. Ecol. Impact Acid Precip.* SNSF Proj. Rept., Norway, 68-74.

Kramer, J., and Tessier, A. 1982. Acidification of aquatic systems: A critique of chemical approaches. *Environ Sci Technol* 16:606-615.

Likens, G.E., Wright, R.F., Galloway, J.N., and Butler, T.J. 1979. Acid rain. *Science* 241:43-52.

Lindmark, G.K. 1982. Acidified lakes: Sediment treatment with sodium carbonate — a remedy? *Hydrobiol.* 92:537-547.

Menz, F.C., and Driscoll, C.T. 1983. An economic evaluation of liming to neutralize acidified Adirondack surface waters. *Wat Res Res* 19:1139-1149.

Miljöaktuellt. 1981. *Statens Naturvårdverkets Tidning*, June.

Pearson, F.H., and McDonnel, A.J. 1975. Use of crushed limestone to neutralize acid wastes. *J Environ Eng Div ASCE* 101:139-158.

Ripl, W. 1981. *Lake Restoration Methods Developed and Used in Sweden*, Wayne, NJ: Atlas Copco Inc.

Schofield, C.L. 1976. Acid precipitation: Effects on fish. *Ambio* 5:228-230.

Schofield, C.L., and Trojnar, J.F. 1980. *Aluminum Toxicity to Brook Trout in Acidified Waters*, New York: Plenum.

Sevaldrud, I.H., Muniz, I.P., and Kalvenes, S. 1980. Loss of fish populations in Southern Norway: Dynamics and magnitude of the problem. In D. Drablos and A. Tollan, eds., *Proc. Int. Conf. Ecol. Impact Acid Precip.*, SNSF Proj. Rept. Norway, 350-351.

Wright, R.F., Dale, T., Gjessing, E.T., Hendrey, G.R., Henriksen, A., Johannessen, M., and Muniz, I.P. 1976. Impact of acid precipitation on freshwater ecosystems in Norway. *Water Air Soil Pollut* 6:483-499.

Wright, R.F., and Henriksen, A. 1978. Chemistry of small Norwegian lakes with special reference to acid precipitation. *Limnol Oceanogr* 23:487-498.

Wright, R.F., and Snekvik, E. 1978. Acid precipitation: Chemistry and fish populations in 700 lakes in southernmost Norway. *Verh Int Ver Limnol* 20:755-765.

16

Summary and Conclusions

We have critically reviewed and summarized procedures for protecting, improving, and restoring eutrophic lakes and reservoirs, and have included a chapter on a rapidly developing and equally serious problem, acidified lakes. A chapter on basic limnology and on methods of study and data analyses that would be used in the typical lake diagnosis and feasibility study has also been included. Our intent has been to describe the procedures we believe are ecologically sound and can bring about long-term control of nuisance vegetation without further exacerbation of the problem. Where possible, we have described how to use each procedure, the causes of successes and failures, any adverse effects, and the costs associated with their application. A separate chapter on algicides and herbicides has not been presented, since this topic is not within the purview of a book on lake restoration. The use of these chemicals poses significant and unanswered questions regarding costs, effectiveness, human health implications, and short-term and long-term impacts on aquatic ecosystems. We do not recommend their use, particularly as the sole treatment of nuisance aquatic plants, unless no other options are possible. We believe that many additional studies of these chemicals are needed, especially on cost-effectiveness compared to other procedures and on their impacts on the ecosystem level of organization. Because this topic is a complex and significant one, it is deserving of a separate critical review.

Techniques for restoration of eutrophic lakes may be divided into those whose purpose is to reduce nutrient concentration in the water column and those whose purpose is to control algal or macrophyte biomass. These categories can be further divided into procedures with demonstrated effectiveness and procedures that require further documentation and research.

TECHNIQUES WITH DEMONSTRATED EFFECTIVENESS

Procedures to Reduce Nutrient Concentration

Nutrient Diversion
Diversion of nutrient income is widely recognized as a highly necessary and usually essential first step to obtaining improvement in lake trophic state.

Uttormark and Hutchins (1980) examined the responses of 23 lakes to diversion and found that, while concentration of nutrients (P in this case) may have declined, in only 30 percent of the cases was there improvement to a lower trophic state. This may have been because diversion was insufficient to lower concentration to the level that would produce mesotrophic lakes. Even so, reduction in algal biomass, absence of blue-green scums, and return of fecal coliform colony counts to the acceptable range will make a diversion a success.

Another reason for failure to significantly improve trophic state is continued release of nutrients from sediments into the water column. This factor, along with morphometric and hydrologic features such as mean depth and flushing rate, is important in determining how much change in concentration there will be after diversion and how long it will take to reach a new, lower concentration. Ultimately, internal P release should decline as older, richer sediments are covered with those of lower nutrient content, and then lake concentration should decline.

Diversion is thus the first step in stopping the process of nutrient enrichment and in lowering the concentration of nutrients. Recovery of the lake to a lower trophic state may be slow, requiring an in-lake treatment such as dilution or sediment removal or watershed controls of nonpoint nutrient sources such as agricultural runoff.

Silt income control may be very significant for preventing the rapid development of shallow water and the spread of nuisance macrophytes. Unless the siltation problem is recent, an extensive littoral area from long-term deposition might require sediment removal or other procedures before macrophyte growth is controlled. Control of income from the watershed is thus, again, an essential first step to improvement in trophic state.

Dilution and Flushing

Dilution reduces the concentration of nutrients in the water column by adding nutrient-poor water. Flushing reduces algal biomass by increasing the loss rate of cells. Both effects occur if large amounts of low-nutrient water are added.

Dilution has been very effective in Moses Lake and Green Lake, Washington (Oglesby, 1969; Welch and Patmont, 1980). Welch (1981) concluded that the best plan is to have a continuous low-rate input of low-nutrient water, coupled with nutrient diversion. Otherwise, high-rate inputs would be required.

While this procedure will bring about lake improvement, its applicability is limited by the general absence, in most watersheds, of nutrient-poor waters that could be diverted into eutrophic lakes.

Phosphorus Inactivation/Precipitation

The purpose of this technique is to lower the concentration of P in the water column by either precipitating it out or preventing its release from sediments (inactivation). Salts of aluminum (aluminum sulfate or sodium aluminate) are added to the lake surface or to the hypolimnion, and P is sorbed to the surface of the aluminum hydroxide floc. In the case of inactivation, sufficient aluminum is added to create a barrier to P release from sediments.

Like the use of herbicides and algicides, this technique has the potential to create toxic conditions. This situation could occur through an increase in soluble aluminum (Al^{+3}), or through a decrease in pH. Kennedy and Cooke (1981; 1982) have described a method for determining the maximum safe dose of aluminum sulfate for a lake, based on lakewater alkalinity and toxicity of dissolved aluminum to trout. An alternative dose procedure for soft-water lakes has been described (Dominie, 1980). These procedures allow for maximum P inactivation without creating toxic conditions.

The P inactivation technique is particularly effective in long-term control of P concentration in the water column. Cooke and Kennedy (1981a; b) have reviewed published case studies of this technique. The longest period of P control to date is 12 years (Garrison and Knauer, 1984). Failures have been due to insufficient dose, polymixes, and insufficient nutrient diversion.

Sediment Removal

Sediment removal is one of the most commonly prescribed techniques for long-term lake improvement. Lake deepening and the removal of toxic materials, macrophytes, and nutrient-rich sediments, are its main purposes. Until recently (Peterson, 1979; 1981; 1982) little published documentation about its effectiveness, environmental impact, and costs was available.

Lake-deepening projects are generally successful, except in instances where sediment income remains high. Similarly, control of nutrient release from sediments can be achieved by removal, although it is often less costly to do so through P inactivation. There is little documentation of macrophyte control by dredging.

There is widespread concern about the negative environmental impact of this procedure. Most negative effects (algal blooms, high turbidity) are of short duration. One common problem is inadequate sizing of the disposal area. Diking in upland areas is the most frequently used means of containing the dredge spoil, and dike failure and contamination of wells have occurred.

Dredging can be expensive, with costs ranging from $0.22 m^{-3} to $13.93 m^{-3}. Removal of contaminated material may exceed $25.00 m^{-3}. One way of lowering costs is to sell uncontaminated dredge material for topsoil dressing.

Procedures to Control Algal and Macrophyte Biomass

Drawdown

Lake-level drawdown is used successfully to control the growth of certain macrophyte species, to consolidate flocculent lake sediments, to provide an opportunity to repair docks and dams, to remove sediment or to install sediment covers, and for fish management (Cooke, 1980a).

Not all aquatic plants are susceptible to the freezing or high-temperature conditions of a winter or summer drawdown. Alligator weed (*Alternanthera philoxeroides*) and *Najas flexilis* (brittle naiad) have been reported to increase in density after drawdown, whereas water lily (*Nuphar* sp.), water hyacinth (*Eichhornia crassipes*) and musk grass (*Chara vulgaris*) usually decrease. For most macrophytes, the published data are too sparse to generalize. A danger may be that susceptible species will be replaced by tolerant ones, leading some investigators to suggest that the best control will be achieved by two to three years of drawdown followed by two years of stable water levels (Lantz et al., 1964; Lantz, 1974). Most drawdowns have been during the winter because this does not interfere with most recreational uses, because there is greater assurance of refill during spring melt and rainfall, and because there is no danger of invasion by terrestrial plants. There are insufficient published data to compare the effectiveness of winter and summer drawdowns.

Drawdown has been effective in fish management. In Louisiana reservoirs, populations of shad and sunfish are removed by winter drawdown and their spawning prevented by summer drawdown (Lantz et al., 1964). In reservoirs with five or more consecutive years of drawdown, fish standing crop and game-fish size and reproduction have increased.

There are negative effects of water-level drawdown. Algal blooms have occurred upon refilling, possibly from nutrient release from rewetted, highly organic sediments. Fish kills in pools of water and great changes in the macroinvertebrate community have been reported.

This low-cost management technique can be effective in macrophyte control and can be beneficial in the sense that it provides management personnel with access to the lake for the purpose of enacting other techniques.

Sediment Covers

Several sheeting materials have been tested for their effectiveness in preventing macrophyte growth (Cooke, 1980b). Impermeable materials such as polyethylene have been found to be effective but troublesome because gases accumulate beneath them and they are difficult to apply. One way to prevent "ballooning" is to cover the sheeting with gravel and silt, but this has usually provided sufficient substrate for plant regrowth. Burlap has been shown to be effective for one growing season, but the material usually decomposes and must be reapplied annually.

One material, a PVC-coated fiberglass screen, has been shown to be very effective in preventing macrophyte growth, comparatively easy to install, does not "balloon." but it is very costly ($140/700 ft^{-2}; $2.15 m^{-2}). Perkins et al. (1980) suggested that the screens control biomass through physical or space limitation. In most instances, the screens must be repositioned annually to remove accumulated silt. Due to its cost, this material appears to be best suited for small areas such as docks, beaches, or waterfronts. New materials, now being tested, may be less costly and more effective.

TECHNIQUES REQUIRING MORE RESEARCH AND DEMONSTRATION

Procedures to Reduce Nutrient Concentration

Hypolimnetic Withdrawal

The object of this technique is to siphon, pump, or discharge nutrient-rich hypolimnetic water from the lake or reservoir. This should reduce the impact of vertical entrainment of nutrients on the epilimnion, and it will remove nutrients from the lake. The siphon used is sometimes called an "Olszewski tube," after its originator (Olszewski, 1961).

There are few reports of the use of this device in North America. Experiences in small European lakes have been described (Bjork, 1974; Gächter, 1976). Total P and internal P release decreased, transparency increased, and fewer blue-greens were present.

This technique deserves further study. It should be dependable and inexpensive. The high P and low dissolved oxygen content of the hypolimnetic water could pose a threat to receiving waters. Perhaps the lake discharge could be used in irrigation. Other possible problems could include disruption of the thermocline, excessive water loss from the lake, and the necessity that the discharge end of the hose be below the surface level of the lake.

Sediment Oxidation

Ripl (1976) has described a procedure for oxidizing the top 15-20 cm of anaerobic lake sediment to reduce internal nutrient release. There are few published data regarding the success of this new procedure and its costs.

Fe (Cl_3) is added first if iron is naturally insufficient. Ferric hydroxide for P binding is formed, and the sulfur in FeS is lost as H_2S. Calcium hydroxide is added next to increase pH to the optimum for denitrification. Finally, calcium nitrate is added to the sediments to oxidize reduced carbon and act as an alternate electron receptor. Reduction of iron^{+3} does not occur until the nitrate is exhausted. The chemicals are added by a harrow. A positive result (lower P, NH_4 decrease, O_2 demand lowered) was found in Lake Lillesjon, Sweden (Ripl and Lindmark, 1978), but costs of chemicals and equipment were high.

Further laboratory and field evaluations of this technique are needed.

Procedures to Control Algae and Macrophyte Biomass

Biological Controls

Biological control methods may be the most promising because they offer the possibility of exerting control of nuisance plants without eradication and without the use of expensive machinery or toxic chemicals. Significant progress has been made in recent years.

Among the most controversial of the biological control methods is introduction of the grass carp, or white amur (*Ctenopharyngodon idella* Val.), a voracious consumer of macrophytes. Several apparently successful introductions have been reported (Mitzner, 1978; Shireman and Maceina, 1981), and the U.S. Army Corps of Engineers and others are conducting large-scale feasibility studies in Florida. The animal, however, was used as a macrophyte control agent in the United States long before adequate scientific studies were conducted on potential adverse impacts. While new data are now being accumulated, there remains no certainty with regard to the effects of these animals on other species, their role in nutrient recycling, whether they transmit fish diseases, and whether or not they will reproduce and infest nontarget waters. Currently the fish is widespread in waters such as the Mississippi River. We will be fortunate if this animal proves to be useful and innocuous in North America, since it was widely released before these questions were asked.

A plant pathogen (*Cercospora rodmanii*) appears to be effective in control of water hyacinth (*Eichhornia crassipes*) and is now being prepared for a large-scale operations test by the U.S. Army Engineers (Freeman, 1977; Freeman et al., 1981).

Monophagous insects have been introduced in southern states for aquatic weed control, following careful screening under quarantine. Two beetles and a moth have been released as controls of water hyacinth (Center, 1981). Their effectiveness is currently under study.

Since the initial report of LaMarra (1975) concerning the nutrient recycling ability of the common carp, others have further demonstrated the eutrophying activities of benthivorous fish (Anderson et al., 1978; Keen and Gagliardi, 1981). The word "biomanipulation" has been used to include lake improvement activities in which fish such as bullheads are removed, or in which the pelagic food web is altered to decrease predation on the large herbivorous zooplankton (Shapiro et al., 1975; Shapiro, 1979). These and other procedures deserve considerable further support and attention. These methods may, for example, be a way to improve lakes where significant nutrient diversion is not possible.

Artificial Circulation

The purpose of this procedure is to prevent thermal stratification, or to destratify a lake. The most effective and least expensive technique is to introduce compressed air at maximum depth, creating a curtain of bubbles and intense mixing. An air-flow rate of 9.4 m^3 min · km^{-2} should be adequate (Lorenzen and Fast, 1977). Pastorok et al. (1981) found that in only 20 percent of the cases reviewed was this flow rate achieved, a factor that may account for some of the reported failures of the technique.

Artificial circulation is included in this section on "techniques requiring more research" because the expected results often do not occur. The principal improvement is increased aeration of the water column and expanded habitat. Undersizing the flow rate may prevent this from happening. Hypothetically, aeration and circulation should inhibit P release from sediments, limit algal productivity by increasing cell residence time in dark waters, decrease pH, and perhaps shift the competitive balance from blue-green algae to greens and diatoms, limit the survival value of blue-green pseudovacuoles (buoyancy), and favor the survival of herbivorous zooplankton. Phytoplankton have decreased in about half the cases examined by Pastorok et al. (1980). Transparency has decreased more often than it increased. Circulation may introduce nutrients to the water column, and in nutrient-limited lakes an increase in biomass may occur. Perhaps this technique might be most effective in lakes where nutrients are not limiting. Additional research is needed.

Hypolimnetic Aeration

This is a technique to aerate the hypolimnion without achieving destratification (Pastorok et al., 1982). It is accomplished by an "air-lift," in which water is lifted up a cylinder by compressed air to the lake surface, aerated, and then returned to the hypolimnion. In some cases, a partial air lift is used.

Successful oxygenation is usually achieved, although metalimnetic oxygen minima may occur with an undersized aerator. Phosphorus concentration may be reduced following the introduction of oxygen, although hypolimnetic levels tend to remain high. There are few data on biological effects, including effects on phytoplankton, zooplankton, and fish (Pastorok et al., 1982). More documentation about the responses of lakes to this treatment is needed.

Harvesting

Harvesting is one of the more controversial procedures, in part because there is little scientific evidence to demonstrate long-term effectiveness. The technique has been incorrectly categorized with herbicides as a cosmetic treatment. Unlike the introduction of toxic materials to lakes, nutrients and oxygen-consuming biomass are removed during a harvest, thus protecting the lake, and there is some evidence that longer-term control of macrophytes can be achieved by using the right cutting frequency and cutting technique. Since harvesting is usually less costly than herbicides, and because removal of nuisance plant biomass can have a protective effect for the lake, the use of this procedure deserves more careful attention and additional experimental work.

ACIDIFIED LAKES

The acidification of lakes through the deposition of acids in precipitation and dry fall is emerging as one of the most serious environmental issues. It is generally agreed that the cause is fossil fuel combustion and transport of air masses containing sulfuric and nitric acids. The ecological effects are just beginning to be known, but they include decrease in lake species diversity and toxicity to fish by increased hydrogen ion and dissolved aluminum concentrations. Soft-water lakes treated with aluminum to remove P or to control P release from lake sediments could become toxic if pH later falls below 6.0 from acid precipitation.

Since lake protection in this case may be difficult or impossible, given the rate of current and projected fossil fuel consumption, restoration of acid-damaged lakes at present includes only the replacement of lost alkalinity. Liming is the principal procedure, and waters restored in this way have been shown to support reintroduced game fish. Since the problem of acidified lakes may increase sharply, much more research on their protection and restoration is needed.

COMBINATIONS OF LAKE RESTORATION TECHNIQUES

We have, for the most part, treated each lake or reservoir restoration or improvement technique in this book separately, as if the use of one precludes the simultaneous use of another. On the contrary, "integrated control methods," similar to those used in agriculture, may be far more effective and less costly than use of any single technique.

Some procedures, when used together, could counter each other or produce a negative effect. Some of these situations are summarized below.

Flushing. Do not use with hypolimnetic withdrawal because the probability of destratification is greatly enhanced and would negate the effectiveness of withdrawal if it occurred. Flushing is not compatible with dilution unless the water used for flushing is very low in nutrients, in which case the techniques are identical. If used with phosphorus inactivation or sediment oxidation, there is risk that chemicals added to the sediments would be removed. Flushing would have to be used with care with sediment removal, since turbidity and nutrients generated by the dredge would quickly be transported through the lake. If these are used together, then barriers around the dredge site would be needed. Flushing and lake-level drawdown would not be used in the same season. The introduction of phytophagous fish must be done only where the fish cannot escape to downstream ecosystems. Flushing would be incompatible with this goal unless barriers to escape were used at the outlet.

Phosphorus Inactivation. Flushing is not compatible with this procedure, as described earlier. Aluminum salts and the reagents used in sediment oxidation could be incompatible, especially if pH is affected and denitrification is impeded. Artificial circulation and hypolimnetic aeration could interfere with the effectiveness of phosphorus inactivation, by mixing the treated sediments with overlying water.

Sediment Oxidation. As noted above, there could be problems with combining use of sediment oxidation and flushing or phosphorus inactivation.

Sediment Removal. Problems with flushing and artificial circulation were described earlier.

Biological Controls. Phytophagous fish should be introduced only where escape downstream is impossible. Hypolimnetic withdrawal, dilution, or flushing would require barriers to escape.

There has been very little published research on the application of two or more techniques in the same lake or reservoir. This is due in part to high costs and also to the desire to learn about one procedure without the confounding effect of additional manipulations. The next major advances in this technology will probably be in the area of combinations of lake restoration techniques. The work of Haag (1985) with herbicides and insects is an example.

The compatibility of some procedures is obvious, and we have mentioned these cases in the text. As examples, drawdown is an excellent way to provide access for repairs and to improve the efficiency of sediment removal and the application of sediment covers. It is also used in fish management, which could be part of a biomanipulation. Harvesting can enhance the effectiveness of 2,4-D applications by allowing access of the chemical to root systems, and plants damaged by a harvester may be more susceptible to plant pathogens. Diversion of nutrient loading is obviously compatible or necessary with every technique.

The entire area of integrated control of nuisance aquatic vegetation is in its infancy. We strongly urge investigators in lake and reservoir management to develop and report about experiments with combinations of techniques, emphasizing not only determinations of effectiveness but also analyses of costs.

CONCLUSION

Our goal has been to develop an up-to-date review, summary, and analysis of the current state of knowledge of the procedures to protect, improve, and restore eutrophic and acidified lakes and reservoirs. An obvious finding, while not always stated explicitly in the text, is the small amount of knowledge we actually have about protection, management, and restoration of our freshwater resources. At present, our solutions seem to be based almost exclusively upon the application of chemicals or the introduction of machinery. In many cases these approaches are effective, but they are also expensive. The developing interest in biological control may greatly add to our arsenal of effective and inexpensive techniques. This avenue may become extremely important in the near future as the demand for clean water increases and the level of tax dollars available for lake restoration decreases.

There are ample reasons to believe that surface water resources of both developed and developing nations will continue to deteriorate with rising demands on agriculture and the inability of governmental units to provide adequate sewage treatment as urban populations increase. We need to greatly increase our knowledge of freshwater ecosystems and of the methods to manage, protect, and restore them. This is vital, not only for recreational purposes, but for human health and economic development.

We urge our colleagues in aquatic ecology to turn their attention to these matters. We need to know far more about the littoral zone and its interactions with sediments and the pelagic zone. We are just beginning to see the significance of the predator-prey system upon the biomass of producers in aquatic communities. Several of the procedures described in this book rely on chemical manipulations of the sediments. Far more knowledge is needed here, including studies of the significance of internal loading. We have very little information about the long-term effectiveness and impacts of these restoration and improvement methods and, finally, we know next to nothing about the efficacy of combined uses of these procedures.

We view this book as a beginning in the area of lake and reservoir restoration. We have been successful with it if we have summarized and evaluated knowledge, if we have provided some guidelines for the users and managers who desire restoration of their lakes, and if we have stimulated additional research and an awareness of the need to know about lake and reservoir protection, management, and restoration.

REFERENCES

Anderson, G., Berggen, H., Cronberg, G., and Gelin, C. 1978. Effects of planktivorous and benthivorous fish on organisms and water chemistry in eutrophic lakes. *Hydrobiologia* 59:9-16.

Bjork, S. 1974. *European Lake Rehabilitation Activities*, Institute of Limnology, Univ. of Lund, Lund, Sweden.

Britt, D.L., and Fraser, J.E. 1983. Effectiveness and uncertainties associated with the chemical neutralization of acidified surface waters. In *Lake Restoration, Protection, and Management*, EPA-440/5-83-001, 96–103.

Center, T.D. 1981. *Release and Establishment of Sameodes albiguttalis for Biological Control of Waterhyacinth*, Tech. Rept. A-81-3, Vicksburg, MS: U.S. Army Corps of Engineers.

Cooke, G.D. 1980a. Lake-level drawdown as a macrophyte control technique. *Water Res Bull* 16:317–322.

Cooke, G.D. 1980b. Covering bottom sediments as a lake restoration technique. *Water Res Bull* 16:921–926.

Cooke, G.D., and Kennedy, R.H. 1981a. *Precipitation and Inactivation of Phosphorus as a Lake Restoration Technique*, EPA-600/3-81-012.

Cooke, G.D., and Kennedy, R.H. 1981b. State-of-the-art summary of phosphorus inactivation as a lake restoration technique. In *Proceedings, Workshop on Algal Management and Control*, Tech. Rept. E-81-7, Las Vegas, NV: U.S. Army Corps of Engineers, 32–56.

Dominie, D.R. II. 1980. Hypolimnetic aluminum treatment of softwater Annabessacook Lake. In *Restoration of Lakes and Inland Waters*, EPA-440/5-81-010, 417–423.

Freeman, T.E. 1977. Biological control of aquatic weeds with plant pathogens. *Aquatic Bot* 3:175–184.

Freeman, T.E., Charudattan, R., Conway, K.E., Cullen, R.E., Martyn, R.D., McKinsey, D.E., Olexa, M.T., and Reese, D.F. 1981. *Biological Control of Aquatic Plants with Pathogenic Fungi*, Tech. Rept. A-81-1, Vicksburg, MS: U.S. Army Corps of Engineers.

Gächter, R. 1976. Lake restoration by bottom water siphoning. *Schweiz Z. Hydrol* 38:1–28.

Garrison, P.J., and Knauer, D.R. 1984. Long-term evaluation of three alum treated lakes. In *Lake and Reservoir Management,* EPA 440/5-84-001, 513–517.

Haag, K.H. 1985. Does herbicide application affect water hyacinth weevils? *Aquatics* 7:13–15.

Keen, W.H., and Gagliardi, J. 1981. Effect of brown bullheads on release of phosphorus in sediment and water systems. *Prog Fish-Cult* 43:183–185.

Kennedy, R.H., and Cooke, G.D. 1981. Aluminum sulfate dose determination and application techniques. In *Restoration of Lakes and Inland Waters*, EPA-440/5-81-010, 405–411.

Kennedy, R.H., and Cooke, G.D. 1982. Control of lake phosphorus with aluminum sulfate. Dose determination and application techniques. *Water Res Bull* 18:389–395.

LaMarra, V.J. Jr. 1975. Digestive activities of carp as a major contributor to the nutrient loading of lakes. *Verh Int Ver Limnol* 19:2461–2468.

Lantz, K.E. 1974. *Natural and Controlled Water Level Fluctuation in a Backwater Lake and Three Louisiana Impoundments*, Baton Rouge, LA: Louisiana Wildlife and Fisheries Comm.

Lantz, K.E., Davis, J.T., Hughes, J.S., and Schafer, H.E. 1964. Water level fluctuation — Its effects on vegetation control and fish population management. *Proc. 18th Ann. Conf. Southeast. Assoc. Game and Fish Comm.* 18:483–494.

Lorenzen, M.W., and Fast, A.W. 1977. *A Guide to Aeration/Circulation Techniques for Lake Management*, EPA-600/3-77-004.

Mitzner, L. 1978. Evaluation of biological control of nuisance aquatic vegetation by grass carp. *Trans Am Fish Soc* 107:135-145.
Oglesby, R.T. 1969. Effects of controlled nutrient dilution on the eutrophication of a lake. In *Eutrophication: Causes, Consequences, Correctives*, Washington, DC: National Academy of Sciences, 483-493.
Olszewski, P. 1961. Versuch einer Ableitung des hypolimnischen Wassers aus einen See. Ergebnisse des ersten Versuchsjahres. *Verh Int Ver Limnol* 14:855-861.
Pastorok, R.A., Ginn, T.C., and Lorenzen, M.W. 1981. *Evaluation of Aeration/Circulation as a Lake Restoration Technique*, EPA-600/3-81-014.
Pastorok, R.A., Lorenzen, M.W., and Ginn, T.C. 1982. *Environmental Aspects of Artificial Aeration and Oxygenation of Reservoirs: A Review of Theory, Techniques, and Experiences*, Tech. Rept. E-82-3, Vicksburg, MS: U.S. Army Corps of Engineers.
Perkins, M.A., Boston, H.L., and Curren, E.F. 1980. The use of fiberglass screens for control of Eurasian water milfoil. *J Aquatic Plant Manage* 18:13-19.
Peterson, S.A. 1979. Dredging and lake restoration. In *Lake Restoration*, EPA-440/5-79-001, 105-114.
Peterson, S.A. 1981. *Sediment Removal as a Lake Restoration Technique*, EPA-600/3-81-013.
Peterson, S.A. 1982. Lake restoration by sediment removal. *Water Res Bull* 18:423-435.
Ripl, W. 1976. Biochemical oxidation of polluted lake sediment with nitrate — A new lake restoration method. *Ambio* 5:132-135.
Ripl, W., and Lindmark, G. 1978. Ecosystem control by nitrogen metabolism in sediment. *Vatten* 2:135-144.
Shapiro, J. 1979. The need for more biology in lake restoration. In *Lake Restoration*, EPA-440/5-79-001, 161-167.
Shapiro, J., LaMarra, V., and Lynch, M. 1975. Biomanipulation — An ecosystem approach to lake restoration. In P. Brezonik and J.L. Fox, eds., *Water Quality Management through Biological Control*, Gainesville, FL: Univ. Florida, 85-96.
Shireman, J.V., and Maceina, M.J. 1981. The utilization of grass carp, *Ctenopharyngodon idella* Val., for hydrilla control in Lake Baldwin, Florida. *J Fish Biol* 19:629-636.
Uttormark, P.D., and Hutchins, M.L. 1980. Input-output models as decision aids for lake restoration. *Water Res Bull* 16:494-500.
Welch, E.B. 1981. The dilution/flushing technique in lake restoration. *Water Res Bull* 17:558-564.
Welch, E.B., and Patmont, C.R. 1980. Lake restoration by dilution: Moses Lake, Washington. *Water Res* 14:1317-1325.

Index

Acidified lakes
  causes, 363-364
  ecological effects, 364-365
  liming costs, 369
  liming effects, 369-374
  liming practices, 365-369
Air lift design, 233, 239-243
Algae. *See* Phytoplankton
Aluminum
  chemistry, 102
  toxicity, 119-122
  toxicity to fish, 120, 367, 370
Aluminum sulfate. *See* Phosphorus inactivation
Annabessacook Lake, Maine, 109, 111-113, 122
Aquascreen, 353-354, 357
Aquashade, 359
Artificial circulation
  costs, 243
  devices and air quantities, 233-243
  effects of
    adverse, 245, 246
    beneficial, 234
    on internal loading, 225, 243
    on pH, 244-245
    on phytoplankton, 226-233, 244-247
    on temperature, 239-241
    on trophic indicators, 243-245
  lakes receiving treatment, 235-238
  summary, 245-247
  theory, 225-233

Bacteria
  effect of alum application, 122
  in lake restoration, 325

Ballinger Lake, Washington, 75, 76, 78, 80
Basin morphometry, 13
Benthos
  description, 21
  effect of sediment removal, 142
  effect of water level drawdown, 270
Biological controls
  biomanipulation, 328-337
  Cerocospora rodmanii, 326
  Ctenopharyngodon idella, 316-324. *See* Grass carp
  cyanophages and bacteria, 325
  Neochetina bruchi, 326-327
  Neochetina eichhorniae, 326-327
  phytophagous fish, 315-325
  plant pathogens and insects, 325-328
  restructuring fish communities, 338-340
  Sameodes albiguttalis, 326
  summary, 340
  Tilapia mossambica, 315-316
  Tilapia zillii, 325
  Vogtia malloi, 328
Biomanipulation, 113, 328-337
  models, 300, 334
Blue-green algae, effect of pH and CO_2, 231-233
Brunsviken Lake, Sweden, 216-220
Burlap sediment cover, 356

Chemung Lake, Ontario, 289, 291, 297, 303, 305
Chlorophyll
  relationship to phosphorus, 19-20
  relationship to transparency, 57
  and trophic state, 35, 38

Clean Lakes Program, 4
Communities of lakes and reservoirs, 20
Copper sulfate, 117, 126, 335, 350
Cyanophages, 325

Dartek, 354-356
Detention basins, 118-119
Diagnosis/feasibility study
    biological determinations, 25
    physical determinations, 23-24
    study design, 23-40
    water chemistry determinations, 24-25
Dilution/flushing
    calculation of dose, 94-96
    case studies, 86-96
    cost, 92, 96, 97
    dilution, definition, 83, 96-97
    effect on phytoplankton, 89-93
    limiting nutrients, 84-86
    precautions, 96-97
    summary, 97-98
    theory, 83-86
Dilutional pumping, 111
Disposal area design, 181-191
Dissolved oxygen
    effect of hypolimnetic withdrawal, 78
    fish, 334-335
    sediment covers, 353
    and temperature, 17
    and zooplankton, 18, 334-335
Diversion. *See* Nutrient diversion
Dollar Lake, Ohio, 102, 109, 114, 115, 122, 201
Drawdown. *See* Water-level drawdown
Dredges
    diver operated, 278, 281-282
    grab bucket, 146-148
    hydraulic, 148-151
    selection of, 160-181
    special purpose, 151-156
Dredging. *See* Sediment removal
Drinking water, 18

East Twin Lake, Ohio, 70, 114, 116, 289, 296-298, 303
Epilimnion, 17
Eutrophication
    definition, 3, 10, 76

effect of basin shape, 13, 16
feedback loops, 13, 335
pictorial model, 21-22
process, 10-22
Export coefficients, 28-29

Fish
    effects on
        of acidification, 364-365
        of alum application, 120
        of hypolimnetic aeration, 213, 220
        of low dissolved oxygen, 18
        of water-level drawdown 263, 268, 270, 339-340
    effects of
        on nutrient regeneration, 192, 337
        on trophic state, 113, 193, 219, 244, 328-337
        rough fish removal, 337-340
        winterkill, 139, 329
Flushing, definition, 83, 96-97
Fly ash, 123-126

Grass carp, 316-324
    effect of
        on nutrients, 317, 319-321
        on other species, 318-319
        on phytoplankton, 317, 320-321
    preferential feeding, 321-322
Green Lake, Washington, 83, 84, 86, 92-94, 97, 98, 148, 355
Groundwater, 31-32

Harvesting
    advantages and disadvantages, 276-277
    Aquamarine, 279
    case studies, 288-297
    control of nutrients, 302-304
    costs, 295, 297-298
    effect on fish, 301-302
    effect on phytoplankton, 300
    effectiveness, 291, 296, 304-306
    environmental effects, 300-302
    and grass carp, 324
    machine costs, 282
    multiple stage, 286-287
    objectives, 275
    options, 285

plant cutter costs, 283
root tillage, 278, 280, 282
single stage, 286
summary, 306-308
types of equipment, 277-288
utilization of plants, 288-289, 307
Herbicides, 6, 324, 327, 350
use with harvesting, 275
use with water-level drawdown, 254, 266
Horseshoe Lake, Wisconsin, 109, 111, 112, 122
Hypolimnetic aeration
case studies, 216-220
costs, 221
effect on trophic state, 216-220
fish management, 220
objectives, 213-214
sizing of system, 214
undesirable side effects, 220
Hypolimnetic oxygen deficit. *See* Oxygen deficit
Hypolimnetic withdrawal
case studies, 75-79
negative effects, 80
theory, 75
Hypolimnion, 17
Hypsograph, 13, 16

Index of basin permanence, 16, 17
Insects
effect of alum application, 120
in lake restoration, 325-328
Integrated control, 275; 276, 324, 326, 327, 383-385
Internal phosphorus loading
effect of alum, 101, 104, 113-117
effect of hypolimnetic withdrawal, 79
models, 36-37
relationship to iron, 18, 64
relationship to lake recovery, 58-60, 64-67, 139-140
Iron, 18, 64, 102, 118, 134, 137, 213, 243

Lake evaluation, 22-40
Lake evaluation index, 39
Lake Mendota, Wisconsin, 18, 61, 66, 101, 102, 296, 297

Lake restoration. *See* specific techniques
combination techniques, 383-385
decision tree, 44
definition, 4
history, 4
procedures proven effective, 377-380
procedures requiring research, 380-383
report, 46
selection of alternatives, 40-46
Lake Sammamish, Washington, 36, 58, 61, 62, 64-66, 68, 69
Lake Trummen, Sweden, 70, 143, 148, 157, 192-193, 198, 204, 205
Lake Washington, Washington, 60-65, 68, 69, 94, 353, 355
Lake Zurich, Switzerland, 61, 67-70
Land use and nutrient loading, 12, 14, 15
Liberty Lake, Washington, 103, 117, 120-122, 304
Lilly Lake, Wisconsin, 142, 148, 155, 194-198, 202
Liming. *See* Acidified lakes
Limiting nutrients, 5, 19, 83, 84, 90, 112
Littoral zone, 20-21
Long Lake, Washington, 114, 117, 122, 140, 204, 268-271, 304

Macrophytes
basin shape, 16
control by sediment removal, 140-141, 144-146
effect of drawdown, 256-262, 266-267
effect of phosphorus inactivation, 117, 121
and eutrophication, 11
light and depth of growth, 143-146
regrowth on sediment covers, 350-351
relationship to internal loading, 45, 141, 268-269, 303-304
response to shading, 358-359
and trophic state, 38-39
Medical Lake, Washington, 111, 112, 120, 122, 198, 201, 204
Metalimnion, 17
Moses Lake, Washington, 83-98
Murphy Flowage, Wisconsin, 263-265

NALMS, 4

N:P ratio, 40
Nitrogen fixation, 19, 90-91, 137
Nitrogen as limiting nutrient, 90-91
Nonpoint sources, 12
Norrviken Lake, Sweden, 57, 58, 60-62, 65, 66, 68, 69
North American Lake Management Society, 4
Nutrient budgets, 26-37
   export coefficients, 27-29
   uncertainty, 33
Nutrient diversion
   case studies, 60-67
   costs, 67-70
   and lake recovery, 60-67
   recovery expected, 56-60
   research needed, 70-71
   theory, 55
Nutrient inactivation. *See* Phosphorus inactivation
Nutrient-loading models, 26-28, 35-36, 56, 60, 84-85, 95
Nutrient sources, 12-15
Nutrient-water budgets
   errors in, 31, 33-34
   ground water, 31, 32
   methods, 26, 29
   point sources, 32
   use in trophic state estimation, 33-36

Okanangan Valley Lakes, British Columbia, 267, 289, 291-294, 297, 305
Oligotrophic, 11
Olszewski tube, 75, 77, 78
Organic matter loading, 11, 13, 26
Oxygen deficit, 18, 39, 80, 214

Pelagic zone, 21
pH
   and dissolved aluminum, 102, 104-106, 109, 113, 121
   and lake acidification, 365-375
pH and CO_2
   and artificial circulation, 244-245
   effect on phytoplankton, 231-233
Phosphatases, 103
Phosphorus
   export coefficients, 27-29

   input-output models, 27, 56, 59, 84-85
   as limiting nutrient, 5, 19
   precipitation, definition, 101-103
   relationship to chlorophyll, 19-20
   seasonal model, 71
   sedimentation, 27-28, 58, 84
   steady state model, 27-28, 56
Phosphorus inactivation
   application method, 103, 109-110
   case studies, 111-118
   chemistry of, 102
   costs, 122-123, 198-205
   definition, 101, 103
   dose determination, 103-108
   effect on phytoplankton, 111-112, 114-117
   and internal loading, 112-117
   negative effects, 119-122
   ponds, 117
   research needed, 122-123
Phosphorus loading, measurement methods, 34
Phytoplankton
   adaptive phosphatase synthesis, 103
   effect on
      of artificial circulation, 226-233, 244-247
      of dilution, 89-93
      of grass carp, 317, 320-321
      of harvesting, 300
      of herbivory, 329-337
      of hypolimnetic withdrawal, 77, 79
      of pH and CO_2, 231-233
      of sediment removal, 141-142, 193, 195, 197
      of water-level drawdown, 270
   light limitation, 226-228
   light and nutrient limitation, 228-230
Plant pathogens, 325-328
Point sources, 12
Ponds, 117-118
Potable water, 3, 10, 18

Reservoirs
   compared to lakes, 10-11
   evaluation, 22-40
   outflow dissolved oxygen, 80
   upstream phosphorus removal, 118

Riplox, 133, 135, 137
Root tillage, 278, 280, 282
Rotenone, 329, 339

Secchi disk
 and chlorophyll, 57
 effect of artificial circulation, 240, 244
 and trophic state, 35-37
Sediment covers
 advantages and disadvantages, 349-350
 application procedures, 357-358
 case studies, 350-356
 properties of materials, 350-356
 summary, 356-357
Sediment oxidation
 case studies, 134-136
 compared to alum, 137
 costs, 134-136
 equipment and dose, 133-134
 theory, 133
Sediment oxygen demand, 134
Sediment removal. *See* Dredges
 case studies, 191-197
 Cleanup system, 154-155
 costs, 197-205
 Darcy-Weisbach formula, 165
 disposal area design, 143, 165, 181-191
 drawdown method, 142
 dredge selection, 157-181
 effect on phytoplankton, 141-142, 193, 195, 197
 environmental effects, 141-143, 194-197
 hypothetical feasibility study, 158-181
 macrophyte control, 140-141, 144-146, 197
 Moody diagram, 165-168
 Mud Cat, 151-152
 Oozer System, 153-155
 Pneuma System, 153
 purposes, 139
 removal depth, 143-146
 Reynolds number, 165, 167-168
 silt curtains, 147-148
 summary, 205-206
 types of dredges, 146-156
 uses of dredged material, 156, 198
Seepage meters, 32
Shading, 358-359

Shagawa Lake, Minnesota, 36, 58, 59, 61, 62, 67-69, 101, 140
Silt curtain, 147-148
Silt loading, 11, 13, 26
Size-selective predation, 329-337
Sodium aluminate, 102, 109, 113
Stormwater, 5, 45

Thermal stratification, 17, 19
Trihalomethanes, 10, 22
Trophic state index, 33-39

Water budget, 26-37
 equation, 30
 errors, 31-34
 evaporation, 32-33
 groundwater, 31-32
 lake level, 33
 methods, 30, 31
 model, 30
 precipitation, 30
Water-level drawdown
 case studies, 253-266
 effect on fish, 253, 268, 339-340
 fish management, 268, 271
 nutrient release, 270
 positive and negative effects, 268-271
 responses of macrophyte species, 256-262, 266-267
 sediment consolidation, 269
 summary, 271-272
 waterfowl management, 269
Watershed
 mapping, 26
 nutrient-water budgets, 26-37
 of reservoirs, 10
West Twin Lake, Ohio, 70, 102, 109, 114, 116, 120, 122, 201, 204, 289, 296
Winterkill, 139, 194, 329, 338

Zooplankton
 biomanipulation, 329-337
 effect on
  of alum application, 121
  of artificial circulation, 244
  of fly ash application, 124, 126
  of low dissolved oxygen, 18, 334-335
  of planktivory, 329-337
 grazing on phytoplankton, 21, 335-336